Second Edition

# SEWER
# PROCESSES

Microbial and Chemical
Process Engineering of
Sewer Networks

Second Edition

# SEWER PROCESSES

## Microbial and Chemical Process Engineering of Sewer Networks

Thorkild Hvitved-Jacobsen
Jes Vollertsen
Asbjørn Haaning Nielsen

CRC Press
Taylor & Francis Group
Boca Raton  London  New York

CRC Press is an imprint of the
Taylor & Francis Group, an **informa** business

CRC Press
Taylor & Francis Group
6000 Broken Sound Parkway NW, Suite 300
Boca Raton, FL 33487-2742

© 2013 by Taylor & Francis Group, LLC
CRC Press is an imprint of Taylor & Francis Group, an Informa business

No claim to original U.S. Government works

Printed on acid-free paper
Version Date: 20130226

International Standard Book Number-13: 978-1-4398-8177-4 (Hardback)

### Library of Congress Cataloging-in-Publication Data

Hvitved-Jacobsen, Thorkild.
    Sewer processes : microbial and chemical process engineering of sewer networks / authors, Thorkild Hvitved-Jacobsen, Jes Vollertsen, Asbjørn Haaning Nielsen. -- Second editon.
        pages cm
    Includes bibliographical references and index.
    ISBN 978-1-4398-8177-4 (hardback)
    1. Sewage--Microbiology. 2. Sewage--Analysis. 3. Sewerage--Design and construction. I. Vollertsen, Jes. II. Nielsen, Asbjørn Haaning. III. Title.

TD736.H85 2013
628.3--dc23                                                                      2013004190

**Visit the Taylor & Francis Web site at**
**http://www.taylorandfrancis.com**

**and the CRC Press Web site at**
**http://www.crcpress.com**

# Contents

# Preface

The first edition of this book was published in 2001. Since then, considerable improvements in the fundamental understanding of sewer processes have taken place. Furthermore, the conceptually formulated WATS (Wastewater Aerobic/anaerobic Transformations in Sewers) sewer process model for prediction and assessment of sewer processes and related adverse effects has been extensively upgraded. These developments have required a substantial extension of the text.

As was the case for the first edition, this book serves a dual purpose. First, it will be of use to students in environmental engineering by enabling them to understand sewer networks from a process engineering point of view and to apply this knowledge in quantitative terms. Second, this book is a practical reference intended to help planners, designers, operators, and consultants working with collection systems comprehend and control the adverse effects of sewer processes. Practicing engineers will find the contents of the book directed to solve problems by adding a process dimension to the design and operation of sewer networks.

Traditionally, books dealing with sewer systems have been devoted to hydraulics and pollutant transport phenomena. In this context, urban drainage and wet weather impacts onto the adjacent environment are in focus. With its concentration on dry weather conditions in the sewer and on the potential adverse effects of in-sewer chemical and microbiological processes, this book is different. It adds a corresponding process-related dimension to the management and engineering of sewers. A well-known example is the generation of hydrogen sulfide and its impacts in terms of concrete corrosion, malodors, and health-related effects. The book provides the reader with knowledge-based information on its formation and fate, and based on this background provides models for prediction and assessment of its occurrence and effects. The general important point is that the sewer is not just a collector and transport system for wastewater but also a chemical and biological reactor with impacts on the system itself, the wastewater treatment plant, and the adjacent environment.

The text offers a fundamental understanding of the chemical and microbiological processes that take place in sewers and quantifies these processes. The process engineering issues of wastewater in sewers are the ultimate objective, and the book provides in this respect the reader with an integrated description of sewer processes in model terms. The text is furthermore useful as an engineering guide to troubleshooting sewer problems.

The organization of the book follows from two perspectives: a general viewpoint with focus on the fundamental principles of chemical and microbiological transformations of wastewater in sewers and a specific viewpoint on the quantitative formulation of sewer processes that is directly applicable for engineers. About 110 figures and 50 tables illustrate the fundamental contents, concepts, and engineering relevance of the text. Furthermore, numerous example problems are included to highlight applications. Chapter 1 offers an overview providing an understanding of the sewer as a process reactor. Chapters 2 and 3 stress chemical and microbiological fundamentals needed to understand the processes in sewers. In Chapter 4, the transfer phenomena between the wastewater phase and the sewer atmosphere are dealt with, particularly in terms of reaeration, odor emission, and impacts of volatile substances. Chapters 5 and 6 investigate the aerobic, anoxic, and anaerobic processes in sewers. Besides hydrogen sulfide-induced corrosion, the major objective of the two chapters is to establish a conceptual understanding of sewer processes and develop corresponding mathematically based formulations. The focal point of Chapter 7 is mitigation directed to control adverse effects of sewer processes, in particular, those related to hydrogen sulfide and volatile organic compounds (VOCs). Chapter 8 deals with the basic characteristics of sewer process modeling and Chapter 9 is on this background devoted to the formulations of the WATS sewer process model. The main subject of Chapter 10 is to quantify and provide information on wastewater compounds and model parameters based on bench scale, pilot scale, and field experiments and directed toward a kinetic description of sewer processes. The text is concluded with Chapter 11 focusing on selected examples on structural and operational measures to improve sewer networks.

The theory and findings of this book have several sources. The first studies on sewer processes were principally carried out 60 to 70 years ago in California, followed by further developments, particularly in Australia, the United Kingdom, and South Africa. The combined scientific and technological understanding acquired through these studies is important and appreciated. The authors of this book have, during the past 30 years, carried out dozens of sewer projects, which have contributed with a conceptually formulated description of sewer processes. This understanding is the basis for the formulation of the WATS sewer process model. The validity of a conceptual understanding of sewer processes has, via the WATS model, been tested through a number of projects in Europe, the Middle East area, North Africa, and the United States.

**Thorkild Hvitved-Jacobsen**
**Jes Vollertsen**
**Asbjørn Haaning Nielsen**
*Aalborg University, Denmark*
*January 2013*

# Acknowledgments

Today's understanding of the sewer as a chemical and biological reactor is relatively new. The authors acknowledge in particular the innovative works 60 to 70 years ago by C.D. Parker, R.D. Pomeroy, and F.D. Bowlus, who made the first contributions to a solid understanding of sewer processes.

Several people have, during the period this book was being developed, in different ways contributed to its contents and present state. The authors are grateful for the contributions received from their former PhD students, in particular by their work that supported the conceptual description of the in-sewer processes. These PhD students, Per Halkjaer Nielsen, Niels Aagaard Jensen, Kamma Raunkjaer, Hanne Lokkegaard, Jes Vollertsen, and Naoya Tanaka, have all contributed with invaluable results to the first edition of this book. The improvements from the first to the present edition of this book are in several chapters attributable to the work of the following PhD students: Chaturong Yongsiri, Asbjørn Haaning Nielsen, Weixiao Yang, Heidi Ina Madsen, Henriette Stokbro Jensen, and Elise Rudelle. Numerous MS students in Environmental Engineering at Aalborg University have likewise contributed with valuable results. Mary Bjerring Rasmussen and Kristian Kilsgaard Ostertoft are also acknowledged for careful and qualified help to produce the figures for this book.

Last, but not least, the first author of this book is grateful to experience that two of his former PhD students, now Professor Jes Vollertsen and Associate Professor Asbjorn Haaning Nielsen, are continuing the work on sewer processes that started 30 years ago at Aalborg University.

# Authors

**Thorkild Hvitved-Jacobsen, MSc,** is professor emeritus at Aalborg University, Denmark. In 2008, he retired from his position as professor of environmental engineering at the Section of Environmental Engineering, Aalborg University, Denmark. His primary research and professional activities concern environmental process engineering of the wastewater collection and treatment systems, including process engineering and pollution related to urban drainage and road runoff. His research has resulted in more than 320 scientific publications in primarily international journals and proceedings. He has authored and coauthored a number of books published in the United Kingdom, the United States, and Japan. He has worldwide experience as consultant for municipalities, institutions, and private firms, and he has produced a large number of technical reports and statements. He has chaired and cochaired several international conferences and workshops within the area of his research, and he has organized and contributed to international courses on sewer processes and urban drainage topics. He is a former chairman of the International Water Association/International Association for Hydraulic Research (IWA/IAHR) Sewer Systems and Processes Working Group. He is also a member of several national research and governmental committees in relation to his research area.

Web sites: http://www.sewer.dk and http://www.bio.aau.dk.

**Jes Vollertsen, PhD,** is a professor of environmental engineering at the Section of Water and Soil, Department of Civil Engineering, Aalborg University, Denmark. His research interests are urban stormwater and wastewater technology, where he combines experimental work on bench scale with pilot scale studies and field studies. He integrates the gained knowledge on conveyance systems and systems for wastewater and stormwater management by numerical modeling of the processes. His research has resulted in more than 190 scientific publications in primarily international journals, conference proceedings, and book chapters. He has authored and coauthored a number of books and reports published in Europe. He is an experienced consultant for private firms and municipalities as well as on litigation support. He has cochaired international conferences, has been a member of numerous scientific committees, and has organized international courses on processes in conveyance systems and stormwater management systems. He is a reviewer for a national research committee in relation to environmental engineering.

Web sites: http://www.sewer.dk and http://www.civil.aau.dk.

**Asbjørn Haaning Nielsen, PhD,** is an associate professor of environmental engineering at the Section of Water and Soil, Department of Civil Engineering, Aalborg University, Denmark. His research and teaching has primarily been devoted to wastewater process engineering of sewer systems and process engineering of combined sewer overflows and stormwater runoff from urban areas and highways. He has extensive experience with chemical analyses of complex environmental samples, particularly relating to the composition of wastewater and sewer gas. He has authored and coauthored more than 60 scientific papers, which have primarily been published in refereed international journals and proceedings from international conferences. He is a committee member of the Danish National Committee for the IWA.

Web sites: http://www.sewer.dk and http://www.civil.aau.dk.

# 1 Sewer Systems and Processes

## 1.1 INTRODUCTION AND PURPOSE

The flows of wastewater originating from households, industries, and runoff from precipitation in urban areas are generally collected and conveyed for treatment and disposal. The systems used for this purpose are named sewer networks or collection systems. A sewer network is thereby defined as the wastewater system located between the sources for generation of wastewater and a wastewater treatment plant, alternatively, a point of discharge into an adjacent receiving water system. A sewer system consists of individual pipes (sewer lines) and a number of installations and structures, such as inlets, manholes, drops, shafts, and pumps, used to facilitate collection and transport. In the European definition, a sewer system is a network of pipelines and ancillary works that convey wastewater from its sources such as a building, roof drainage system, or paved area to the point where it is discharged into a wastewater treatment plant or directly into the adjacent environment (BS EN 752-1, 1996).

The efficient, safe, and cost-effective collection and transport of wastewater and runoff water have been identified as key criteria to be observed. In this context, the word "safe" means that public health, welfare, and environmental protection have high priority. The demand for solutions toward more sustainable water management in the cities is furthermore a challenge.

A sewer network is subject to great variability in terms of its performance. During dry weather periods, the flow rates reflect the behavior of the community in the upper part of a sewer system, often with a flow rate variability of about a factor of 10 over day and night. In sewer pipes receiving both municipal wastewater and urban stormwater runoff, i.e., the combined sewer networks, the flow rates during extreme rainfall events are often increased by a factor of 100–1000 compared with the average dry weather flow. It is clear that efforts in both research and practice have been devoted to developing systems and procedures for the design of collection systems and their operation under varying conditions. During the past 20 to 30 years, emphasis has been placed on drainage phenomena in terms of flow conditions for the sewer network and integrated solutions comprising the treatment plant performance and receiving water impacts during wet weather periods. Urban drainage has thereby been an important issue in both research and practice.

The focus of this book is on the dry weather aspects of the sewer network and, in this respect, its design and performance from a microbial and chemical process standpoint. Wastewater includes substances with a pronounced chemical and biological reactivity, and the sewer network is therefore—in addition to being a collection and conveyance system—also a reactor for transformation of wastewater. In this respect, the point is that several substances from this transformation have a severe impact on the sewer network and its surroundings. Concrete corrosion, odor nuisance, human health impacts, and effects on a treatment plant located downstream are important examples.

During wet weather conditions, quite different problems related to the pollution and process performance of the sewer network arise, and the impacts on the adjacent environment are potentially severe. The authors of this book are aware of this situation. In their book, *Urban and Highway Stormwater Pollution: Concepts and Engineering*, Hvitved-Jacobsen et al. (2010) pose a "wet-weather parallel" to the present book.

Because of the basic requirements of wastewater collection and conveyance, sewer networks are traditionally dealt with from a physical point of view, i.e., the hydraulics and sewer solids transport processes are addressed. From this point of view, design and operational principles have been developed, to a great extent supported by numerical procedures and an ever-increasing capacity of computers. Under wet weather conditions, the hydraulics and solids transport phenomena in a sewer play a central role. Because of dilution and reduced residence time of wastewater and runoff water in the sewer, the chemical and microbiological processes in the network itself are often of minor importance. Not surprisingly, interests devoted to urban drainage have focused on the physical behavior of the sewer, "sometimes"—although at an increasing rate—taking into account the chemical and biological impact of wastewater and runoff water discharged to the receiving environment.

In contrast, under dry weather conditions, which may occur more than 95% of the time, chemical and biological sewer processes may exert pronounced effects on the sewer performance and on the interaction between the sewer and subsequent treatment processes. Possibly, because researchers' and operators' interests have been devoted to wet weather conditions, the dry-weather biological and chemical performance of a sewer, i.e., the sewer as a "chemical and biological reactor," have been of minor concern. Or, at least, it has not been dealt with in terms of a detailed understanding of the chemical and biological processes and their quantification based on the underlying fundamental phenomena. In this respect, it is interesting—but also a bit depressing to note—that sewers and treatment plants have been very differently managed. It is, however, apparent that the sewer cannot be neglected as a chemical and biological process system. These processes may exert severe impacts on the sewer itself, the treatment plant, the environment, and the humans in direct or indirect contact with the sewer.

Textbooks dealing with collection systems have normally been devoted to planning, design, operation, and maintenance focusing on the physical processes. This book is different, in that it will primarily be concerned with chemical and biological processes in sewers under dry weather conditions, and will emphasize and quantify the microbiological aspects. There are several examples that illustrate the importance of these sewer processes and call for their control and consideration in practice. The impact of sulfide (hydrogen sulfide) produced under anaerobic conditions in wastewater is probably the most widely known example. Sulfide is a serious health hazard for humans and is a malodorous compound that may also create severe corrosion problems in the sewer network. On the other hand, anaerobic conditions may produce and preserve those easily biodegradable substrates that enhance advanced wastewater treatment in terms of improved conditions for denitrification—nitrogen removal—and biological phosphorus removal. In an aerobic sewer, removal of these easily biodegradable organic substances and production of less biodegradable particles (e.g., microorganisms) may occur. Aerobic conditions may, therefore, improve conditions for in-sewer treatment of the wastewater and result in a positive interaction with subsequent mechanical and physicochemical treatment processes. These few examples show that a sewer is not just a collection and conveyance system but also a process reactor that must be considered an integral part of the entire urban wastewater system.

In conventional design and management practice, treatment of wastewater is considered to take place entirely within the treatment plant, whereas a sewer network serves the sole purpose of collecting and conveying wastewater from its sources to treatment. The concept of considering the sewer as a process reactor also serves the purpose of breaching this rather rigid understanding of a sewerage system. When considering the processes "starting at the sink," a number of basic aspects for improved engineering can be more directly and correctly taken into account. Furthermore, it is important that more holistic approaches expressed in terms of sustainability, public health, environmental protection, and enhancing the standard of living for the general population, can be considered. Figure 1.1 illustrates that sewers, treatment plants, and receiving water systems should not, from a process point of view, be viewed as stand-alone units.

It is the purpose of this book to provide a fundamental basis for understanding sewer processes and demonstrate how this knowledge can be applied for design, operation, and maintenance of collection systems. The overall criteria of a sewer network to observe efficient, safe, cost-effective, and sustainable collection and

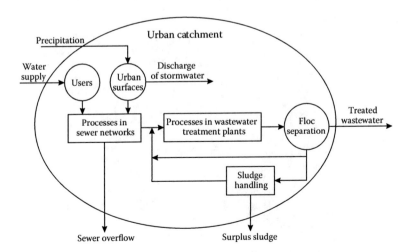

**FIGURE 1.1** Integrated sewerage process system and its interactions with the surroundings.

conveyance of wastewater are still valid; however, they must be expanded with a process dimension.

Specific aspects of microbiology and chemistry will be dealt with in this book whenever considered relevant for the understanding of the in-sewer processes. Although a fundamental basis in applied microbiology, and chemistry is beneficial when reading this book, it is the authors' experience that the book in itself will provide sufficient basic knowledge to understand sewer processes and make use of this knowledge in practice.

In particular, the text will focus on the microbiological processes in the sewer, but also the chemical processes, especially the physicochemical processes. The environmental engineering relevance is the ultimate goal when deciding to which extent such details will be included. Hydraulics and solid-transport phenomena, sewer construction details, materials, and traditional sewer design and management will only shortly be dealt with and primarily when relevant for the microbial and chemical processes.

## 1.2 SEWER DEVELOPMENTS IN A HISTORICAL PERSPECTIVE

### 1.2.1 EARLY DAYS OF SEWERS

Sewer networks belong to the urban environment and may therefore, in principle, date back to the days of the urban revolution starting about 7000 BC when the first urban settlements were established. Several ancient civilizations, particularly those located in the Middle East, developed sewers and drainage systems to remove either wastewater from houses or surface runoff in populated areas (Burian et al. 1999; Bertrand-Krajewski 2005).

An example is Mohenjo-Daro, a city settlement of the Indus Valley Civilization (today located in West Pakistan). Buildings from the period 2500–2000 BC show bathing and latrine facilities and a sewer system equipped with a grit chamber (Figure 1.2). It is assumed that the grit chamber was important for a proper function of the sewer pipe located downstream.

**FIGURE 1.2**   View of a sewer with a grit chamber, Mohenjo-Daro, now located in West Pakistan, 2500–2000 BC (Source: Picture courtesy of Dr. Michael Jansen.)

### 1.2.2  SEWERS IN ANCIENT ROME

From the golden age of ancient Rome, where water consumption per capita was in the same order of magnitude as present-day levels in the developed part of the world, sewers became both needed and common. During the reign of Emperor Augustus, it is known that his son-in-law, Agrippa—while sailing a boat—visited Rome's main sewer system, Cloaca Maxima, about 30 years B.C. (Figure 1.3). Apparently, he was not satisfied with its performance because he ordered it to be flushed, which was done by simultaneously diverting water from seven aqueducts. It is not known if the reason for this action was the sediment problems or malodors. However, what we know is that in ancient Rome, it was considered a punishment and a degrading work to clean sewers. Although this example indicates that the sewer network receives wastewater from households, these systems were typically constructed with the main purpose of conveying stormwater runoff from urban paved areas, protecting them against flooding. However, the contents of solid waste as a constituent of the runoff might also originate from wastes dumped at the streets. Sporadically, similar drainage systems were also in use in Europe during the 16th and 17th centuries. Typically, it was prohibited to discharge wastes from households into these storm drains.

### 1.2.3  SEWERS IN MIDDLE AGES

In general, the knowledge on sewer construction and management gained in Ancient Rome got lost and the European cities of the Middle Ages had typically at best a rudimentarily developed drainage infrastructure. It was not until about the seventeenth century, that large European—and to some extent, American cities—started

**FIGURE 1.3**  A tract of Cloaca Maxima from about 4 B.C. (Source: Picture courtesy of Dr. John Hopkins.)

developing underground sewer networks as we know them today. As late as the period of the French revolution—a little more than 200 years ago—the total length of the sewer network in Paris was only 26 km. In 1887, the total length was extended to 600 km.

### 1.2.4  SEWER NETWORK OF TODAY UNDER DEVELOPMENT

The sewer network we know today is a relatively newly invented infrastructure of cities. Not until the early days of the Second Industrial Revolution, starting around 1830 in Europe and America, did it become common to construct underground wastewater collection systems. Often, the sewers developed as "wild mixed systems" with a main purpose of fast removal of wastes and wastewater from the growing cities.

London and Paris were among the first to develop efficient sewer networks, but other European, Australian, and American cities followed rapidly. The first sewers developed from the storm drains, which were then allowed to receive waterborne wastes from flush toilets, in principle converting these drains into combined sewers. A major reason for collecting the wastewater in underground systems was the enormous problem of the unpleasant smell from the open sewers, cesspools, and privies, and the requirements for space in the streets of densely populated cities. Therefore, design procedures and construction details facilitating the efficient transport of the solid constituents of wastewater became central, an aspect that remains valid even today. As an example, in 1906 the 1177-km sewer system in Paris was equipped with 4369 small ancillary works with the sole purpose of flushing the pipe section located downstream. The so-called man-entry sewer is basically also a "construction detail" for the efficient removal of sediments that otherwise could cause blocking.

### 1.2.5  SANITATION: HYGIENIC ASPECTS OF SEWERS

The knowledge on the hygienic aspects of sewers and the corresponding human impacts in terms of water-borne diseases developed slowly from the late eighteenth century during the next 100 years. It was definitely in contrast to the rapidly developing urbanization that took place during the same period. The lack of an efficient infrastructure for wastes in growing cities and new settlements caused severe pandemics. There were numerous attempts to improve the understanding of how human wastes could affect health and the development of diseases. Several correct observations and solidly performed investigations on this relationship failed to find general acceptance. Based on studies of the environmental conditions in prisons, the famous French chemist Antoine Lavoisier concluded that the quality of water supply and sewerage affected the health of the inmates. Unfortunately, Lavoisier was beheaded in 1794 during the French revolution. The occurrence of pandemics of cholera and numerous local outbreaks from the beginning of the nineteenth century until about 1860 caused more than 5 million deaths, particularly in growing European cities with faulty sewerage infrastructure. An illustrative example is the cholera epidemic in Copenhagen, Denmark, during the months of June to August in 1853. A total of 4737 inhabitants died in less than 3 months, which at that time comprised almost 5% of the population in Copenhagen.

A major step toward understanding the impact of wastewater contamination of drinking water sources took place in 1848 when an English physician, John Snow, observed that outbreaks of cholera in London occurred within geographically concentrated areas where people received their water supply from the very same water well. Then, he correctly concluded that a substance "materia morbus" was excreted by cholera-infected humans and transported into the drinking water systems, typically local underground wells. At that time, however, his considerations were totally rejected by physicians who continued to insist that diseases such as cholera and "black death" were caused by foul air. During a cholera epidemic in the Soho area of London in 1854, John Snow continued his studies and identified the outbreak as the public water pump on Broad Street (now Broadwick Street). However, the water pump was not finally closed until 1866—proving that the adoption of new knowledge takes time. The significance of John Snow's accomplishment is that he, based on solid statistical facts, identified the link between cholera infection and water polluted with human excreta, i.e., he identified polluted water as the vector for the cholera disease. However, it was not until 1883 when the German doctor, Robert Koch, isolated *Vibrio cholerae*, that the microbial cause of the disease was finally identified. For his work as the founder of bacteriology, Robert Koch received the Nobel Prize in physiology and medicine in 1905.

The concept of a deliberate separation of wastewater and drinking water was, however, not generally accepted and might, in most cases—compared with the need for space in the narrow streets—not have been a principal initiating factor in the establishment of underground sewers. Later, during the twentieth century, it became quite clear that a sewer is a "technical hygienic and sanitary installation" that efficiently reduces epidemic diseases. This characteristic is still valid and is a major reason why underground sewer networks are still expanding, even in developing countries under conditions of limited financial resources. The need for sanitation is clearly shown by the fact that the World Health Organization (WHO) reported that around 2005, more than 1.1 billion people lack access to drinking water from an improved source and that 2.6 billion people do not have basic sanitation.

It is generally recognized that the sewer network is a sanitary installation that effectively hinders wastewater in being a vehicle for dissemination of infectious diseases. It is in this respect interesting that the *British Medical Journal* (*BMJ*), during the period between January 5 and January 14, 2007, invited its readers (mainly medical doctors from countries all over the world) to submit on its web site a nomination for the top medical breakthrough since 1840, the year the journal was launched. *BMJ* posted 15 areas of medical advances as potentials for nomination, of which sanitation garnered the top vote—ahead of, for example, antibiotics, anesthesia, vaccines, and discovery of the DNA structure (Hitti 2007).

### 1.2.6 SEWER AND ITS ADJACENT ENVIRONMENT

The polluting impact onto the environment of the waste stream from sewers is historically a relatively recent concern. In 1889, the following excerpt was included in an article on the advantages of a separate sewer system (Manufacturer and Builder 1889): "A theoretically perfect sewer would be one in which all the sewage would be carried rapidly to its outfall outside the city, so that no time would be given for decomposition." Although the environment outside the city is not considered a problem, it is apparently problematic if "decomposition" occurs within the city area. In the same article, it is referred to as a problem that the combined sewers result in "obstructions form a series of small dams in the sewer, and in dry weather the sewage stands in a succession of pools along the sewers, decomposing and sending volumes of sewer gas out of every crevice through which it can escape." It is, however, not further discussed what "sewer gas" is. In 1889, details relating to the chemical and microbial processes in sewers were unknown among constructors and operators of sewers.

Until the middle of the twentieth century, the sewage collected in cities was typically discharged into the adjacent environment without any type of treatment, resulting in problems such as bacterial contamination, malodors, dissolved oxygen depletion, and fish kills in downstream receiving waters. Even today, such problems are well known, and other problems such as eutrophication and toxicity of heavy metals and organic micropollutants including, for example, chemical substances originating from pharmaceutical drugs have been added. The "end-of-pipe" solution to these problems in terms of wastewater treatment was not introduced in several countries until after World War II. Although wastewater treatment plants—with different levels of treatment—are now common worldwide, the process of further development of treatment and reuse of resources from the waste streams is still in progress. We still suffer from the aftereffects of a missing integrated development of the urban wastewater system by a narrow distinction between the sewer as a collecting and conveying system for wastewater and the treatment plant as a pollutant reduction system.

The older parts of the cities in Europe and the United States are typically served by combined sewer networks with outfalls from where the excess wet weather flows are more or less untreated and routed into an adjacent receiving water system. Although separate sanitary and storm sewers have dominated construction for the past 50 to 100 years, numerous old combined systems are still in operation, often being upgraded and equipped with basins to detain wet-weather discharges of untreated water.

### 1.2.7   Hydrogen Sulfide in Sewers

Today's sewerage in terms of a combined wastewater conveyance in sewers followed by its treatment is, as described in Sections 1.2.5 and 1.2.6, the solution for problems associated with both sanitation and environmental effects. However, the way we construct sewer networks may lead to in-sewer anaerobic process-related problems, primarily in terms of the formation of hydrogen sulfide and volatile organic compounds. The corresponding problems appear as concrete and metal corrosion degrading the sewer network, health-related impacts on the sewer personnel, and malodors observed in the adjacent environment.

It is likely that anaerobic processes in terms of production of hydrogen sulfide and other malodors have always been associated with conveyance of human wastes. "Sewer processes" are in this respect not a new "invention." However, it was not until the beginning of the twentieth century that hydrogen sulfide formation and concrete corrosion were finally identified in terms of a cause–effect relationship (Olmsted and Hamlin 1900).

The scientific and technical aspects of chemical and biological sewer processes focusing on hydrogen sulfide formation and its impacts and control were not deliberately dealt with until about 1930 (Bowlus and Banta 1932; Pomeroy 1936; Parker 1945a, 1945b). Knowledge on the sulfide problem was the result of the work of prescient scientists and practitioners and particularly developed in the Los Angeles area in California, the United Kingdom, and Australia. Although this knowledge was published internationally, it was not generally known among constructors and operators of sewer networks worldwide. Even in the late twentieth century, several structures were constructed that reflected their builders' lack of knowledge on the nature of chemical and biological processes in sewers, and corresponding wrong network constructions led to fatal disruptions caused by hydrogen sulfide formation. Even today it is frequently seen that inappropriate sewer constructions and operational mistakes are due to an incomplete understanding of the process.

It is, however, important to notice that since about the 1960s, problems related to hydrogen sulfide in terms of deterioration of sewer networks and concerns for the toxicity and odor nuisance have became clear to several municipalities and network stakeholders. Investigations were performed in several countries to gain knowledge on the formation and control of hydrogen sulfide. A number of models for prediction of sulfide formation and guidelines for solving the sulfide problems in gravity sewers and pressure mains were developed. These models and guidelines have been widely used for sewer design and for implementation of appropriate control methods to prevent the generation of sulfide or its effects (cf. Chapters 6 and 7).

### 1.2.8 FINAL COMMENTS

The development of sewers that has taken place is the result of 100 to 150 years of enormous investments. All over the world, it has left us with a sewer and treatment plant infrastructure that will be in use for an unknown length of time. We will still see developments in terms of technical improvements and sustainable solutions. However, as a general trend, we will not see the present wastewater collection and treatment concept replaced by, for example, centralized collection of "solid" human excreta or on-site solutions. It might have been a realistic option for implementation and further development 150 years ago. Not now!

## 1.3 TYPES AND PERFORMANCE OF SEWER NETWORKS

Sewer network characteristics in terms of design, use of materials, and operation affect sewer processes, and, what is important, knowledge on sewer processes can be actively applied in the design and operation of a sewer system. As an example, the type of sewer determines, to a great extent, if aerobic or anaerobic processes dominate. Furthermore, the flow regime in the water phase and ventilation of the sewer atmosphere may affect the gas phase buildup of odorous, corroding, and toxic volatile substances produced by microbiological processes in the water phase.

Sewers can be classified into different categories. The three major ways of classification refer to (1) which type of sewage is collected, (2) which type of transport mode is applied, and (3) the size and function of the sewer. These three different categories of sewers divide sewers into groups with different characteristics in terms of wastewater collection and transport. In addition, it is also extremely relevant to consider these aspects when addressing sewer processes.

### 1.3.1 TYPE OF SEWAGE COLLECTED

There are three main types of sewer networks that refer to the sources of the sewage: sanitary sewers, storm sewers, and combined sewers. Wastewater of domestic, commercial, and industrial origin is conveyed in both sanitary sewers and combined sewers, whereas the storm sewers in principle only transport runoff water from urban surfaces and roads. Each of the three types of sewers has, in terms of different flow and pollutant characteristics, specific properties related to sewer processes. This book does not cover the specific characteristics related to stormwater runoff. These aspects are, in terms of processes and pollution, addressed in a "parallel" book by the same authors (Hvitved-Jacobsen et al. 2010).

### 1.3.2 TRANSPORT MODE OF SEWAGE COLLECTED

There are two main types of sewer networks that refer to the transport mode: gravity sewers and pressure sewers (Figure 1.4). A gravity sewer is designed with a sloping bottom and the flow occurs by gravitation. In contrast, the driving force for flow in a pressure sewer is pumping. A pressure sewer is therefore also named a pumping sewer or a force main. In terms of processes, it is important that the water surface in a gravity sewer is most of the time exposed to a gas phase (sewer atmosphere), whereas this is clearly not the case in a pressure sewer. The exchange of volatile compounds between the water phase and the gas phase in a gravity sewer, e.g., molecular oxygen resulting in reaeration of the water phase, is crucial. In contrast, anaerobic conditions in wastewater of pressure sewers are typically occurring.

### 1.3.3 SIZE AND FUNCTION OF SEWER

A sewer system is a network and the sewage flows typically from small collecting sewers in the upstream part of a catchment through larger and larger sewers until it reaches the wastewater treatment plant. Different terms are used to characterize a sewer in this respect. The small-diameter sewers in the upper part of a catchment are often called lateral sewers. A number of lateral sewers typically discharge into

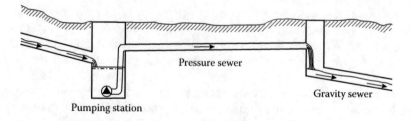

**FIGURE 1.4** Principle of wastewater flow from a partly filled gravity sewer into a full flowing pressure sewer followed by a gravity sewer.

a trunk sewer that thereby serves a larger catchment area. The main structure of the sewer network is the intercepting sewer that may receive sewage from several trunk sewers and diverts the sewage to treatment and disposal. The size and function of a sewer affect sewer processes in different ways. As an example, the water phase processes are more dominating compared with surface related processes—the biofilm processes—in large-diameter sewers.

The classification of sewers in terms of which type of sewage is collected will in the following be further explained and focused on.

*Sanitary sewer networks.* Sanitary sewers—often identified as separate sewers—are designed to collect and transport the daily wastewater flow from residential areas, commercial areas, and industries to a wastewater treatment plant. Typically, the wastewater transported in these sewers has a relatively high concentration of more or less biodegradable organic matter and microorganisms and is therefore biologically active. Wastewater in these sewers is, from a process point of view, a mixture of biomass (especially heterotrophic bacteria) and substrates for this biomass.

The flow in sanitary sewers is controlled by either gravity (gravity sewers) or pressure (pressure sewers) exerted by pumps installed in pumping stations (Figure 1.4). In a partially filled gravity sewer, transfer of oxygen across the air–water interface (reaeration) is possible, and aerobic heterotrophic processes in the wastewater may proceed. On the contrary, pressurized systems are full flowing and do not allow for reaeration of the water phase. In these sewer types, anaerobic processes in the wastewater will, therefore, generally dominate.

Among other parameters, the residence time in a sewer network affects the degree of transformation of the wastewater. The residence time depends on the size of the catchment, the distance to the wastewater treatment plant located downstream, and specific sewer characteristics, such as pipe diameter and slope of a gravity sewer pipe. The residence time is often relatively high in a pressure sewer, principally during nighttime hours when a reduced flow of wastewater is typical.

*Storm sewer networks.* Storm sewers or stormwater sewers are constructed for collection and transport of stormwater (runoff water) originating from impervious or semipervious urban surfaces such as streets, parking lots, and roofs and from roads and highways. Surface waters typically enter these networks through inlets located in street gutters. The storm sewers are in operation during and after periods with rainfall—and in cold climate areas, also during snowmelt periods. Typically, a storm sewer diverts the runoff water into adjacent watercourses with no or limited chemical and biological treatment although treatment of the runoff in, e.g., detention and retention ponds or other types of facilities, is becoming more common.

In the context of this book, focusing on the dry weather performance of sewers, the storm sewer network will only be given limited attention. The quality aspects of storm sewer networks in terms of design, performance, and impacts are central subjects of Hvitved-Jacobsen et al. (2010).

*Combined sewer networks.* The combined sewers collect, mix, and transport the flows of municipal wastewater and urban runoff water. In terms of the processes that proceed in the combined sewers, these systems generally perform like sanitary sewers during dry weather periods. However, because of their ability to serve runoff purposes, combined sewer networks are designed differently compared with separate sewers and include constructions such as overflow structures and detention basins to manage large quantities of wet weather flows. These ancillaries may influence a number of process details. Furthermore, the combined systems are subject to a higher degree of variability in the processes compared with the sanitary sewers because of the frequent shift in the flow conditions. Combined sewer networks are constructed as either gravity sewer lines or pressure pipes—or as a combination of both of these types.

The main characteristic features of the three different types of sewer networks are depicted in Figure 1.5.

The three types of sewer networks represent the main types. In practice, sanitary sewers often exist in catchments partially operated by separate sewers, i.e., the sanitary sewers may to some extent receive runoff water. Furthermore, sanitary sewage may also be discharged into storm sewers. Such illicit connections are frequently seen. Other alternative sewer systems include, for example, the vacuum sewers that are typically small systems and are operated locally.

## 1.4 SEWER AS A REACTOR FOR CHEMICAL AND MICROBIAL PROCESSES

Sewer processes proceed in complex systems, i.e., in one or more of the five phases: the suspended water phase, the biofilm, the sewer sediments, the sewer atmosphere, and at the sewer walls. Furthermore, exchange of substances across the interfaces takes place. Processes that proceed in the sewer system also affect other parts of the urban system, e.g., the urban atmosphere with malodorous substances. Furthermore, wastewater treatment plants and local receiving waters receive not just those substances discharged into the sewer but also products that are the result of the sewer processes (Figures 1.1 and 1.6).

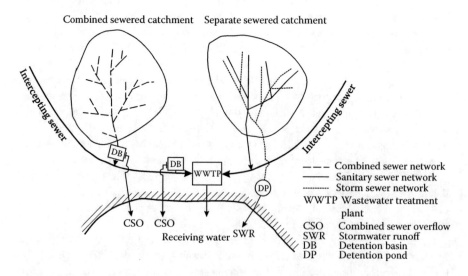

**FIGURE 1.5** Outline of sewer networks in separate sewer catchments and combined sewer catchments.

The conditions for reduction and oxidation of substances—the redox conditions—are in general central for understanding which chemical and biological processes can proceed. For the sewer as a chemical and biological reactor, the redox conditions therefore play a central role. In principle, a redox reaction proceeds by transferring electrons at the atomic or molecular scale from one compound to another compound. By this transfer of electrons, oxidation and reduction, respectively, of the involved compounds will proceed.

The microbial system in wastewater of sewers is dominated by heterotrophic microorganisms, which degrade and transform wastewater components. The redox conditions are determined by the availability of the electron acceptor, i.e., the substance that receives electrons in a redox reaction. Examples of important electron acceptors in a sewer are dissolved oxygen ($O_2$), nitrate $\left(NO_3^-\right)$, and sulfate $\left(SO_4^{2-}\right)$, determining whether aerobic, anoxic, or anaerobic processes, respectively, may occur. By the transfer of electrons, reduction of the central element of these three compounds to water ($H_2O$), molecular nitrogen ($N_2$), and hydrogen sulfide ($H_2S$), respectively, takes place. The importance of the processes for the sewer and the surroundings is not just caused by the removal and transformation of organic substrates—the electron donor—but is also a result of transformation of the electron acceptors exemplified by the formation of hydrogen sulfide from sulfate.

The design characteristics and operation mode of a sewer network determine, to a great extent, which redox conditions prevail. Table 1.1 gives an overview of sewer system characteristics important for the process conditions. Primarily, aerobic and anaerobic conditions arise, whereas anoxic conditions in principle only

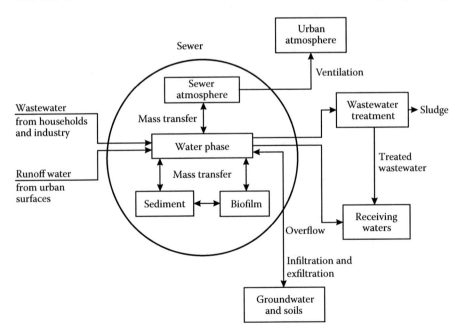

**FIGURE 1.6** Outline of wastewater flows related to a sewer system showing locations for potential occurrence of sewer processes and the receiving environment.

exist if nitrate—or oxidized inorganic nitrogen substances—is artificially added to the wastewater. The extent of reaeration, which is closely related to the design and operation of the sewer, is a fundamental process that determines if aerobic or anaerobic conditions exist. Under aerobic conditions, degradation of easily biodegradable organic matter is a dominating process. If dissolved oxygen or nitrates are not available, strictly anaerobic conditions occur, and sulfate is typically the external electron acceptor, resulting in the formation of hydrogen sulfide, a phenomenon well known among practitioners dealing with collection systems. Furthermore, fermentation occurring under anaerobic conditions plays a major role for formation of malodors.

It should briefly be noted that the redox processes shown in Table 1.1 are the so-called respiration processes, i.e., they require the participation of an external electron acceptor. In contrast, organic matter can, in fermentation processes, undergo a balanced series of oxidative and reductive reactions, i.e., organic matter reduced in one step of the process is oxidized in another.

In addition to the redox conditions affected by the design characteristics of the sewer network as outlined in Table 1.1, several other sewer characteristics influence the process conditions. The following examples illustrate the close relations between design and operation characteristics and the resulting conditions for the sewer processes:

Turbulence and flow of wastewater affect reaeration and release of odorous and corrosive substances into the sewer atmosphere.

Ventilation of the sewer system releases odorous and toxic substances into the urban atmosphere, and if—as an example—hydrogen sulfide is released to the surroundings, the corroding effect of this substance is reduced in the network itself.

The hydraulic mean depth of the water phase, i.e., the cross-sectional area of the water volume divided by the width of the water surface, affects the process conditions for reaeration and thereby the potential occurrence of aerobic transformation of organic matter.

The wastewater velocity and shear stress at the sewer walls affect the buildup of sewer biofilms and deposits.

---

## TABLE 1.1

## Electron Acceptors and Corresponding Conditions for Microbial Redox Processes (Respiration Processes) in Sewer Networks

| Process Conditions | External Electron Acceptor | Typical Sewer System Characteristics |
| --- | --- | --- |
| Aerobic | + Oxygen | Partly filled gravity sewer |
| | | Aerated pressure sewer |
| Anoxic | – Oxygen | Pressure sewer with addition of nitrate |
| | + Nitrate or nitrite | |
| Anaerobic | – Oxygen | Pressure sewer |
| | – Nitrate and nitrite | Full-flowing gravity sewer |
| | + Sulfate | Gravity sewer with low slope and deposits |
| | ($+CO_2$) | |

The relation between design and operation characteristics and process conditions is emphasized to draw attention to the fact that knowledge on sewer processes can actively be used to design sewers to observe specific process requirements. What is important in this respect is, of course, a quantification of relevant knowledge on sewer processes addressing relevant aspects of engineering sewer networks in terms of planning, design, and management. Knowledge that can be expressed for prediction and simulation in terms of models is of specific importance. In this way, it should, as an example, not be unrealistic to "design" wastewater for intended purposes. Applying sewer design characteristics that enhance the formation of low molecular organics and thereby improve nutrient removal in wastewater treatment is an example.

Wastewater characteristics play an important role for the nature and the course of the sewer processes. A number of parameters such as temperature and pH and quality characteristics, for example, biodegradability of the organic matter and the amount of active biomass, are crucial for the outcome of the transformations. Microbial transformations—and thereby biochemical transformations—characterize the sewer environment in terms of wastewater quality. On the other hand, the physicochemical characteristics, e.g., diffusion in the biofilm and transfer of substances across the water–air interface, play an important role and are therefore an integral aspect of microbial transformations. The hydraulics and the sewer solids transport processes have also a pronounced impact on the sewer performance. These physical processes, however, are dealt with in books on hydraulics and are, therefore, only included in this book when directly related to the chemical and biological processes. The following section is, however, included to give a brief introduction to the physical processes.

## 1.5   WATER AND MASS TRANSPORT IN SEWERS

The need for collection and conveyance of wastewater from its many sources to treatment and discharge are the fundamental cause why the sewer infrastructure is established. The hydraulics of the water phase with its contents of constituents is therefore traditionally the basis for design and management of sewer networks. In addition, it must be addressed that the sewer processes need to be included to observe safe, efficient, environmentally acceptable, and sustainable transport of the sewage.

Details of sewer hydraulics and corresponding procedures for design of sewers are, as previously mentioned, not a central subject of this book. Knowledge on basic flow characteristics of the water phase and its constituents is, however, needed to fully understand the nature of sewer processes. Such fundamental aspects will therefore briefly be dealt with in this section. It is in this respect important that traditional sewer hydraulics does not include transport processes such as mass transport across the air–water interface and the water–biofilm surface. Neither is the movement of the sewer atmosphere and ventilation considered in traditional hydraulic design of sewers. Such mass transport processes play a central role when dealing with sewer processes and are therefore definitely an objective of this work to deal with.

Several textbooks on different hydraulic subjects are available and relevant for sewer design. The books published by Hauser (1995) and Chanson (2004) are two examples of work that deal with basic and general hydraulics. Comprehensive details on water and mass transport in sewers are found in the work of, for example, Ashley et al. (2004) and ASCE (2007).

The physical transport processes are, in terms of chemical and microbial processes in sewers, generally important as a basis to determine where the pollutants occur. Transport-related aspects concern both soluble and particulate forms of the constituents. An important aspect relate to sediment deposition in sewers that may cause both physical blocking and hydrogen sulfide problems. Scouring of sediments in combined sewers caused by increased flow during extreme rainfall runoff may induce the release of pollutants into the water phase and discharge of these substances into adjacent receiving waters.

## 1.5.1 ADVECTION, DIFFUSION, AND DISPERSION

A fundamental characteristic of flowing waters is related to the mode of movement defined as either being laminar or turbulent—and a transition state between these two types of flow regimes. Laminar flow generally exists at low flow velocities, whereas at increased flow, the movement of the water changes from a calm to a whirling motion. When laminar flow exists, it is possible to determine the velocity field in time and place as a transport of the water and its associated constituents in just one direction and with a constant flow velocity over time. In turbulent flow, however, the velocity profile of the fluid varies, caused by a mutual exchange of the water elements. Contrary to laminar flow, the turbulent flow velocity is therefore determined by two terms: a mean value and a component that varies stochastically. At turbulent flow, the velocity vector therefore changes both magnitude and direction.

In the following sections, a number of fundamental hydraulic phenomena that are important for the movement of soluble as well as particulate pollutants are briefly defined and described:

* Advection
* Molecular diffusion (Fick's first law)
* Dispersion (eddy diffusion)

### 1.5.1.1 Advection

Advection—also named convective transport—describes a mode of transport where a constituent is transported by the net flow of the water phase. Advection thereby describes a situation where no spreading of a constituent takes place. Advection is quantified in terms of a flux. The flux, $J$, of a constituent is defined as a transport, i.e., as an amount of a constituent transported per unit of time and per unit of cross sectional area. It is basically a vector, i.e., it includes both the magnitude of the phenomenon and a direction. The magnitude of the flux—also named the flux rate is:

$$J = Cu \qquad (1.1)$$

where:

$J =$ flux (rate) of a constituent (g m$^{-2}$ s$^{-1}$)
$C =$ constituent concentration (g m$^{-3}$)
$u =$ net flow velocity of water (m s$^{-1}$)

Advection describes a mode of flow where all soluble and particulate constituents are exposed to a uniform velocity, i.e., the flux vector is equal for all constituents in the water phase, soluble species, as well as suspended particles.

### 1.5.1.2 Molecular Diffusion

Diffusion is a disordered movement of the constituents that takes place at the molecular scale. The molecular scale movement of a constituent is caused by a temperature-induced mutual impact of the molecules. This temperature-induced movement is also known as Brownian movement. The net movement of the constituents always (on average) takes place from a high concentration to an area of lower concentration.

At the macroscopic level, molecular diffusion is expressed by Fick's first law of diffusion. The temperature-dependent driving force for the flux of a constituent is the concentration difference per unit of distance, i.e., the concentration gradient:

$$J = -D\frac{dC}{dx} \tag{1.2}$$

where:

$D =$ molecular diffusion coefficient (m$^2$ s$^{-1}$)
$x =$ distance (m)

The magnitude of the diffusion coefficient depends on both the properties of the constituent that is transported and the fluid (water). The temperature has an influence on the magnitude of $D$, but it is, by definition, independent of the mode of transport because it fundamentally describes a phenomenon that takes place at the molecular scale.

Molecular diffusion is a phenomenon that causes a slow movement of a constituent, and the molecular diffusion coefficient for small molecules is typically in the order of $10^{-9}$ m$^2$ s$^{-1}$. The characteristic distance of travel versus time is:

$$x = (2Dt)^{0.5} \tag{1.3}$$

where $t$ denotes time (s).

Molecular diffusion is important when dealing with, e.g., transport in biofilms and transport across interfaces (e.g., the air–water interface). However, within the water phase it is generally exceeded by several orders of magnitude by dispersion, which is described in the following section.

### 1.5.1.3  Dispersion

Dispersion of a constituent describes a phenomenon related to its spreading in a fluid (water). In contrast to molecular diffusion, dispersion is a movement at the macroscopic scale and is caused by flow velocity variations in time and place. Dispersion is therefore related to turbulent conditions and typically exceeds molecular diffusion by several orders of magnitude. In practice, it is always occurring in sewer flows. Dispersion is also known as eddy diffusion.

Dispersion is a random process that by nature is quite different compared with molecular diffusion. However, dispersion on average also concerns transport of constituents from areas of high concentration to areas of low concentration. The empirical description of dispersion therefore follows a similar form as shown by Fick's first law (Equation 1.2):

$$J = -\varepsilon \frac{dC}{dx} \qquad (1.4)$$

where $\varepsilon$ is the dispersion coefficient ($m^2\ s^{-1}$).

Contrary to values for $D$, it is basically not possible to produce tables that give values of $\varepsilon$ for specific compounds. The reason is that the actual flow pattern—more than the type of compound—determines the magnitude of the dispersion coefficient. Determination of $\varepsilon$ therefore basically requires flow measurements followed by a model calibration procedure.

### 1.5.2  Hydraulics of Sewers

The following is a rather brief introduction to the hydraulics of sewers. Further details are found in the work Yen (2001) and WEF (2007).

As described in Section 1.3 and illustrated in Figure 1.4, the flow regime in gravity sewers and pressure sewers are fundamentally different. For both types of sewers, it is crucial that they are designed and constructed to manage varying flow rates of the sewage over time. The pressure sewer must therefore be designed with sufficient pipe and pump capacity to observe this variability. In the gravity sewer, the capacity is determined by the gravity force as the driving force for the water flow and the resistance in terms of friction. The one-dimensional unsteady state flow in a gravity sewer pipe is expressed by the two so-called Saint-Venant equations (Equations 1.5 and 1.6) that are formulated here for open-channel flow (cf. Figure 1.7).

The continuity equation:

$$\frac{dQ}{dx} + b\frac{dy}{dt} = q \qquad (1.5)$$

The momentum equation:

$$\frac{dQ}{dt} + \frac{d}{dt}\left(\frac{Q^2}{A}\right) + gA\frac{dy}{dx} = -gAS_f \qquad (1.6)$$

where:

$Q =$ volumetric flow rate (m³ s⁻¹)
$x =$ distance downstream channel (m)
$b =$ channel width at the water surface (m)
$y = z + h$ water level coordinate (m)
$z =$ vertical bottom coordinate (m)
$h =$ water depth (m)
$t =$ time (s)
$q =$ lateral (constant) inflow of water (m³ m⁻¹ s⁻¹, i.e., m² s⁻¹)
$A =$ cross-sectional area of the water volume (m²)
$g =$ gravitational acceleration (m s²)
$S_f =$ friction slope, approximately the slope of water surface (–)

Equations 1.5 and 1.6 express in principle the mass balance of the water volume between two cross-sections of the channel and Newton's second law (force = mass × acceleration), respectively. The Saint-Venant equations cannot be solved explicitly, and numerical techniques must therefore be used.

The sewer flow is often described under more simplistic conditions compared with the Saint-Venant equations, and several formulas exist for the calculation of flow in both full and partly filled pipes and in open channels. Several semiempirical models that were developed some 100 to 150 years ago are still relevant and widely used. These formulas are typically based on rather few and central parameters describing the physical system, and often expressed in terms of a power function. One of these empirical formulas, the so-called Manning formula (Equation 1.7) is widely applied in engineering practice and is here briefly described for open-channel flow.

The Manning formula is expressed for calculation of the mean flow velocity and is valid under steady state and uniform flow conditions, i.e., at constant flow and

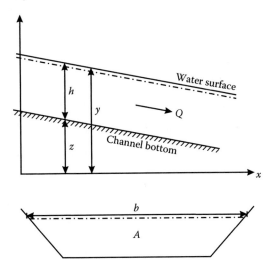

**FIGURE 1.7**  Longitudinal and cross-sectional characteristics of an open channel (cf. Equations 1.5 and 1.6).

water depth. Under these conditions, the Manning formula for open-channel flow can be formulated as follows:

$$u = \frac{Q}{A} = \frac{1}{n} R^{2/3} I^{1/2} \tag{1.7}$$

where:
     $u =$ average flow velocity (m s$^{-1}$)
     $Q =$ flow rate (m$^3$ s$^{-1}$)
     $A =$ cross-sectional area of the channel (m$^2$)
     $n =$ Manning coefficient of roughness (s m$^{-1/3}$)
     $R = A/P =$ hydraulic radius (m)
     $P =$ wetted perimeter of the channel (m)
     $I =$ slope of the water surface or the bottom (m m$^{-1}$)

The Manning coefficient of roughness, $n$, is a parameter that expresses the resistance to the flow. A pipe or channel with rather smooth concrete surfaces has a Manning coefficient of roughness of 0.012–0.014 s m$^{-1/3}$ in contrast to a value about 0.02 s m$^{-1/3}$ for rough surfaces.

The description of the friction in pipe flows is expressed by the so-called Colebrook–White equation (Equation 1.8; Colebrook and White 1937). This implicit equation describes the relationship between the Reynolds number, $Re$, the roughness of the pipe, $\varepsilon$, and the friction factor, $f$.

$$\frac{1}{\sqrt{f}} = -2\log\left(\frac{\varepsilon}{3.7D} + \frac{2.51}{Re\sqrt{f}}\right) \tag{1.8}$$

where:
     $f =$ the Darcy friction factor (–)
     $\varepsilon =$ roughness height (m)
     $D =$ pipe diameter (m)
     $Re =$ Reynolds number $= Du/\upsilon$ (–)
     $u =$ flow velocity (m s$^{-1}$)
     $\upsilon =$ kinematic viscosity of water (m$^2$ s$^{-1}$)

For laminar flow, i.e., $Re < 2000$, the Darcy friction factor $f = 64/Re$.

It should finally be noted that the use of simple empirical formulas for calculation of the flow in pipes and channels are now widely replaced by computer models, often formulated in terms of varying degrees of simplification of the Saint-Venant equations. It is, however, important to recall that the impact of sewer processes generally occur under dry weather conditions, i.e., at approximately steady state and uniform flow. A rather simple description of the hydraulics in sewers is therefore, in this case, typically sufficient.

### 1.5.3 Mass Transport in Sewers

The flow of water in sewer pipes is the carrier of soluble constituents, colloidal matter, and suspended particles. The variability in flow has a crucial impact on the transport of these different forms of constituents. It is particularly important that solids may settle and accumulate in a sewer. Solids deposited can under high flow conditions be eroded, resuspended in the water phase, and transported to a new site, where under less turbulent and relatively quiescent conditions it can settle and be deposited.

It is evident that sewers must be designed to transport sewer solids and that accumulation of deposits be avoided. Deposits of solids may increase the overall roughness of a pipe and reduce the cross section, thereby reducing the hydraulic capacity of the sewer network, potentially leading to flooding. At higher flows, large quantities of deposits may be eroded and resuspended resulting in a changed quality of the flow. In combined sewers, in particular, it may result in increased pollutant loads into receiving waters from overflow structures. Furthermore, the pollutants associated with the deposits may result in deleterious effects, e.g., in terms of hydrogen sulfide problems. Self-cleansing sewers must therefore be designed. Otherwise, flushing of a sewer reach is needed. Procedures for design and operation of self-cleansing sewers are therefore central for their overall function (Ashley et al. 2004). Depending on the sewer type and local practice, selection of a minimum velocity (e.g., 0.6–0.9 m s$^{-1}$) of the flow or a minimum shear stress (e.g., 1–2 N m$^{-2}$) exerted on the sewer walls may be selected to observe self-cleansing conditions. Such conditions are typically required with a given frequency of occurrence. A sufficient pipe slope is therefore central for a well-functioning gravity sewer.

As already dealt with in Sections 1.5.1.1 and 1.5.1.3, advection (convective transport) caused by the mean flow and dispersion due to turbulent diffusion are the two major processes that determine transport of a constituent at the macroscopic scale in a well-designed sewer. The mass balance of a constituent uniformly mixed over the cross section of the channel is, according to Equations 1.1 and 1.4 in time ($t$) and place ($x$), formulated as follows:

$$\frac{d(AC_x)}{dt} + \frac{d(QC_x)}{dx} = \frac{d}{dx}\left(\varepsilon A \frac{dC_x}{dx}\right) + qC_q - s \tag{1.9}$$

where:

$C_x$ = constituent concentration at distance $x$ (g m$^{-3}$)
$\varepsilon$ = dispersion coefficient (m$^2$ s$^{-1}$)
$C_q$ = constituent concentration in the lateral inflow (g m$^{-3}$)
$s$ = sink or process rate term for the constituent per unit length (g m$^{-1}$ s$^{-1}$)

As an example, and if a first-order reaction in the water phase proceeds, the sink term can be formulated as follows:

$$s = AkC_x \tag{1.10}$$

where $k$ is the first-order reaction rate constant in the water phase (s$^{-1}$).

Because, Equation 1.9 cannot be solved explicitly, numerical techniques must therefore be used.

## 1.6  SEWER PROCESS APPROACH

A sewer network and any related treatment plant have traditionally been considered separate units for the management of wastewater. Consequently, they have been separately designed and operated, typically by engineers and other practitioners with different professional background. The link between these two units has typically just been expressed in terms of the wastewater flow from the sewer as an input parameter for the treatment plant. Two different and separate functions have been dealt with: the sewer system must collect and convey the wastewater to the treatment plant, and the treatment plant must reduce the load of pollutants into the receiving water according to the quality standards set. Consequently, sewers are often just considered simple input systems for the quantity of the water flows at the boundaries where they are connected with either wastewater treatment plants or overflow structures discharging untreated wastewater into adjacent watercourses during rainfall. This traditional approach to sewer performance is not appropriate and needs considerable improvement.

Against this background, a sustainable and integrated dimension of wastewater management is needed. The safe and efficient collection and conveyance of wastewater to treatment and disposal are of course still an essential concern. The consideration of sewer processes as an element in the design and operation of sewers will give a new dimension to the overall objective of wastewater management and contribute to improved sustainability. Therefore, the different technical elements of the sewer system must be considered holistically by taking into account the following aspects:

- The type of sources of the wastewater
- The sewer as a physical, chemical, and biological reactor for the wastewater compounds being transported, and their potential deteriorating impacts onto the system itself
- The interaction between the sewer network and the treatment plant
- The consequences for the local receiving waters
- The impact of the sewer air on the surrounding urban environment in terms of, e.g., malodors

The fact that the sewer is a reactor for chemical and biological processes has only played a limited role when considering the function of a sewer. It is remarkable that the quality of wastewater is typically defined (and monitored) at the point of inlet to a treatment plant or at a point of discharge, totally omitting why and how this quality has been created. The fact that wastewater is subject to transformations in a sewer network—often supported by a relatively long hydraulic retention time—has generally not been considered and included in the entire management of wastewater. A similar way of neglecting microbial and chemical processes in the other urban subsystems—the treatment plant and the local receiving waters—where wastewater also appears, would be quite unthinkable.

Heterotrophic bacterial processes dominate the transformations of wastewater components in the sewer. There are similarities with the corresponding processes in biological treatment plants. It is, however, from the very beginning, important to emphasize that transformations of wastewater under sewer conditions and in activated sludge, biofilm systems or anaerobic digesters develop differently. The processes in sewers, treatment plants, and receiving waters must be dealt with as an entity. However, they must also be considered with their own specific characteristics and conditions.

The conveyance aspects of a sewer network in terms of capacity, pollution control, structural integrity, and cost are beyond dispute. However, a new approach is needed to include the chemical and biological processes in a sewer more actively in the design and operation procedures. The first step is to give those sanitary and environmental engineers and technicians dealing with sewers a comprehensive understanding that covers the relevant scientific and engineering approaches of sewer processes.

The first edition of this book was published in 2002. Since then, the basic theoretical knowledge on sewer processes and results in terms of implementation of this knowledge in practice have developed. In particular, it is important that knowledge has developed to a level where the sewer processes now are integrated in computer models with different objectives, e.g., in terms of specific network design and integrated sewer management, and planning aspects at catchment scale. In spite of this, it must be realized that because the chemical and biological process aspects, until now, have played a limited role in the engineering of collection systems, there are still "dark spots" where further research and development are needed. This book is therefore not able to give complete answers to all relevant questions, but it hopefully gives the readers updated and valuable knowledge on the sewer as a chemical and biological reactor.

The objective of the book is clear: to add a chemical and microbial dimension to design and management of sewer networks with an overall purpose of improved sustainability for the system itself and the surroundings it impacts.

## REFERENCES

ASCE (2007), *Gravity Sanitary Sewers—Design and Construction*, ASCE (American Society of Civil Engineers) *Manuals and Reports on Engineering Practice No. 60* and WEF (Water Environment Federation) *Manual of Practice No. FD-5*, ASCE and WEF, Reston, VA, USA, p. 422.

Ashley, R.M., J.-L. Bertrand-Krajewski, T. Hvitved-Jacobsen, and M. Verbanck (eds.) (2004), *Solids in Sewers—Characteristics, Effects and Control of Sewer Solids and Associated Pollutants*, Scientific and Technical Report No. 14. IWA Publishing, London, p. 340.

Bertrand-Krajewski, J.-L. (2005), Sewer systems in the 19th century: diffusion of ideas and techniques, circulation of engineers, 4th IWHA (International Water History Association) Conference, Paris, France, December 1–4, 2005, p. 14.

Bowlus, F.D. and A.P. Banta (1932), Control of anaerobic decomposition in sewage transportation, *Water Works Sewerage*, 79(11), 369.

BS EN 752-1 (1996), Drain and sewer systems outside buildings—Part 1. Generalities and definitions.

Burian, S.J., S.J. Nix, S.R. Durrans, R.E. Pitt, C.-Y. Fan, and R. Field (1999), The historical development of wet-weather flow management, USEPA/600/JA-99/275, p. 23.

Chanson, H. (2004), *Hydraulics of Open Channel Flow*, Elsevier, Oxford, UK, p. 650.

Colebrook, C.F. and C.M. White (1937), Experiments with fluid friction in roughened pipes, *Proc. R. Soc. Lond., Ser. A Mathem. Phys. Sci.*, 161(906), 367–381.

Hauser, B. (1995), *Practical Hydraulics Handbook*, CRC Press, Boca Raton, FL, p. 368.

Hitti, M. (2007), Greatest medical advance: sanitation, *Br. Med. J.*, January 18, 2007.

Hvitved-Jacobsen, T., J. Vollertsen, and A.H. Nielsen (2010), *Urban and Highway Stormwater Pollution: Concepts and Engineering*, CRC Press, Boca Raton, FL, p. 347.

Manufacturer and Builder (1889), The separate sewer system, *Manuf. Builder*, 21(9), 210.

Olmsted, F.H. and H. Hamlin (1900), Converting portions of the Los Angeles outfall sewer into a septic tank, *Eng. News*, 44(19), 317–318.

Parker, C.D. (1945a), The corrosion of concrete 1. The isolation of a species of bacterium associated with the corrosion of concrete exposed to atmospheres containing hydrogen sulphides, *Aust. J. Expt. Biol. Med. Sci.*, 23, 81–90.

Parker, C.D. (1945b), The corrosion of concrete 2. The function of *Thiobacillus concretivorus* (nov. spec.) in the corrosion of concrete exposed to atmospheres containing hydrogen sulphide, *Aust. J. Expt. Biol. Med. Sci.*, 23, 91–98.

Pomeroy, R. (1936), The determination of sulfides in sewage, *J. Sewage Works*, 8(4), 572.

WEF (2007), Gravity sanitary sewer design and construction, WEF (Water Environment Federation) Manual of Practice No. FD-5, Reston, VA, USA, p. 422.

Yen, B.C. (2001), Hydraulics of sewer systems, in: L.W. Mays (ed.), *Stormwater Collection Systems Design Handbook*, McGraw-Hill, New York, 6.1–6.113.

# 2 In-Sewer Chemical and Physicochemical Processes

In the context of this book on sewer processes, it is important to understand that a sewer consists of five main parts, also referred to as phases:

1. A water phase including suspended particles
2. Biofilms attached to the submerged solid parts of the sewer
3. Sewer sediments also referred to as deposits
4. A sewer atmosphere or headspace
5. Sewer walls exposed to the sewer atmosphere including a moisture layer and adhered substances

The chemical and biological processes proceed within these phases, at their internal surfaces such as particle surfaces of the suspended water phase and at the sewer walls. Furthermore, mass transfer, i.e., exchange of substances, takes place between the different phases, e.g., transfer of volatile substances from the water phase into the sewer atmosphere.

Because of the system-related process conditions, it is evident that a sewer becomes complex in terms of the processes that proceed. However, basic and solid understanding—and corresponding application—of chemical and biological knowledge is the key for a conceptual description of process-related phenomena leading to generally valid mathematical expressions for prediction. In this respect, it is crucial that observations of specific phenomena typically cannot be transferred to a new case without taking into account dissimilarities in system and process conditions.

The chemical and microbial processes are, in the context of engineering, the sewer system subjects integrated in the entire book. However, there are fundamental aspects of understanding that are convenient to treat specifically. The objective of this chapter is, in this respect, to highlight the fundamental chemical and physicochemical aspects of general importance for sewer systems and in-sewer processes. The contents of this chapter are selected with this issue in mind, and our focus is on chemical reactions (redox reactions), chemical equilibrium, the stoichiometry of the reactions (i.e., mass balance aspects), and the kinetics concerning the reaction rates of the processes. The thermodynamics of the chemical reactions—i.e., the energy-related changes that take place during the transformations—takes a central role as a determining factor for the course of the processes. In brief, the thermodynamics is a description of the driving forces for the processes.

Although microbial processes play a dominating role when dealing with sewer processes, it is crucial to understand that they are, by nature, chemical transformations that proceed in biological systems. Often these chemical processes, i.e., changes in chemical substances initiated by living organisms, are referred to as biochemical processes. To a high degree, it is the chemical process-related characteristics that establish the basis for the description and prediction of sewer processes. Therefore, there are good reasons to deal with the basic aspects of the chemical processes and apply this knowledge on the in-sewer microbial systems.

This chapter is not a comprehensive discourse on general chemistry but written to support the understanding of the fundamentals of chemistry in the context of in-sewer processes. Books with subjects in general and applied chemistry and chemical engineering are legion and not generally referred to here. However, details of chemistry related to aquatic environmental systems, and therefore also relevant for sewer systems, are given by Snoeyink and Jenkins (1980), Stumm and Morgan (1996), and Speece (2008).

## 2.1 REDOX REACTIONS

Redox reaction is an abbreviated name for reduction–oxidation reaction. It is a chemical reaction or process where one of the reacting components is reduced and another reacting component is correspondingly oxidized. By this process, the latter component donates electrons to the component that is reduced. By definition, electrons are transferred from an electron donor—the component that is oxidized—to an electron acceptor that is hereby reduced. Oxidation and reduction are thereby words for processes identical with "donating electrons" and "receiving electrons," respectively. These aspects are the basics behind the "headings" of this section. The major part of chemical and biochemical processes in sewers is redox reactions.

### 2.1.1 CHEMICAL EQUILIBRIUM AND POTENTIAL FOR REACTION

A key aspect of process engineering involves understanding the basics of equilibrium for a chemical reaction and its potential to proceed. In the following discussion, these fundamentals will be outlined. Further details can be found in texts referred to in the introduction portion of this chapter.

Chemical equilibrium describes a state in which no net changes in the components involved takes place. A chemical equilibrium is dynamic, meaning that the formation of products ($P$) for a reversible reaction proceeds with the same rate as the backward reaction in terms of the formation of reactants ($R$). The following exemplifies the chemical equilibrium of a simple, reversible chemical reaction that takes place within a homogenous system, i.e., a one-phase system such as a water phase:

$$r_1R_1 + r_2R_2 \Leftrightarrow p_1P_1 + p_2P_2 \tag{2.1}$$

where:

$R_i$ = reactants, i.e., reacting components, here $i = 1$ and $i = 2$
$P_i$ = products, here $i = 1$ and $i = 2$
$r_i$ and $p_i$ = stoichiometric coefficients, i.e., the ratio between $R$ and $P$ components
    (–), here $i = 1$ and $i = 2$

Chemical equilibrium in a homogenous system corresponding to Equation 2.1 is at constant temperature expressed by the following equation:

$$K = \frac{[P_1]^{p_1}[P_2]^{p_2}}{[R_1]^{r_1}[R_2]^{r_2}} \tag{2.2}$$

where:

$K$ = equilibrium constant (unit depends on the values of the stoichiometric coefficients)

$[R_i]$ and $[P_i]$ = concentrations, i.e., concentrations at equilibrium, of reactants and products, respectively (mol L$^{-1}$)

It should be noted that equilibrium as expressed by Equation 2.2 is based on the concentrations of the constituents. The equation is theoretically only valid in very dilute aqueous systems, where the interactions between the constituents can be neglected. In more concentrated solutions, it may be needed to account for these interactions by introducing the activity coefficient defined as the ratio between the "active concentration," i.e., the activity, $a$, of the substance, and its concentration, $c$, determined using analytical methods.

Equation 2.2 is valid for a wide range of equilibrium descriptions in both natural and technical systems. In the case of acid–base reactions, $K$ is referred to as the acid and base dissociation constant, $K_a$ and $K_b$, respectively, and often in tables given as $pK_a = -\log K_a$ and $pK_b = -\log K_b$. Table 2.1 shows selected equilibrium constants for chemical equilibria relevant for sewer systems.

---

**TABLE 2.1**

**Selected Equilibrium Constants at 25°C for Chemical Equilibria Relevant for Sewer Systems**

| Chemical Equilibrium Acid Form ⇔ Base Form + H$^+$ | $pK_a = -\log K_a$ |
| --- | --- |
| Sulfuric acid: $H_2SO_4 \Leftrightarrow HSO_4^- + H^+$ | −3 |
| Bisulfate: $HSO_4^- \Leftrightarrow SO_4^{2-} + H^+$ | 1.9 |
| Ferric ion: $Fe(H_2O)_6^{3+} \Leftrightarrow Fe(H_2O)_5(OH)^{2+} + H^+$ | 2.2 |
| Acetic acid: $CH_3COOH \Leftrightarrow CH_3COO^- + H^+$ | 4.7 |
| Carbon dioxide: $H_2CO_3 \Leftrightarrow HCO_3^- + H^+$ | 6.3 |
| Hydrogen sulfide: $H_2S \Leftrightarrow HS^- + H^+$ | 7.1 |
| Ammonium ion: $NH_4^+ \Leftrightarrow NH_3 + H^+$ | 9.3 |
| Bicarbonate ion: $HCO_3^- \Leftrightarrow CO_3^{2-} + H^+$ | 10.3 |
| Water: $H_2O^a \Leftrightarrow OH^- + H^+$ | 14 |

a The activity of water is defined as $a_{H_2O} = 1$, although the molar concentration of diluted solutions is approximately $1000/18 = 55.55$.

The equilibria shown in Table 2.1 are all pH-dependent. For pH values in wastewater of sewers typically varying between 7.0 and 8.0, the acid form and the base form for both carbon dioxide and hydrogen sulfide may exist. At such typical pH values, the acid form will relatively dominate if the $pK_a$ value for the equilibrium is high, whereas the base form dominates at low values of $pK_a$. It must furthermore be noted that the occurrence of other substances may influence what is actually found. As an example, sulfide ($HS^-$) may react with heavy metals such as $Fe^{2+}$ and form precipitates of sulfide, and bicarbonate $\left(HCO_3^-\right)$ may react with $Ca^{2+}$ and form precipitates of limestone.

Chemical equilibrium as expressed by Equation 2.2 is a question of establishing thermodynamic stability in terms of energy changes for the substances involved. It is the potential work that a system—here, a chemical reaction—can do onto its surroundings that determines in which direction the reaction proceeds or if equilibrium is established. The numerical value of this work is thermodynamically given by Gibbs' free energy, $G$ (cf. Section 2.1.3). An illustration of the changes in Gibbs' free energy that can take place for a chemical or biochemical reaction is outlined in Figure 2.1.

As shown in Figure 2.1, the minimum value of $G$, $G_{eq}$, corresponds to a composition of $R$ and $P$ that results in chemical equilibrium. All spontaneous reactions will proceed in a direction of reducing $G$, i.e., $\Delta G$ is negative for spontaneous reactions. The conditions for the reaction $R \rightarrow P$ is therefore thermodynamically expressed as follows:

$\Delta G < 0$

If $\Delta G$ for the reaction $R \rightarrow P$ is negative, the reaction has a potential to proceed. The potential for work, $G$, of the process is reduced when the starting point in

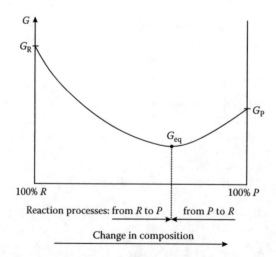

**FIGURE 2.1** Illustration of the changes in Gibbs' free energy, $G$, for a chemical or biochemical process for reacting components ($R$) to form products ($P$) or the opposite: for $P$ to form $R$.

Figure 2.1 is to the left of the point corresponding to $G_{eq}$. The process is defined as spontaneous.

**$\Delta G = 0$**
Equilibrium exists, i.e., no net transformation of the components will take place.

**$\Delta G > 0$**
The reaction $R \rightarrow P$ has no potential to proceed when the mixture of $R$ and $P$ corresponds to a composition to the right of the point for $G_{eq}$. The reverse process $P \rightarrow R$ is spontaneous.

The thermodynamic conditions as described for a chemical or biochemical reaction and equilibrium are generally valid, although here they are exemplified for a simple homogeneous reaction. The conditions therefore also apply for heterogenous reactions between substances that occur in multiphase systems (e.g., reacting across interfaces).

Finally, it must be stressed that it is important to distinguish between the potential for a chemical or biochemical reaction to proceed and the reality: does it really seem to proceed under the actual conditions? The latter question concerns—if the thermodynamic requirements are fulfilled—the process kinetics in terms of the magnitude of the reaction rate. The consequence is that a thermodynamically unstable compound may easily exist if the reaction rate for its transformation is low. The kinetics for chemical and biochemical reactions is the subject of Section 2.2.

### 2.1.2 Redox Reactions in Sewers

It is the energy conversions in terms of the thermodynamic properties of the reacting and produced chemical substances that fundamentally determine if a chemical reaction proceeds or not (cf. Sections 2.1.1 and 2.1.3). Regardless of the fact that a chemical reaction proceeds in an abiotic environment or a biotic environment, i.e., initiated by living organisms and often named a biochemical reaction, the basic thermodynamic properties are the same. What might be different is—as an example—the enzyme initiated rate by which the transformation takes place. Although the living organisms (microorganisms) play a central role in the sewer environment for which processes in practice will proceed, it is equally important to notice that the fundamental chemical reaction-related properties are not affected by their presence. These basic thermodynamic properties of chemical (biochemical) reactions do not conflict with the fact that the biological system is crucial for sewer processes. In particular, the presence of the heterotrophic biomass in the wastewater phase, in biofilms, and in sediments is central for the course of the biochemical processes.

A central aspect of biochemical processes is the fact that the heterotrophic biomass uses organic matter for two fundamental purposes:

1. Organic matter (the substrate) is the carbon source for microbial growth and therefore needed for the formation of new cells.

2. Organic matter is the energy source needed for growth-related processes and for maintaining life of the existing biomass.

The anabolic processes transform substances resulting in the production of new biomass, whereas the catabolic processes provide energy needed for production of new cell biomass as well as maintenance of the basic functions of the living biomass (cf. Figure 2.2).

The energy accumulated in the organic matter is made available for the microorganisms by catabolism, i.e., a "degradation" process performed by oxidation of the organic matter (substrate). The organic matter is, thereby, the electron donor. The corresponding reduction of an external electron acceptor takes place when dissolved oxygen (DO) (aerobic process), nitrate (anoxic process), or sulfate (anaerobic process) is present. Moreover, metals can act as external electron acceptors, e.g., iron ($Fe^{3+}$) and manganese ($Mn^{4+}$).

These energy-producing reactions are termed respiration processes. They require the presence of an external compound that can serve as the terminal electron acceptor of the electron transport chain. However, under anaerobic conditions, fermentation processes and methanogenesis that do not require the participation of an external electron acceptor can also proceed. In this case, the organic substrate undergoes a balanced series of oxidative and reductive reactions, i.e., organic matter reduced in one step of the process is oxidized in another.

The basic understanding of the microbial processes (biochemical processes) in wastewater is based on the fact that substrate utilization for growth of biomass takes place parallel to its removal for energy purposes by an electron acceptor. Figure 2.3 shows the general concept of a biochemical redox process and examples where an external electron acceptor is involved.

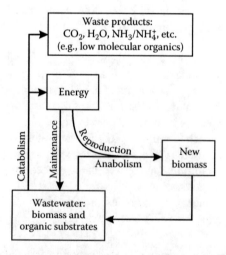

**FIGURE 2.2**  Main pathways of organic matter in wastewater of a sewer system.

General concept:

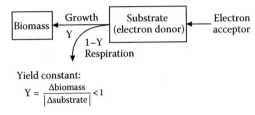

Yield constant:

$$Y = \frac{\Delta biomass}{|\Delta substrate|} < 1$$

Examples:

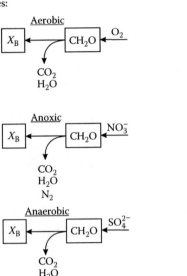

**FIGURE 2.3** Concept of a redox reaction in terms of wastewater transformations in sewer systems showing microbial biomass, $X_B$, and organic substrate, $CH_2O$, relations. The examples show aerobic, anoxic, and anaerobic transformations involving an external electron acceptor.

### 2.1.3 REDOX REACTIONS AND THERMODYNAMICS

#### 2.1.3.1 Nature of Redox Reactions

The microbial catabolic processes, which proceed in wastewater, provide the biomass with energy (cf. Figures 2.2 and 2.3). These processes include two major steps:

1. Oxidation of organic matter as the electron donor
2. Reduction of an electron acceptor

A redox process with organic matter consists basically of transfer of electrons from the electron donor (the organic matter) to the relevant electron acceptor, i.e., from

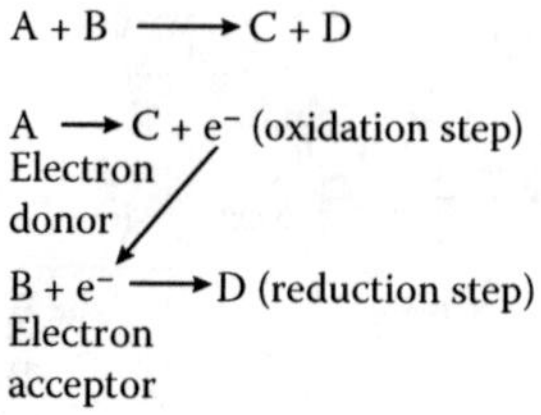

$$A + B \longrightarrow C + D$$

$A \longrightarrow C + e^-$ (oxidation step)
Electron
donor

$B + e^- \longrightarrow D$ (reduction step)
Electron
acceptor

**FIGURE 2.4**  Oxidation and reduction steps and corresponding electron transfer of a redox process.

the oxidation step to the reduction step. Figure 2.4 outlines the basic concept of a redox process involving the chemical components A, B, C, and D. The oxidation of component A (electron donor) to component C produces an electron. This electron is transferred to the reduction step, thereby resulting in transformation of component B (electron acceptor) to component D.

The examples referred to—and as also shown in Figure 2.3—concern organic matter as electron donors and electron acceptors relevant for wastewater transformations in sewers. It is, however, crucial to stress that Figure 2.4 is generally relevant and that the concept is also valid for redox reactions involving transformations of inorganic substances.

### 2.1.3.2   Redox Reactions and Thermodynamics

The energy transformations related to redox processes in wastewater follow the general rules of thermodynamics (Stumm and Morgan 1996; Atkins and de Paula 2002). A central thermodynamic energy function in this respect is the Gibbs' free energy, $G$, that defines the state and the potential for a change in state of a redox reaction. $G$ is a measure of the driving force (the work-producing potential) of a redox process. For naturally occurring processes, the change in Gibbs' free energy ($\Delta G$) is therefore negative, i.e., the redox process loses work potential when it proceeds (cf. Figure 2.1). At constant temperature and pressure, $\Delta G$ equals the maximum work that can be produced by the redox process. $\Delta G$ is therefore also a measure of the tendency for a redox process to proceed.

The change in Gibbs' free energy is related to two other thermodynamic state functions: enthalpy and entropy. Without going into details, the change in enthalpy of a redox reaction is equal to the heat transferred to the process at a constant pressure. In brief, the change in entropy is a consequence of the change in composition and is, for reversible reactions, related to the heat transferred to the process and the absolute temperature. The relation between the three thermodynamic state functions, $\Delta G$, $\Delta H$, and $\Delta S$, is:

$$\Delta G = \Delta H - T\Delta S \qquad (2.3)$$

where:
  $G$ = Gibbs' free energy (kJ mol$^{-1}$)
  $H$ = enthalpy (kJ mol$^{-1}$)

$T$ = temperature (K)

$S$ = entropy (kJ mol$^{-1}$)

The majority of redox processes are exothermic, i.e., heat producing and consequently $\Delta H < 0$, in their natural direction of reaction. Often, they are so highly exothermic that the term $T\Delta S$ has little influence on the magnitude of $\Delta G$. In the early days of thermodynamics, it was therefore wrongly believed that all exothermic processes would proceed spontaneously and that $\Delta H$ was a measure of this tendency.

As shown in Figure 2.4, the characteristic feature of a redox reaction is the transfer of electrons from one atom to another. Therefore, the Gibbs' free energy equals the work potential that is lost by transfer of electrons from the oxidation step to the reduction step. The difference in electron potential between these two half-reactions is therefore directly related to $\Delta G$ for the redox reaction:

$$\Delta G° = -nF\Delta E° = -nF(E°_{red} - E°_{ox}) \tag{2.4}$$

where:

$G°$ = Gibbs' free energy at standard conditions, i.e., 25°C, pH 7, 1 atm and 1 molar concentration of relevant components (kJ mol$^{-1}$)

$n$ = number of electrons transferred according to the reaction scheme (–)

$F$ = Faraday's constant equal to 96.48 (kJ mol$^{-1}$ V$^{-1}$)

$\Delta E°$ = redox potential at standard conditions of electron acceptor, $E°_{red}$, minus standard redox potential of electron donor, $E°_{ox}$ (V)

Standard redox potentials ($E°$) for a large number of half-reactions relevant for chemical and biochemical reactions are shown in CRC (2009). Selected standard redox potentials for half-reactions of particular importance for biochemical processes are in order of magnitude in an "electron tower" version shown in Figure 2.5.

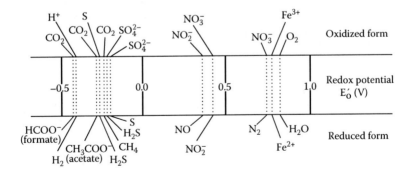

**FIGURE 2.5** Order of magnitude of standard potentials for selected redox pairs of importance for microbial processes, partly after Madigan et al. (1997). The redox potentials are given at standard conditions, i.e., 25°C, pH 7, 1 atm, and 1 molar concentration of relevant compounds. The "electron tower" is shown horizontally.

The following important results can be drawn from Figure 2.5 in accordance with Boon (1995):

- Aerobic conditions correspond to a redox potential of about 800 mV.
- Anoxic conditions correspond to a redox potential of about 400 mV.
- Anaerobic conditions correspond to a redox potential less than about −200 mV.

For simplicity reasons, only the redox pairs—and not the half-reactions—are shown in Figure 2.5. As an example, the equilibrium half-reaction for oxygen between the oxidized form ($O_2$) and the reduced form ($H_2O$) is:

$$1/4 O_2 + H^+ + e^- \Leftrightarrow 1/2 H_2O \tag{2.5}$$

The so-called Nernst's equation can be used to account for external conditions different from standard values. The formulation of this equation will not be covered in this book, but interested readers are referred to texts on electrochemistry or books such as those written by Stumm and Morgan (1996), Atkins and de Paula (2002), and Hvitved-Jacobsen et al. (2010).

Referring to Figures 2.4 and 2.5, it should be noted that two half-reactions are combined to form a redox reaction. When combining such two half-reactions, reduction and oxidation will take place at the highest redox potential and at the lowest redox potential, respectively, with a corresponding transfer of electrons. Referring to Equation 2.4, it should furthermore be noted that the difference in redox potentials of the two half-reactions is related to the "tendency" for the redox reaction to proceed, i.e., the energy-related aspect, and not the kinetics in terms of a reaction rate. The thermodynamic details of a redox reaction are discussed in Example 2.1.

### Example 2.1: Oxidation of Organic Matter

This example concerns the oxidation (degradation) of organic matter exemplified by acetate. For simplicity reasons, this example will not include external conditions resulting in redox potentials different from standard values, and redox potentials as shown in Figure 2.4 are therefore directly referred to.

Oxidation of acetate (the electron donor) is possible under aerobic, anoxic, and anaerobic conditions, i.e., with molecular oxygen ($O_2$), nitrate ($NO_3^-$), and sulfate ($SO_4^{2-}$), respectively, as electron acceptors. Referring to Equation 2.4 and Figure 2.5, it is readily seen that as long as $O_2$ is available and dominating, a relatively high redox potential also results in a corresponding relatively high value of $\Delta G$ compared with anoxic and anaerobic transformations. It is only after $O_2$ is depleted from the system that nitrate and sulfate will take over as electron acceptors for the oxidation of acetate. Figure 2.5 also shows that $Fe^{3+}/Fe^{2+}$ with a relatively high redox potential can act as an electron acceptor system in microbial reactions for oxidation of acetate.

Traditionally, aerobic, anoxic, and anaerobic conditions are the terms applied when dealing with processes in wastewater systems (cf. Table 1.1). However, it must be noted that when using redox potentials, further details and quantitatively expressed information can be included to determine which chemical and bio-chemical processes might proceed.

The redox potential and corresponding energy release for oxidation (deg-radation) of organic matter will be exemplified by both aerobic and anaerobic oxidation of acetate. At aerobic conditions, the redox pair $O_2/H_2O$ is therefore combined with the redox pair $CO_2/CH_3COO^-$ (cf. Figure 2.4). Corresponding to the equilibrium half-reaction for $O_2/H_2O$ as shown in Equation 2.5, the equilib-rium half-reaction for $CO_2/CH_3COO^-$ can be established. The details of how to arrange such half-reactions will be dealt with in Section 2.1.4.

$$2CO_2 + 5H_2O + 8e^- \Leftrightarrow CH_3COO^- + 7OH^-$$

The oxidation of carbon in acetate to carbon in $CO_2$ will take place at the high-est redox potential, and an equivalent reduction of oxygen in $O_2$ to oxygen in $H_2O$ correspondingly proceeds at the highest redox potential. For the entire reaction, $\Delta E°$ can be directly calculated based on the redox potentials for each of the two half-reactions:

$$\Delta E° = E_{red}° - E_{ox}° = 0.82 - (-0.29) = 1.11\,V$$

The energy transformation in terms of Gibbs' free energy for the oxidation of ace-tate can be calculated as follows. From the half-reaction for $CO_2/CH_3COO^-$, it is seen that 8 mol of electrons per mole of acetate are transferred from the oxidation step to the reduction step. According to Equation 2.4, the energy transformation per mole of acetate at standard conditions is:

$$\Delta G° = -nF\Delta E° = -nF(E_{red}° - E_{ox}°) = -8 \times 96.48 \times 1.11 = -856.7\,kJ\,mol^{-1}$$

The negative value of $\Delta G°$ shows that the oxidation of acetate proceeds sponta-neously and that the reaction produces "work" by correspondingly reducing the work potential of the system.

In principle, the magnitude of $\Delta G°$ for aerobic degradation of organic matter (here, exemplified by acetate) can be compared with a corresponding anaerobic transformation applying the redox pairs $SO_4^{2-}/H_2S$ as electron acceptor:

$$SO_4^{2-} + 10H^+ + 8e^- \Leftrightarrow H_2S + 4H_2O$$

The redox potential for this half-reaction is −0.22 V. Compared with the cor-responding value for $O_2/H_2O$, it is clear that the difference in redox potential between $SO_4^{2-}/H_2S$ and $CO_2/CH_3COO^-$ is relatively low. The $\Delta E°$ and $\Delta G°$ for the anaerobic redox reaction are:

$$\Delta E° = -0.22 - (-0.29) = 0.07\,V$$

$$\Delta G° = -nF(E_{red}° - E_{ox}°) = -8 \times 96.48 \times 0.07 = -54.0\,kJ\,mol^{-1}$$

A direct comparison with the amount of Gibbs' energy released under aerobic conditions demonstrates that the energy efficiency for anaerobic degradation of organic matter (here, exemplified by acetate) is rather low. As will be further dealt with in Chapter 3, the degradation pattern under anaerobic conditions is far more complex, as acetate is often an end-product for degradation of organic substrates and therefore accumulates under such conditions.

Finally, it is also important to note that the calculations in this example for simplicity reasons have been performed at standard conditions. The real situation in a sewer is in terms of concentration levels for the relevant substances and temperature difference.

Basically, Table 1.1 and Figure 2.5 express the same process-related characteristics in terms of which processes are relevant and might proceed. However, the information in Figure 2.5 is energy-related and, therefore, also expresses the fundamental of the processes: to proceed if Gibbs' free energy is released. Table 1.1 more pragmatically refers directly to the availability of the relevant electron acceptors. Microbiologically, and as shown in Example 2.1, aerobic processes are preferred because they correspond to the highest amount of energy release.

### 2.1.3.3   Redox Reactions and Phase Changes

In continuation of what is discussed in this section concerning the thermodynamics of redox processes, it is important that the following three parameters—in addition to temperature and pressure—are main external factors for characterizing process-related aspects. In particular, these three factors determine which chemical and biochemical processes are thermodynamically possible and which phases of the chemical substances are stable or unstable:

1. The acid–base characteristics, i.e., $pH = -\log(C_{H^+})$
2. The concentration, $C$, or $pC = -\log C$ (for soluble species), of a substance
3. The redox potential, $E$, or $p\varepsilon = -\log E$

The importance of these three factors on process-related aspects are often illustrated in diagrams, typically as $C$–pH ($pC$–pH) diagrams and $E$–pH ($p\varepsilon$–pH) diagrams. Correspondingly, $E$–$C$ ($p\varepsilon$–$pC$) diagrams can also be constructed; however, they are less used. In the following, the first two types of diagrams will be illustrated with sulfur as a relevant example for sewer systems.

The $E$–pH ($p\varepsilon$–pH) diagrams are named Pourbaix diagrams, which map out the phases of chemical compounds under equilibrium conditions (Pourbaix 1974). Figure 2.6 shows an example of such a diagram for the binary sulfur and oxygen system in water at 1 atm, 25°C, and with the sum of the concentrations of all soluble sulfur-containing compounds equal to 0.1 mM (0.1 mmol $L^{-1}$ or 3.2 g S $m^{-3}$). Sulfur is selected as an example because of its position as a central and important process-related element in sewers.

The lines in Figure 2.6 correspond to equilibrium between the phases of the sulfur compounds that exist at each side of the line. As an example, the equilibrium at pH < 2 between $H_2S(aq)$ and elemental sulfur, $S(s)$, is:

$$S(s) + 2H^+ + 2e^- \Leftrightarrow H_2S(aq) \tag{2.6}$$

Equation 2.6 shows that both pH and the redox potential, $p\varepsilon$, have an influence on the equilibrium. In the corresponding $E$–pH diagram (Figure 2.6), this equilibrium is

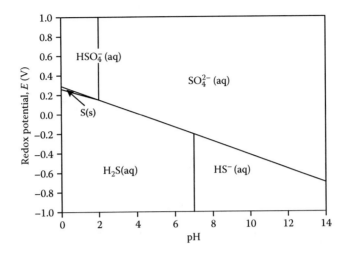

**FIGURE 2.6** Pourbaix diagram ($E$–pH diagram) for the binary sulfur and oxygen system at 1 atm, 25°C, and a total concentration of soluble sulfur compounds equal to 0.1 mM (0.1 mmol L$^{-1}$ or 3.2 gS m$^{-3}$).

therefore shown as a sloping line. In contrast, the equilibrium between $H_2S$ and $HS^-$ is only dependent on $H^+$ and will therefore in the diagram appear as a vertical line.

As an example, when moving from this line in the Pourbaix diagram in the direction of lower redox potential, the balance in Equation 2.6 between the two compounds is shifted to the right side, i.e., both phases still exist, although $H_2S(aq)$ becomes more and more dominating. This phenomenon of shift between phases is also illustrated in a $C$–pH diagram (cf. Figure 2.7). This figure shows, as an example, the equilibrium between the aqueous phases of $H_2S$ and $HS^-$.

Referring to Figure 2.6, where the equilibrium line for $H_2S/HS^-$ is shown at pH 7, it is readily seen in Figure 2.7 that both sulfide species also exist at other pH values, although in different relative quantities.

The equilibrium distribution between the different sulfide species can be quantified in mathematical terms. The pH-dependent equilibrium between $H_2S$ and $HS^-$ as shown in Figure 2.7 is mathematically expressed as follows:

$$f_{H_2S} = (10^{pH-pK_a} + 1)^{-1} \qquad (2.7)$$

where:

$f_{H_2S}$ = fraction of $H_2S$ in a mixture between $H_2S$ and $HS^-$ (–)
$K_a$ = equilibrium constant for $H_2S(aq) \Leftrightarrow HS^- + H^+$ (mol L$^{-1}$)
$pK_a = -\log K_a$ (cf. Table 2.1)

Figure 2.6 does not include sulfur compounds such as sulfite and thiosulfite. It thereby indirectly indicates that these species are not thermodynamically stable under the current conditions. Such external conditions in terms of, for example, pressure, temperature, and level of concentration, different from those valid for Figure

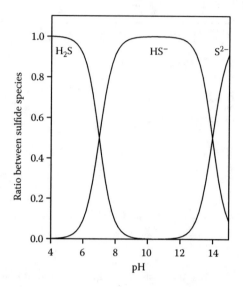

**FIGURE 2.7**  $C$–pH diagram for $H_2S(aq) \Leftrightarrow HS^- + H^+$ and $HS^- \Leftrightarrow S^{2-} + H^+$.

2.6, result in changes of stability for the compounds. As an example, the area where elemental sulfur, $S(s)$, is thermodynamically stable widens considerably when the concentration of soluble sulfur compounds increases. Furthermore, it is important that thermodynamically unstable compounds may easily exist for long periods because of the low reaction rates for their transformations. Compounds such as thiosulfite and elemental sulfur are therefore often found, although they, according to an $E$–pH diagram, are not stable.

### 2.1.4  STOICHIOMETRY OF REDOX REACTIONS

Stoichiometry refers to the balancing of a chemical reaction that, in principle, is a mass balance for the components included in a reaction. As an example, stoichiometry of redox reactions plays a central role when setting up conceptually based sewer process models.

The catabolic energy-producing processes constitute an important part of redox reactions in sewers. In this respect, organic matter is typically the electron donor that undergoes oxidation (Figures 2.2 and 2.4).

Concerning the reduction step of the redox reaction, the heterotrophic microorganisms may use different electron acceptors (Figure 2.3). If oxygen is available, it is the terminal electron acceptor, and the process proceeds under aerobic conditions. In the absence of oxygen, and if nitrates are available, nitrate becomes the electron acceptor, i.e., the redox process proceeds under anoxic conditions. If DO, nitrate, or ferric iron is not available, strictly anaerobic conditions exist, and sulfate is a potential external electron acceptor. Table 1.1 and Figure 2.5 give an overview of the process conditions related to sewer systems.

The stoichiometry of redox reactions will, in the following discussion, be dealt with in terms of basic elements, procedures, and examples.

### 2.1.4.1 Oxidation Level

An important parameter for the determination of stoichiometry of chemical reactions is the oxidation level (OX) of a chemical substance defined as follows:

The oxidation level is an imaginary charge of an element (atom) in a stable state of a molecule relative to the corresponding low stability of the free atom.

The oxidation level is not a fundamental theorem; however, it is useful for practical purposes when balancing chemical reactions. The concept of OX is related to the formation of a stabilized electron configuration around the atoms being part of a molecule compared with the (typically unstable) electron configuration of the single, free atom. The oxidation level is thereby also related to the formation of covalent bonds in terms of shared electrons between atoms in a molecule.

The major elements of organic matter relevant for the transformations of wastewater in sewers are C, O, H, N, and S. The classical understanding of the stability of these elements is associated with the configuration of the electrons around each of the atoms resulting in the formation of stable molecules. Carbon is the central element that shows specific—and unique—characteristics, which will also be dealt with in this chapter. It is important to note that the following refers to the elements C, O, H, N, and S as constituents of organic matter (the electron donor) and not to these elements as part of an electron acceptor. The oxidation levels of the relevant electron acceptors will, however, also be discussed.

To understand what is behind the definition of the oxidation level, OX, and to apply this understanding correctly, it is crucial to realize that a redox process is an exchange of electrons between the reacting compounds (cf. Figure 2.4). The understanding of the willingness for chemical substances to undergo a transformation fundamentally has its origin in the electron structure of the elements and the corresponding degree of stability.

From a pragmatic point of view, the electrons around the nucleus of an element are organized in shells. The shell number, $n$, starting with 1, refers to the order in which the electrons around the nucleus occupy "sites" when successively increasing the atomic number. The maximum number of electrons, $N$, that can exist in a shell is given by the following formula:

$$N = 2n^2 \tag{2.8}$$

Focusing on the first three shells, and thereby also including the elements that are the most interesting for organic matter, the maximum number of electrons in each of these shells is therefore:

Shell #1: maximum of two electrons
Shell #2: maximum of eight electrons
Shell #3: maximum of 18 electrons (also stable with eight electrons)

In each shell, the electrons are organized in orbits that can be interpreted as a space of the shell where a maximum of two electrons with opposite spin can occupy a position. The shells and the orbits will, with increasing atomic number, be filled with electrons in the order of lowest energy, i.e., according to a maximum of stability of each element. The configuration of the first 18 elements of the periodic system includes the atoms C, O, H, N, and S, which are the most central for the microbial processes in sewers (Table 2.2).

Referring to Table 2.2, all elements have a tendency to reach maximum of stability, i.e., to obtain the same structure as the noble gases (i.e., He, Ne, and Ar). Taking hydrogen as an example, it will reach a stable configuration with two electrons in the first shell, and the same orbit. By being surrounded by two electrons, H equals a structure corresponding to He. H can approach this structure by supplying its single electron to an electronegative atom, i.e., an atom with the ability to attract electrons. This electronegative atom will share one of its electrons with H in the same orbit. The hydrogen atom will thereby resemble He, and the oxidation level of hydrogen, $OX_H$, is equal to +1. When H shares electrons with another atom, the two electrons in the orbit form a bond that ties the two interacting atoms together in the molecule that is formed. The two electrons in the orbit are thereby situated in a stable configuration. In other words, the reaction—i.e., the formation of the molecule—is completed at a lower energy level compared with the energy of the two single atoms involved in the reaction. In a corresponding manner, O lacks two electrons to resemble Ne and will, as an electronegative atom, attract two electrons resulting in $OX_O = -2$.

As shown in Table 2.2, both O and N have two electrons in the first shell that is thereby stabilized corresponding to the structure of He. Stabilization in the four

## TABLE 2.2
### Electron Configuration of the First 18 Elements of the Periodic System

| Atomic Number | Element | No. of Electrons | | |
| --- | --- | --- | --- | --- |
| | | First Shell, Max 1 Orbit | Second Shell, Max 4 Orbits | Third Shell, Max 9 Orbits |
| 1 | H, hydrogen | 1 | | |
| 2 | He, helium | 2 | | |
| 3 | Li, lithium | 2 | 1 | |
| 4 | Be, beryllium | 2 | 2 | |
| 5 | B, boron | 2 | 3 | |
| 6 | C, carbon | 2 | 4 | |
| 7 | N, nitrogen | 2 | 5 | |
| 8 | O, oxygen | 2 | 6 | |
| 9 | F, fluorine | 2 | 7 | |
| 10 | Ne, neon | 2 | 8 | |
| 11 | Na, sodium | 2 | 8 | 1 |
| 12 | Mg, magnesium | 2 | 8 | 2 |
| 13 | Al, aluminum | 2 | 8 | 3 |
| 14 | Si, silicon | 2 | 8 | 4 |
| 15 | P, phosphorous | 2 | 8 | 5 |
| 16 | S, sulfur | 2 | 8 | 6 |
| 17 | Cl, chlorine | 2 | 8 | 7 |
| 18 | Ar, argon | 2 | 8 | 8 |

orbits of the second shell requires eight electrons. Stabilization of O and N corresponds to an electron configuration where each of these atoms obtains a structure comparable with Ne. The O and N atoms require an additional supply of two and three electrons, respectively, to establish the configuration of Ne, i.e., $OX_O = -2$ and $OX_N = -3$. Sulfur is an atom with a stabilized first and second shell and a configuration in the third shell equal to the second shell for oxygen, i.e., with six electrons. By accepting two more electrons, S will approach Ar in terms of stability. The oxidation level of S is, therefore, the same as for O, i.e., $OX_S = -2$.

As an example, Figure 2.8 shows both a simple (plane) and a more advanced (tetrahedral) picture of the electronic structure around the oxygen atom. In addition to the existing six electrons in the outer orbits, oxygen requires another two electrons to be stable, i.e., according to the definition of OX for an electronegative atom: $OX_O = -2$.

Except for carbon, C, that—as will be shown in the following—has specific characteristics, the oxidation levels for the most important elements as constituents of organic matter are constant:

$$OX_H = +1$$

$$OX_O = -2$$

$$OX_N = -3$$

$$OX_S = -2$$

For all practical purposes, it is important to take note of the following:

- The oxidation levels for each of the elements H, O, N, and S in organic matter remain unchanged by interacting in the microbiological processes that proceed in sewer systems.
- $\Sigma n OX_E = 0$ for the elements as part of a molecule, where $n$ is the number of atoms for each of the elements, $E$.

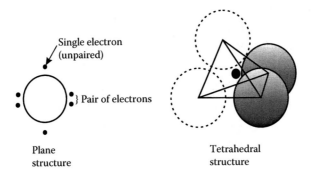

**Plane structure**

**Tetrahedral structure**

**FIGURE 2.8**  Atomic structure for oxygen (O) with two paired (two electrons per orbit) and two unpaired electrons in the outer four orbits.

Taking $H_2O$ as an example, this statement results in the following:

$$n_H OX_H + n_O OX_O = 2(+1) + 1(-2) = 0$$

When observing these two basic statements, the following two examples show the two extremes of $OX_C$ for carbon:

$$\text{Methane, } CH_4: \quad OX_C = -4$$

$$\text{Carbon dioxide, } CO_2: \quad OX_C = +4$$

The variability in $OX_C$ is interesting. Basically, it occurs because of the four unpaired electrons in the second shell, which results in the maximum of reactivity in four spatial directions. The electron configuration in the second shell provides carbon with the potential to establish bonding with elements having both positive and negative OX values and thereby approaching the stability of Ne. It is not strange that it was carbon, which became the key element of all living organisms! In principle, silicon (Si) has in its third shell the same configuration, although at a higher energy level. This structure, however, resulted in Si being the central element of the inorganic world in terms of its occurrence in the Earth's crust.

The varying oxidation level of carbon, $OX_C$, occurs because of a fixed $OX_O$ value closely related, and linear to, the COD value of the C compound in question. The following equation establishes this relationship:

$$OX_C = 4 - 1.5 \frac{COD}{TOC} \tag{2.9}$$

where:
   COD = chemical oxygen demand (g $O_2$ mol$^{-1}$ or g $O_2$ g$^{-1}$)
   TOC = total organic carbon (g C mol$^{-1}$ or g C g$^{-1}$)

Example 2.2 exemplifies the use of Equation 2.9 for the determination of $OX_C$.

### Example 2.2: Determination of $OX_C$ for Acetate

Formula for acetate $CH_3COOH$
   Chemical reaction for determination of COD:

$$CH_3COOH + 2O_2 \rightarrow 2CO_2 + 2H_2O$$

$$COD = 2 \times 32 = 64 \text{ g } O_2 \text{ mol}^{-1}$$

$$TOC = 2 \times 12 = 24 \text{ g C mol}^{-1}$$

$$OX_C = 4 - 1.5 \frac{64}{24} = 0$$

Example 2.2 shows that $OX_C$ for acetate is 0, which is also the case for all carbohydrates with the average formula $CH_2O$. Other organic compounds typical for wastewater may have different $OX_C$ values, with proteins often ranging from 0 to −0.4, whereas lipids typically show $OX_C$ values between −1 and −2. As shown in Equation 2.9, an organic substance with a high $OX_C$ value means that the potential for further oxidation and thereby energy production is relatively low. The ultimate high value is $OX_C = +4$ for $CO_2$ corresponding to $COD = 0$.

### 2.1.4.2 Electron Equivalent of a Redox Reaction

As shown in Figure 2.4, electrons are transferred from the oxidation step to the reduction step of the redox reaction. The number of electrons exchanged is the basis for the stoichiometry of a redox process. This fact is crucial when establishing a mass balance, as required when modeling sewer processes (cf. Chapters 8 and 9). The OX value is a key element in the determination of this number.

When determining the transfer of electrons, it should be realized that if organic matter as an electron donor is oxidized, i.e., degraded, no changes in the OX values for the elements H, O, and N will take place in the oxidation step when the following products are formed: $CO_2$, $H_2O$, and $NH_3/NH_4^+$.

The number of electrons exchanged in a redox reaction with organic matter as electron donor is, therefore, determined only by the change in the oxidation level for carbon. The unit for this exchange (electron equivalent, e-eq) is:

$$1 \text{ e-eq} = N_A \qquad (2.10)$$

where $N_A$ (Avogadro's number) $= 6.023 \cdot 10^{23}$, i.e., the number of electrons exchanged per equivalent of the chemical compound (−).

The electron equivalent number (e-eq) for a redox process where organic matter is the electron donor, is therefore

$$\text{e-eq} = n_C |\Delta OX_C| \qquad (2.11)$$

where $n_C$ denotes the number of C atoms per mole (mol$^{-1}$).

The use of Equation 2.11 is exemplified in Section 2.1.4.3.

### 2.1.4.3 Balancing of Redox Reactions

The balancing of redox reactions based on the transfer of electrons from the oxidation step to the reduction step follows a procedure based on the calculation of $\Delta OX_C$ and e-eq. These steps are for half-reactions as follows:

1. Balance of the number of electrons (e$^-$)
2. Balance of the charge
3. Balance of H
4. Control of balance for O

The use of this procedure for half reactions is exemplified (cf. Examples 2.3 through 2.6). The balance for an organic matter with a simple formula $CH_2O$ (C/H/O ratio as for a carbohydrate) is shown in Example 2.3.

## Example 2.3: Balancing of the Oxidation Step
## for an Electron Donor, $CH_2O$

For practical purposes, the balance of the half-reaction will be exemplified without taking into account Avogadro's number, $N_A$.

1. Balance of the number of electrons:

$$OX = 0 \quad OX = +4$$

$$\downarrow \qquad \downarrow$$

$$nCH_2O \rightarrow CO_2 + e^-$$

$$\text{e-eq} = n_C \, |\Delta OX_C| = 1 \times 4 = 4$$

$$\Rightarrow n = \frac{1}{4}$$

$$\frac{1}{4}CH_2O \rightarrow \frac{1}{4}CO_2 + e^-$$

2. Balance of charge:

$$\frac{1}{4}CH_2O + OH^- \rightarrow \frac{1}{4}CO_2 + e^- \text{ (balance with } H^+ \text{ or } OH^- \text{ depending on the pH value)}$$

3. H balance:

$$\frac{1}{4}CH_2O + OH^- \rightarrow \frac{1}{4}CO_2 + \frac{3}{4}H_2O + e^- \text{ (balance with } H_2O)$$

4. Control of O balance:

$$\frac{1}{4} + 1 = \frac{5}{4} \qquad \frac{2}{4} + \frac{3}{4} = \frac{5}{4}$$

Examples 2.4 through 2.6 concern the stoichiometry of half-reactions for electron acceptors relevant for sewer processes. Example 2.4 shows the procedure for an aerobic process step, Example 2.5 for an anoxic half-reaction, and Example 2.6 for an anaerobic half-reaction. Again, it should be noted that although the oxidation levels for the elements O, N, and S remain constant as constituents of organic matter when it is an electron donor, it is not the case when these elements act as electron acceptors.

## Example 2.4: Reduction of Electron Acceptor, $O_2$ (Aerobic Half-Reaction)

1. Balance of the number of electrons:

$$OX = 0 \quad OX = -2$$

$$\downarrow \quad\quad\quad \downarrow$$

$$nO_2 + e^- \rightarrow H_2O$$

$$\text{e-eq} = n_O |\Delta OX_O| = 2 \times 2 = 4$$

$$\Rightarrow n = \frac{1}{4}$$

2. Balance of charge:

$$\frac{1}{4}O_2 + H^+ + e^- \rightarrow H_2O \text{ (balance with } H^+ \text{ or } OH^-)$$

3. H balance:

$$\frac{1}{4}O_2 + H^+ + e^- \rightarrow \frac{1}{2}H_2O \text{ (balance with } H_2O)$$

4. Control of O balance:

$$\frac{1}{4}O_2 + H^+ + e^- \rightarrow \frac{1}{2}H_2O$$

$$\frac{1}{4} \times 2 = \frac{1}{2} \quad\quad \frac{1}{2} \times 1 = \frac{1}{2}$$

## Example 2.5: Reduction of Electron Acceptor, $NO_3^-$ (Anoxic Half-Reaction)

1. Balance of the number of electrons:

$$OX = +5 \quad OX = 0$$

$$\downarrow \quad\quad\quad \downarrow$$

$$nNO_3^- + e^- \rightarrow N_2$$

$$\text{e-eq} = n_N |\Delta OX_N| = 1 \times 5 = 5$$

$$\Rightarrow n = \frac{1}{5}$$

$$\frac{1}{5}NO_3^- + e^- \rightarrow \frac{1}{10}N_2$$

2. Balance of charge:

$$\frac{1}{5}NO_3^- + \frac{6}{5}H^+ + e^- \rightarrow \frac{1}{10}N_2$$

3. H balance:

$$\frac{1}{5}NO_3^- + \frac{6}{5}H^+ + e^- \rightarrow \frac{1}{10}N_2 + \frac{6}{10}H_2O$$

4. Control of O balance:

$$\frac{3}{5} = \frac{6}{10}$$

## Example 2.6: Reduction of Electron Acceptor, $SO_4^{2-}$ (Anaerobic Half-Reaction)

1. Balance of the number of electrons:

$$OX = +6 \qquad OX = -2$$

$$\downarrow \qquad\qquad \downarrow$$

$$nSO_4^{2-} + e^- \rightarrow H_2S$$

$$\text{e-eq} = n_s|\Delta OX_s| = 1 \times 8 = 8$$

$$\Rightarrow n = \frac{1}{8}$$

$$\frac{1}{8}SO_4^{2-} + e^- \rightarrow \frac{1}{8}H_2S$$

2. Balance of charge:

$$\frac{1}{8}SO_4^{2-} + \frac{5}{4}H^+ + e^- \rightarrow \frac{1}{8}H_2S$$

3. H balance:

$$\frac{1}{8}SO_4^{2-} + \frac{5}{4}H^+ + e^- \rightarrow \frac{1}{8}H_2S + \frac{1}{2}H_2O$$

4. Control of O balance:

$$\frac{4}{8} = \frac{1}{2}$$

Examples 2.3 through 2.6 show how half-reactions are balanced by using the exchange of electrons as the central element of the procedure. Example 2.7 shows how a redox reaction is completed by combining two half-reactions (cf. the outline in Figure 2.4).

## Example 2.7: Final Balancing of Redox Reactions

Example 2.3 is used as an example of microbial oxidation of an electron donor (organic matter) under anoxic conditions, i.e., with reduction of $NO_3^-$ as electron acceptor (cf. Example 2.5).
    Oxidation of $CH_2O$, Example 2.3:

$$\frac{1}{4}CH_2O + OH^- \rightarrow \frac{1}{4}CO_2 + \frac{3}{4}H_2O + e^-$$

Reduction of $NO_3^-$, Example 2.5:

$$\frac{1}{5}NO_3^- + \frac{6}{5}H^+ + e^- \rightarrow \frac{1}{10}N_2 + \frac{6}{10}H_2O$$

These two half-reactions can be added directly because the number of electrons produced equals the number of electrons consumed. If it is not the case, equalization must be done as the preliminary step. After multiplication with four and realizing that $H^+ + OH^- \rightarrow H_2O$, the final equation showing the stoichiometry of the redox reaction is as follows:

$$CH_2O + \frac{4}{5}H^+ + \frac{4}{5}NO_3^- \rightarrow CO_2 + \frac{7}{5}H_2O + \frac{2}{5}N_2$$

This equation is rather simple and can therefore be relatively easily overseen. The procedure as described may therefore, for similar simple reactions, not be needed by experienced persons. However, more complex redox reactions will certainly require a well-defined stepwise procedure to establish stoichiometry and a corresponding mass balance.

## 2.2 KINETICS OF MICROBIOLOGICAL SYSTEMS

A central subject of kinetics concerns the reaction rate of chemical and biochemical processes, including the description of rate expressions in homogeneous as well as in heterogeneous systems, i.e., one-phase and multiphase systems, respectively. Because microbial processes combine the activity of living organisms with transport of chemical reactants and products between a bulk water phase, across a cell wall and inside a cell, all microbiological processes are—by definition—heterogeneous. For practical purposes, however, microbial processes in a suspended wastewater phase are considered homogeneous, whereas processes in sewer biofilms are considered heterogeneous.

The metabolism of microorganisms is rather complex and occurs typically in several steps and with enzymes. However, to be used in practice, the metabolic pathways followed by anabolic and catabolic processes need, in terms of the related kinetics, to be described in a relatively simple manner. A major task of this book is therefore to quantify the kinetics of the governing steps of these processes in a way that—when combined—will allow the description of a complex series of reactions. The simple formulations of chemical kinetics will be dealt with in this section. Chapters 3 through 6 will exemplify how the relatively simple kinetic expressions can be combined and used for the description of the kinetics of more complex systems, e.g., when formulating models for computers.

Although this section focuses specifically on the kinetics of biochemical processes, it is crucial to realize that the formulations are equally valid for all types of chemical reactions.

Further details concerning the kinetics of chemical and biological processes in terms of both basic theory and applications relevant for this text can be found in a number of books (e.g., Snoeyink and Jenkins 1980; Stumm and Morgan 1996; Atkins and de Paula 2002).

### 2.2.1 KINETICS OF HOMOGENEOUS REACTIONS

As already stressed, processes in the wastewater phase can, from a practical point of view, be considered homogeneous. The reaction rates may depend on the concentration of relevant reactants and can often, in wastewater, be described as either zero-order (0-order) or first-order (1-order) reactions.

#### 2.2.1.1 Zero-Order Reaction

A 0-order reaction is defined as a reaction that is independent of the concentration of the reactants, i.e., the reaction rate is proportional to a constant multiplied by the concentration of a reactant raised to the power of zero:

$$\frac{dC}{dt} = -k_0 C^0 = -k_0 \tag{2.12}$$

where:
   $C$ = concentration of the reactant (g m$^{-3}$)
   $t$ = time (typically, h or day)
   $k_0$ = 0-order rate constant (g m$^{-3}$ h$^{-1}$ or g m$^{-3}$ day$^{-1}$)

If the initial concentration of $C$ for $t = t_0$ is $C_0$, Equation 2.12 yields, by integration from $t_0$ to $t$:

$$C - C_0 = -k_0 (t - t_0) \tag{2.13}$$

In a microbial system, a 0-order reaction will approximately proceed under conditions where a biomass or substrate concentration is high compared with the changes in the concentration. Such conditions are not typical in the wastewater phase of a sewer. However, a 0-order reaction may proceed when factors, such as surface area available for adsorption, limit the reaction rate.

### 2.2.1.2 First-Order Reaction

A 1-order reaction depends on the concentration of a reactant raised to the power of 1, i.e.,

$$\frac{dC}{dt} = -k_1 C^1 = -k_1 C \tag{2.14}$$

where:

$C$ = concentration of the reactant (g m$^{-3}$)

$t$ = time (typically h or day)

$k_1$ = 1-order rate constant (h$^{-1}$ or day$^{-1}$)

If the initial concentration of $C$ for $t = t_0$ is $C_0$, Equation 2.14 yields, by integration from $t_0$ to $t$:

$$C = C_0 e^{-k_1(t-t_0)} \tag{2.15}$$

It is characteristic for a 1-order reaction that the time required for 50% reduction of the reactant concentration is a constant, i.e., when $C = 0.5C_0$. According to Equation 2.15, the so-called half-life for the reaction is:

$$t_{1/2} = \frac{-\ln 0.5}{k_1} = \frac{0.6931}{k_1} \tag{2.16}$$

where $t_{1/2}$ denotes the half-life of the reaction (h or day).

As shown in Equation 2.15, a 1-order reaction is equivalent to an exponential change of a component. A great number of microbial processes follow, under sewer conditions, approximately this type of kinetics. A 1-order reaction is, therefore, the traditional way to describe microbial transformations of organic matter in wastewater. An example of a process expected to follow the 1-order kinetics is shown in Example 2.8.

### Example 2.8: First-Order Kinetics for BOD Removal Under Aerobic Conditions in Wastewater of a Sewer

This example concerns the degradation of wastewater transported for 4 h under aerobic conditions in a half-full intercepting gravity sewer. It is assumed that the transformation of the organic matter only proceeds in the wastewater phase and follows 1-order removal kinetics.

Wastewater and systems characteristics are given as follows: $BOD_5$ is 200 g $O_2$ $m^{-3}$; temperature is 20°C; pipe diameter is 0.3 m; 1-order removal rate of $BOD_5$ is 0.05 $h^{-1}$. According to Equation 2.14, the $BOD_5$ value after 4 h is:

$$BOD_5(t = 4h) = BOD_5(t = 0) \, e^{-0.05 \cdot 4} = 164 \text{ g } O_2 \text{ m}^{-3}$$

The reduction in the $BOD_5$ value equals about 18% during aerobic transport in the sewer.

### 2.2.1.3   *n*-Order Reactions

Although zero-order and first-order descriptions are the most commonly used as sewer process rate expressions, other types of reaction orders exist. The general formulation of a homogeneous reaction between two reacting components, A and B, is.

$$r = k(C_A)^a(C_B)^b \tag{2.17}$$

where:

$r$ = reaction rate, i.e., the rate of change of concentration, $dC/dt$
$k$ = rate constant (unit depends on the expression of the reaction rate, $r$, in terms of $a$ and $b$)
$C_A$ and $C_B$ = concentrations of compounds A and B, respectively (typically in units of g $m^{-3}$ or mmol $L^{-1}$ (mM))
$n = a + b$ = reaction order (–)

It should be noted that $n$, $a$, and $b$ need not be integral numbers.

Aerobic chemical and biological oxidation of sulfide in wastewater of sewers is an example of two reacting substances, DO, $O_2$, and sulfide, $H_2S$ and $HS^-$, that in principle follow a reaction scheme according to Equation 2.17 (cf. Section 5.5.1.2). Another example is the half-order reaction, $r = kC^{0.5}$, that is relevant for partly penetrated biofilms (cf. Section 2.2.2.1).

Typically, the reaction order—and the rate expression—of a chemical or biological process, is empirically determined based on experimental techniques approaching sewer conditions. It is, therefore, also often the case that the reaction order, $n$, differs from the more or less "theoretical" values.

### 2.2.1.4   Growth Limitation Kinetics

Two different types of limitation for growth of a biomass exist: rate limitation and mass limitation. Both cases refer to biomass-initiated chemical reactions. Basically, growth limitation concerns the description of the limiting conditions for the development of biomass but is here extended to include substances transformed or produced by the activity of biomass species.

#### 2.2.1.4.1   Rate Limitation

Rate limitation refers to the fact that there might exist some kind of "bottleneck" for the production of either a specific species of biomass or the consumption of a substrate, i.e., a substance required for the activity of the biomass. In mathematical terms, rate limitation can be expressed in terms of $dX/dt$ or $dC/dt$, where X and C are the concentration of the active biomass and the concentration of a substrate for the biomass, respectively. The cause of rate limitation is legion. As an example, the bottleneck may be due to limitation in the rate of transport (diffusion) of a substrate across the cell membrane of a microorganism. Under optimal external conditions for an organism, rate limitation results in a maximum value of the growth rate or a maximum consumption rate of a substrate.

### 2.2.4.1.2  Mass Limitation

Mass limitation refers to mass balance aspects. It means that if a certain amount of a substrate is available—and if needed for the activity (growth) of the biomass—this amount determines what can be totally produced. Therefore, mass limitation refers to what proceeds in a closed system with no exchange of mass with the surroundings.

It is crucial to clearly distinguish between the two different concepts of growth limitation for biomass. It seems quite logical. However, in practice, it is often seen that if this is not considered, fatal misunderstanding and wrong conclusions will be the result.

Example 2.9 illustrates the difference between rate limitation and mass limitation.

### Example 2.9: Sulfate Limitation for Sulfide Formation Under Anaerobic Conditions in a Sewer

In sewers, the activity of sulfate-reducing bacteria can, under anaerobic conditions, result in the production of hydrogen sulfide. If the total concentration of sulfate is 100 g $SO_4$-S m$^{-3}$, a simple mass balance results in a potential amount of 100 g $H_2$S-S m$^{-3}$ sulfide to be produced (it is hereby assumed that the production of sulfide from sulfur-containing proteins can be neglected). Therefore, the *mass limitation* for this system, in terms of a maximum amount of sulfide that can be produced, is 100 g $H_2$S-S m$^{-3}$. The time it will take to produce this amount of sulfide is not taken into account in case of mass limitation.

The time aspect—i.e., the kinetics—is, in contrast, the core characteristic of *rate limitation*. It is important to note that in this case investigations have revealed that the maximum rate of sulfide production in sewers by sulfate-reducing bacteria occurs at a concentration of sulfate in the order of 3–5 g $SO_4$-S m$^{-3}$ depending on site- and species-specific aspects (Nielsen and Hvitved-Jacobsen 1988). Sulfate concentrations above this level do *not* increase the rate of sulfide production. This is assumed to be the effect of rate limitation of sulfate diffusion into the biofilm formed by the sulfate-reducing bacteria growing at the sewer walls. *Rate limitation* of sulfide formation, therefore, only occurs at sulfate concentrations below 3–5 g $SO_4$-S m$^{-3}$. Above this level, sulfide formation will proceed with a more or less constant rate according to a 0-order reaction as long as other external conditions do not cause limitation.

A main purpose of dealing with growth kinetics for microbial systems in sewers is to develop mathematical expressions for the forecast of changes in biomass and substrates, i.e., phenomena that are important for the design and operation of sewer systems. The understanding of microbial growth kinetics is based on Monod (1949), who divides microbial growth into six well-defined phases (cf. Figure 2.9).

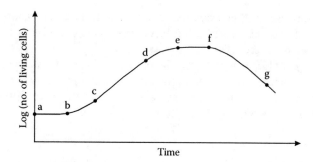

**FIGURE 2.9**   Different phases of microbial growth according to Monod (1949).

In brief, and as shown in Figure 2.9, the different phases of microbial growth are:

- A *lag phase* (a–b), where the microorganisms adapt to the ambient environment.
- An *acceleration phase* (b–c), where the growth rate increases.
- An *exponential growth phase* (c–d). As long as microbial growth occurs under nonlimited rate conditions, exponential growth may take place, and changes in biomass and substrate are maximal. According to Equations 2.14 and 2.15, exponential microbial growth follows Equations 2.18 and 2.19, corresponding to a maximal and constant value of the specific growth rate, $\mu_{max}$:

$$\frac{dX}{dt} = \mu_{max} X \tag{2.18}$$

where:

$X$ = concentration of active microbial biomass (g m$^{-3}$)

$t$ = time (typically h or day)

$\mu_{max}$ = maximum specific growth rate, i.e., the growth rate at nonlimited rate conditions (h$^{-1}$ or day$^{-1}$)

In an integrated form, Equation 2.18 is expressed as:

$$X = X_0 e^{\mu_{max}(t-t_0)} \tag{2.19}$$

- A *phase for declining growth* (d–e) with a negative rate of acceleration for the growth caused by growth rate limitation.
- A *stationary phase* (e–f), where the microorganisms obtain their maximal population size. The substrate in this phase is sufficient for maintaining the population size; however, it does not allow further increase.
- An *endogenous phase* (f–g) corresponding to death and inactivation of the biomass, e.g., caused by environmental conditions for reduced survival or activity.

Equations 2.18 and 2.19 are expressed under conditions corresponding to exponential growth. However, they can be reformulated corresponding to limitations in the growth rate. These two equations are therefore also the basis for description of the kinetics when the growth of the biomass takes place under conditions where substrate or other external environmental factors are limiting the growth rate. Substrate-limited growth in terms of reduced availability of either the electron donor or the electron acceptor—or both of these substances—is common in wastewater of sewer systems. Based on the concept of Michaelis–Menten's kinetics valid for enzymatic processes, Monod (1949) formulated, in operational terms, the relationship between the actual and the maximal specific growth rate constants and the concentration of a limiting substrate.

$$\mu = \mu_{max} \frac{S}{S + K_S} \tag{2.20}$$

where:
$\mu$ = specific growth rate ($h^{-1}$ or $day^{-1}$)
$S$ = concentration of substrate ($g\ m^{-3}$)
$K_S$ = saturation constant ($g\ m^{-3}$)

If two or more substrates are growth limiting, the ratio $S/(S + K_S)$ is included in Equation 2.20 for each of the compounds corresponding to Equations 2.31 and 2.32.

Based on Equations 2.18 and 2.20, the general formulation of microbial growth kinetics for one limiting substrate is:

$$\frac{dX}{dt} = \mu_{max} \frac{S}{S + K_S} X \tag{2.21}$$

The relationship between $\mu$ and $S$ as depicted in Figure 2.10 quantifies the importance of a substrate concentration in terms of its impact on the growth rate. As seen in Equation 2.20, $\mu = 1/2\ \mu_{max}$ results in $K_S = S$. For this reason, $K_S$ is also named the "half-saturation constant." Equation 2.20 and the corresponding curve shown in Figure 2.10 are named the Monod expression and the Monod curve, respectively.

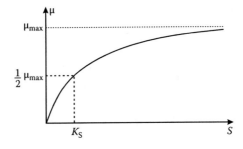

**FIGURE 2.10** The Monod curve.

## 2.2.2  KINETICS OF HETEROGENEOUS REACTIONS

In contrast to the kinetics of homogeneous reactions discussed in Section 2.2.1, this section deals with the kinetics of heterogeneous reactions, i.e., reactions in multiphase systems. The biofilm kinetics and the kinetics of hydrolysis are the two major types of heterogeneous kinetics for sewer processes.

### 2.2.2.1  Biofilms and Biofilm Kinetics

A biofilm—in sewers often named slime—is a concentrated layer of microorganisms that adheres to a solid surface. The thickness of a biofilm varies considerably. The thickness of a biofilm developed under anaerobic conditions is typically rather smooth and less than 500 µm, whereas aerobic biofilms typically are fluffy and may range from 1 to 2 mm up to a few centimeters. Biofilms are typically characterized by the following factors:

- Water content ranging from 70% to 90%
- Organic matter contents from 50% to 90% of the total dry matter contents
- Relatively high contents of carbohydrates and proteins
- Varying cell biomass contents from 10% to 90% of total organic matter contents and a corresponding (high and low, respectively) content of extracellular polymeric substances surrounding the cells

Further details on biofilms in sewer systems are dealt with in Section 3.2.7. General and specific information on biofilms can be found in the work of Characklis and Marshall (1990).

As previously mentioned, all biologically initiated reactions are, in terms of their kinetics, heterogeneous. However, for practical reasons, the processes in the suspended water phase are considered homogeneous and therefore, in terms of their kinetics, included in Section 2.2.1. Processes in biofilms, however, proceed by exchange of electron donors and electron acceptors with the surrounding bulk water phase. As a consequence, these processes proceed in a two-phase system including a water phase and a "solid" biofilm and are, therefore, heterogeneous.

The kinetics of the uptake of soluble substrate by the bacteria in a biofilm is traditionally described by a combination of mass transport across the water–biofilm interface, transport in the biofilm itself, and the corresponding relevant biotransformations. Transport through the stagnant water layer at the biofilm surface is described by Fick's first law of diffusion (cf. Figure 2.11). Fick's second law of diffusion and Michaelis–Menten (Monod) kinetics are used for describing the combined transport and transformations in the biofilm itself (Williamson and McCarty 1976a, 1976b; Harremoës 1978). The development of equations for these transport and transformation phenomena is dealt with in the following discussion.

A detailed description of biofilm kinetics is given by Harremoës (1978), Characklis and Marshall (1990), and Henze et al. (1995). In the following, an overview of biofilm kinetics relevant for sewer processes will be given.

As depicted in Figure 2.11, mass transport of substrate from the bulk water phase proceeds through a stagnant fluid boundary layer (liquid film) and into a biofilm followed by a combined diffusion and utilization of the substrate in the biofilm.

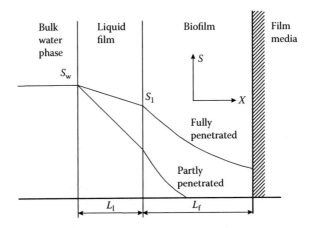

**FIGURE 2.11** Substrate profiles exemplified for a biofilm with fully penetration and partly penetration of a single substrate, respectively.

The diffusion of the substrate from the bulk water phase across the stagnant liquid film is described by Fick's first law (cf. Figure 2.11):

$$J = -AD_w \frac{\partial S}{\partial x} = -AD_w \frac{S_1 - S_w}{L_1} \tag{2.22}$$

where:
$J$ = flux of substrate in the $x$-direction (g s$^{-1}$, g h$^{-1}$, or g day$^{-1}$)
$A$ = cross-sectional area through which the flux is occurring (m$^2$)
$D_w$ = molecular diffusion coefficient of a substance in water (m$^2$ s$^{-1}$, m$^2$ h$^{-1}$, or m$^2$ day$^{-1}$)
$S$ = substrate concentration (g m$^{-3}$)
$x$ = distance (m)
$\partial S/\partial x$ = concentration gradient of substrate in the $x$-direction (g m$^{-3}$ m$^{-1}$)

The biofilm is considered homogeneous, having a well-defined thickness, and the transport of substances in the biofilm is determined by molecular diffusion. As this transport process is relatively slow, it is normally the limiting process for the transformations in the biofilm. If Fick's first law, Equation 2.22, is applied for the transport in the biofilm, the following is valid:

$$\frac{\partial J}{\partial x} = -A \times D_f \frac{\partial^2 S}{\partial x^2} \tag{2.23}$$

where $D_f$ is the molecular diffusion coefficient of a substance in the biofilm (m$^2$ s$^{-1}$, m$^2$ h$^{-1}$, or m$^2$ day$^{-1}$).

If no reaction proceeds in the biofilm, the flux of substrate, $J$, into the film is constant (equal to 0) and, consequently, $\partial J / \partial x = 0$:

$$D_f \frac{\partial^2 S}{\partial x^2} = -\frac{1}{A} \frac{\partial J}{\partial x} = 0 \qquad (2.24)$$

Equation 2.24, which concerns a situation without processes in the biofilm, can be extended to include transformation of a substrate, an electron donor (organic matter), or an electron acceptor, e.g., DO. If the reaction rate is limited by just one substrate and if the reaction is proceeding under steady-state conditions, i.e., with a fixed concentration profile, the differential equation for the combined transport and substrate utilization follows the Monod kinetics. This situation is expressed by Equation 2.25 and illustrated in Figure 2.11. Equation 2.25 states that the molecular diffusion under steady-state conditions is equal to the bacterial uptake of the substrate. Equation 2.25 is an example of Fick's second law.

$$D_f \frac{d^2 S}{dx^2} = k_{0f} \frac{S}{K_S + S} \qquad (2.25)$$

where:

$K_S$ = saturation constant (g m$^{-3}$)
$K_{0f}$ = 0-order rate constant per unit volume in the biofilm (g m$^{-3}$ s$^{-1}$, g m$^{-3}$ h$^{-1}$, or g m$^{-3}$ day$^{-1}$)

The 0-order rate constant, $k_{0f}$, is a maximal volume-based reaction rate for substrate transformation in a biofilm:

$$k_{0f} = q_{max} X_f \qquad (2.26)$$

where:

$q_{max}$ = maximal uptake rate of substrate in the biofilm (h$^{-1}$ or day$^{-1}$)
$X_f$ = bacterial density (concentration) in the biofilm (g m$^{-3}$)

Equation 2.25 is nonlinear and has no analytical solution although analytical approximations exist (Harremoës 1978; Henze et al. 1995).

The biological processes in biofilms are either described by 1-order or 0-order kinetics. However, the 0-order reaction is of specific importance for sewer biofilms as is also the case for treatment processes of wastewater in biofilters. The saturation constant, $K_S$, is normally insignificant compared with the substrate concentration, and the biofilm kinetics (cf. Equation 2.25), can therefore be considered 0-order. As shown in Figure 2.11, two different conditions exist: the biofilm is either fully penetrated or partly penetrated, corresponding to either a fully process-effective or a partly process-effective biofilm. The distinction between these two situations can be expressed by means of a dimensionless constant, $\beta$, called the penetration ratio

(Harremoës 1978). For each of these two situations, the flux of substrate across the biofilm surface can, for typical sewer flow regimes—i.e., when neglecting the transport through the stagnant liquid film—be calculated (cf. Equations 2.28 and 2.30):

Fully penetrated biofilm:

$$\beta = \sqrt{\frac{2D_f S_w}{k_{0f} L_f^2}} > 1 \tag{2.27}$$

$$r_a = k_{0f} L_f \tag{2.28}$$

Partly penetrated biofilm:

$$\beta = \sqrt{\frac{2D_f S_w}{k_{0f} L_f^2}} < 1 \tag{2.29}$$

$$r_a = \sqrt{2D_f k_{0f}} \; S_w^{0.5} = k_{1/2} S_w^{0.5} \tag{2.30}$$

where:
  $S_w$ = substrate concentration in the bulk water phase (g m$^{-3}$)
  $r_a$ = biofilm surface flux (g m$^{-2}$ s$^{-1}$, g m$^{-2}$ h$^{-1}$, g m$^{-2}$ day$^{-1}$)
  $k_{1/2}$ = 1/2-order rate constant per unit area of biofilm surface (g$^{0.5}$ m$^{-0.5}$ s$^{-1}$, g$^{0.5}$
      m$^{-0.5}$ h$^{-1}$, g$^{0.5}$ m$^{-0.5}$ day$^{-1}$)
  $L_f$ = biofilm thickness (m)

Equations 2.28 and 2.30, for the calculation of the biofilm surface flux, $r_a$, are interesting because they relate the 0-order reaction in the biofilm to the substrate conditions in the bulk water phase. As seen from these two equations, a fully penetrated biofilm is of 0-order with respect to the bulk water phase substrate concentration, and a partly penetrated biofilm is of 1/2-order. Sewer biofilms are typically relatively thick with a high process rate. Such biofilms are, therefore, normally partly penetrated by substrate and of 1/2-order with respect to the substrate conditions in the wastewater phase. Figure 2.12 depicts the theoretical transition from 1/2-order to 0-order kinetics in a biofilm.

The redox reactions taking place in a sewer biofilm may require that diffusion of both electron donor and electron acceptor be considered. According to Equation 2.25, the steady-state mass balance for these two components is:

$$D_{f,ox} \frac{d^2 S_{ox}}{dx^2} = k_{0f,ox} \frac{S_{ox}}{K_{S,ox} + S_{ox}} \frac{S_{red}}{K_{S,red} + S_{red}} \tag{2.31}$$

$$D_{f,red} \frac{d^2 S_{red}}{dx^2} = k_{0f,red} \frac{S_{ox}}{K_{S,ox} + S_{ox}} \frac{S_{red}}{K_{S,red} + S_{red}} \tag{2.32}$$

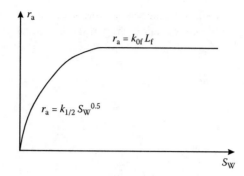

**FIGURE 2.12** Biofilm kinetics with respect to bulk water phase. Typically, conditions corresponding to 1/2-order kinetics exist in sewer biofilms.

These two second-order nonlinear differential equations have no explicit solution and should therefore be solved numerically. The limiting substrate for the biofilm transformations is the one that penetrates the shortest distance into the biofilm. Equations 2.31 and 2.32 are, thereby, reduced to an equation corresponding to Equation 2.25. If the limiting substrate cannot be identified, approximations based on Equation 2.30 can be developed.

### 2.2.2.2 Kinetics of Hydrolysis

Hydrolysis is an enzymatic process by which complex organic compounds that cannot directly be used as substrates by the bacterial biomass can be broken down into simple molecules. Further details of this process are discussed in Section 3.2.3. A complex particulate hydrolyzable substrate, $X_S$, is thereby converted into soluble or colloidal compounds, $S_S$. These relatively small molecules can be transported from the bulk water phase across the bacterial cell wall and into the cell where the substrate utilization processes proceed.

A simple and generally accepted approach of the kinetics for the hydrolysis is a 1-order description (Henze et al. 1995):

$$\frac{dX_S}{dt} = k_H X_S \tag{2.33}$$

where:
$X_S$ = concentration of hydrolyzable substrate (g COD m$^{-3}$)
$k_H$ = 1-order hydrolysis constant (day$^{-1}$)

However, in some situations, it is assumed that the particulate substrate is covered with bacteria that excrete exoenzymes, i.e., enzymes active outside the bacterial cell. Because it is assumed that enzymes are present in excessive amounts, the hydrolysis rate is constant per unit area available for hydrolysis. The kinetics model is therefore surface-related:

$$\frac{dC_S}{dt} = k_A \frac{A}{V} \tag{2.34}$$

where:

$C_S$ = concentration of substrate (g COD m$^{-3}$)
$k_A$ = surface-based hydrolysis constant (g COD m$^{-2}$ day$^{-1}$)
$A$ = surface area available for hydrolysis (m$^2$)
$V$ = volume (m$^3$)

A combination of the concepts behind Equations 2.33 and 2.34 is applied in the activated sludge model for the description of the kinetics of the hydrolysis processes (Henze et al. 1987). This combined concept, originally proposed by Dold et al. (1980), includes a saturation type of expression and a heterotrophic biomass with a maximum capacity for hydrolysis:

$$\frac{dX_S}{dt} = k_n \frac{X_S/X_{Hw}}{K_X + X_S/X_{Hw}} X_{Hw} \tag{2.35}$$

where:

$k_n$ = hydrolysis rate constant (day$^{-1}$)
$X_{Hw}$ = heterotrophic active biomass in the water phase (g COD m$^{-3}$)
$K_X$ = saturation constant for hydrolysis (–)

Equation 2.35 describes the two extremes where either availability of hydrolyzable substrate or biomass limits the hydrolysis. If hydrolyzable substrate is available in excess, i.e., $X_S > X_{Hw}$, the biomass density, $X_{Hw}$, is considered proportional to the enzymatic activity, and a situation equivalent to Equation 2.34 is the result, i.e., the rate of hydrolysis is 1-order with respect to $X_{Hw}$. In contrast, if $X_S \ll X_{Hw}$, Equation 2.35 equals Equation 2.33, and the rate of hydrolysis is 1-order with respect to $X_S$. Situations between these two extremes are appropriately described based on Equation 2.35.

The pathways for transformation of carbohydrate, protein, and lipids as depicted in Figure 3.8, outline a potential basis for the kinetic description of hydrolysis. In this way, it is possible to take into account the specific characteristics, e.g., process rates, for the different groups of organic matter. Christ et al. (2000) use this approach for kinetics and model expressions for anaerobic hydrolysis of organic wastes. However, taking into account the level of knowledge on sewer processes, such details are at present not generally applicable.

## 2.3 TEMPERATURE DEPENDENCY OF MICROBIAL, CHEMICAL, AND PHYSICOCHEMICAL PROCESSES

Physical, chemical, and biological processes are in general, temperature dependent. This temperature dependency is often, in a mathematically expressed quantification of a process, included in a "constant." Such constants appear in the description of, for example, the reaction rate of chemical or biological processes or diffusion- and transport-related processes of a compound. In the following, the focus is on microbial processes, but the derived temperature expressions are typically valid for chemical and physical processes as well.

The temperature in a sewer depends on a number of different conditions, e.g., local climate, source of wastewater, and system characteristics. The microbial system developed in a sewer is typically subject to annual temperature variations and, to some extent, a daily variability. Different microbial communities will develop under different temperature conditions, and process rates relevant for the activity of microorganisms may vary considerably with temperature. Long-term variations affect which microbial population type will develop in a sewer, whereas short-term variations have impacts on the rate of microbial processes in the cell itself as well as on the diffusion rate of substrates.

An important and often overlooked consequence of the impact of temperature is caused by the fact that different populations of bacteria may develop in sewers in the temperate climate zones compared with those in warm climates. The rate of a biochemical reaction determined in a specific case is therefore not generally transferable to different climate conditions. It is clear that such aspects require local experimental investigations to provide data that are valid for prediction and use in modeling.

The temperature dependency of a reaction rate of a microbial process and a chemical or physicochemical reaction can be described based on the fundamental formulation of the Arrhenius equation:

$$\frac{d \ln k}{dT} = \frac{E_a}{RT^2} \tag{2.36}$$

where:
  $k$ = reaction rate constant (unit depends on the reaction order)
  $T$ = temperature (K)
  $E_a$ = activation energy of the reaction (J mol$^{-1}$)
  $R$ = universal gas constant ($R$ = 8.314 J mol$^{-1}$ K$^{-1}$)

The integration of Equation 2.36 between temperatures $T_1$ and $T_2$ corresponding to the rate constants $k_1$ and $k_2$, respectively, results in:

$$\ln \frac{k_2}{k_1} = \frac{E_a(T_2 - T_1)}{RT_2T_1} \tag{2.37}$$

For practical purposes, a temperature coefficient, $\alpha$, is introduced:

$$\alpha = \exp\left( \frac{E_a}{RT_2T_1} \right) \tag{2.38}$$

where $\alpha$ is the temperature coefficient (–).

According to Equations 2.37 and 2.38, the temperature dependency of the reaction rate is:

$$k_2 \quad k_1 \qquad^{T_2\ T_1} \tag{2.39}$$

The magnitude of $\alpha$ varies with the system including the biological community that it may concern and should be determined from experiments. However, a typical value of $\alpha$ is 1.07 for a water phase microbial process, approximately corresponding to a doubling of the reaction rate with a 10°C increase in temperature. The temperature coefficient for biofilm processes is typically considerably lower and equal to about 1.03. The reason is that the reaction rate in the biofilm is limited by the diffusion of a reacting component and not by the reaction itself.

Within a narrow temperature interval typical for sewer networks, the temperature coefficient, $\alpha$, is approximately constant, because the product between $T_1$ and $T_2$ varies only slightly. However, if $\alpha$ cannot be considered constant, more than one value of $\alpha$ may be needed to describe the temperature dependency, with each value of $\alpha$ corresponding to a subinterval.

Equations 2.36 through 2.39 are valid for the temperature dependency of a specific chemical reaction or biochemical process. These equations are not relevant for the description of annual changes of the kinetics if different microbial populations develop.

The Arrhenius equation, normally expressed in a simple version by Equation 2.39, is not the only way to express a description of a process-related influence of temperature. The van't Hoff equation is equally a relevant thermodynamic based option in this case (cf. Section 4.1.2.2).

## 2.4  ACID–BASE CHEMISTRY IN SEWERS

The traditional description of the acid–base chemistry can be found in any textbook dealing with chemistry or physical chemistry and in books with special emphasis on aquatic chemistry (e.g., Snoeyink and Jenkins 1980; Stumm and Morgan 1996).

The basic description of the acid–base chemistry is considered well known and is therefore not a subject of this book. However, it is important to recall the characteristics that play a role for sewer processes. The two parameters, pH and alkalinity, are particularly of importance. As an example, pH controls the emission of hydrogen sulfide to the sewer atmosphere and thereby corrosion and odor problems. Furthermore, in most wastewaters, the so-called carbonate system plays a central role as a pH buffer, i.e., it is a system that determines the magnitude of pH changes. Further details in terms of the acid–base chemistry and its relation to the emission of volatile compounds are dealt with in Section 4.5.

### 2.4.1  Carbonate System

The term "carbonate system" refers to a set of principles for the description of the interactions between inorganic carbon compounds and what is ultimately established in terms of homogeneous and heterogeneous equilibria between the following components:

- Gas phase: $CO_2(g)$
- Water phase: $CO_2(aq)$, $H_2CO_3$, $HCO_3^-$, $CO_3^{2-}$ (and corresponding cations)
- Solid phase: carbonates, e.g., limestone, $CaCO_3$, and dolomite, $CaMg(CO_3)_2$

The carbonate system is important in most aquatic systems including living organisms. The following characteristics of the carbonate system are particularly relevant:

- The carbonate system is a pH buffer, i.e., it maintains a pH value of relatively limited variability.
- The carbonate system provides inorganic carbon for the autotrophic processes. In the sewer itself, autotrophic processes (sulfide oxidation) are primarily of interest at the sewer walls exposed to the gas phase. However, autotrophic processes (nitrification) are highly important in the suspended water phase of wastewater treatment plants.
- The carbonate system determines the exchange of $CO_2$ across the air–water interface and precipitation or dissolution of the solid forms of carbonate at the water–solid interface.
- The carbonate system plays a central role in the global climate.

The carbonate system is under equilibrium conditions illustrated in Figure 2.13. Limestone ($CaCO_3$)—as an example of a solid carbonate—exists in equilibrium with carbonate in the water phase. The water phase is open to the overlying atmosphere with $CO_2$-equilibrium across the air–water interface.

Figure 2.13 shows equilibrium in the carbonate system. However, in most cases relevant for sewers, the equilibrium across the two interfaces is typically not being established, and a net exchange of a compound depending on the magnitude of the transfer rates takes place.

The basic description of the carbonate equilibrium system corresponding to Figure 2.13 follows from Henry's law (air–water equilibrium), the acid–base

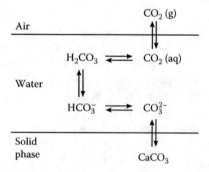

**FIGURE 2.13** Illustration of the carbonate system with both homogenous and heterogeneous equilibria.

equilibria within the water phase, and the description of the precipitation–dissolution equilibrium, i.e., the water–solid phase equilibrium. The equations for these equilibria will be outlined in the following discussion.

The values of the equilibrium parameters in the equations for the carbonate system are, in the following, exemplified at 25°C and 1 atm total pressure corresponding to a partial pressure of $CO_2(g)$ of about 0.00039 atm. The concentrations in the water phase are given in units of mol $L^{-1}$ (also named M). Further details concerning the carbonate system are found in several books (e.g., Snoeyink and Jenkins 1980; Stumm and Morgan 1996; Nazaroff and Alvarez-Cohen 2001).

### 2.4.1.1 Air–Water Equilibrium

The equilibrium between $CO_2(g)$ and $CO_2(aq)$ follows Henry's law (cf. Section 4.1.2):

$$p_{CO_2(g)} = H_{CO_2} C_{CO_2(aq)} \tag{2.40}$$

where:
$p_{CO_2(g)}$ = partial pressure of $CO_2(g)$ (atm)
$H_{CO_2}$ = Henry's law constant for $CO_2$ = 29.4 atm L mol$^{-1}$
$C_{CO_2(aq)}$ = equilibrium concentration of $CO_2(aq)$ (mol $L^{-1}$)

### 2.4.1.2 Water Phase Equilibria

The equilibrium between $CO_2(aq)$ and $H_2CO_3$ follows the general expression for chemical equilibrium in a homogeneous system (cf. Section 2.1.1):

$$\frac{C_{H_2CO_3}}{C_{CO_2(aq)}} = K_m = 1.58 \times 10^{-3} \tag{2.41}$$

where:
$C_{H_2CO_3}$ = equilibrium concentration of $H_2CO_3$ (mol $L^{-1}$)
$K_m$ = equilibrium constant (–)

The total concentration of the nondissociated carbonic acid in the water phase is defined as:

$$C_{H_2CO_3^*} = C_{CO_2(aq)} + C_{H_2CO_3} \tag{2.42}$$

Carbonic acid dissociates in two steps. The corresponding acid–base equilibria and equilibrium expressions are:

$$H_2CO_3^* \Leftrightarrow H_2O + CO_2 \Leftrightarrow H^+ + HCO_3^- \tag{2.43}$$

As shown by the numerical value of $K_m$, the main part of the nondissociated carbonic acid exists as $CO_2(aq)$. The term $H_2CO_3^*$ will therefore not be shown in the following chapters of this book, but will be replaced by $CO_2(aq)$.

$$\frac{C_{H^+}C_{HCO_3^-}}{C_{CO_2(aq)}} = K_{S1} = 4.47 \times 10^{-7} \tag{2.44}$$

The second dissociation step is:

$$HCO_3^- \Leftrightarrow H^+ + CO_3^{2-} \tag{2.45}$$

$$\frac{C_{H^+}C_{CO_3^{2-}}}{C_{HCO_3^-}} = K_{S2} = 4.68 \times 10^{-11} \tag{2.46}$$

where:
$C_{H^+}$ = equilibrium concentration of $H^+$ (mol $L^{-1}$)
$C_{HCO_3^-}$ = equilibrium concentration of $HCO_3^-$ (mol $L^{-1}$)
$C_{CO_3^{2-}}$ = equilibrium concentration of $CO_3^{2-}$ (mol $L^{-1}$)
$K_{S1}$ = equilibrium constant (mol $L^{-1}$)
$K_{S2}$ = equilibrium constant (mol $L^{-1}$)

### 2.4.1.3   Water–Solid Equilibrium

The equilibrium between carbonate in the solid state (e.g., limestone) and carbonate in the water phase follows an equilibrium expression where the activity—the effective concentration—of the solid material per definition is equal to 1:

$$CaCO_3 \Leftrightarrow Ca^{2+} + CO_3^{2-} \tag{2.47}$$

$$C_{Ca^{2+}}C_{CO_3^{2-}} = L_{CaCO_3} = 5 \times 10^{-9} \tag{2.48}$$

where:
$C_{Ca^{2+}}$ = equilibrium concentration of $Ca^{2+}$ (mol $L^{-1}$)
$C_{CO_3^{2-}}$ = equilibrium concentration of $CO_3^{2-}$ (mol $L^{-1}$)
$L_{CaCO_3}$ = equilibrium constant, also named solubility product of $CaCO_3$ (mol$^2$ $L^{-2}$)

### 2.4.1.4   General Physicochemical Expressions

In addition to Equations 2.40 through 2.48, expressing the equilibrium of the carbonate system, Equations 2.49 through 2.53 are essential to consider for aquatic systems.

The dissociation of water and corresponding equilibrium expression are, with an activity of water approximately equal to 1 for dilute solutions:

$$H_2O \Leftrightarrow H^+ + OH^- \tag{2.49}$$

$$C_{H^+} C_{OH^-} = K_w = 10^{-14} \tag{2.50}$$

where:

$C_{OH^-}$ = equilibrium concentration of $OH^-$ (mol $L^{-1}$)
$K_w$ = equilibrium constant for ionization of water (mol$^2$ $L^{-2}$)

Equation 2.50 is the pragmatic—but also theoretically sound—basis for the definition of the pH scale. Defining $pH = -\log C_{H^+}$ and $pOH = -\log C_{OH^-}$, the pH scale at 25°C is determined by:

$$pH + pOH = -\log K_w = 14 \tag{2.51}$$

The requirement of electroneutrality (considering a case with an aqueous solution of compounds that only originate from water and the carbonate system including metal ions associated with the carbonate ions) implies that the balance between anions and cations be observed:

$$C_{H^+} + C_{Me^{n+}} = C_{OH^-} + C_{HCO_3^-} + 2C_{CO_3^{2-}} \tag{2.52}$$

where $Me^{n+}$ denotes the metal ion having the charge $n$.

Typically, the metals are calcium, magnesium, and iron, together with alkali metals, e.g., sodium and potassium.

Furthermore, the total concentration of inorganic carbon in the water phase is:

$$C_T = C_{H_2CO_3^*} + C_{HCO_3^-} + C_{CO_3^{2-}} \tag{2.53}$$

## 2.4.2 ALKALINITY AND BUFFER SYSTEMS

For aquatic systems, there is a need to express in quantitative terms the potential for a change in pH, i.e., either in order to maintain its value in a rather constant level or to quantify the extent of an increase or a decrease in pH. In wastewater of sewer networks, several factors may contribute to a change in pH, and even relatively small changes in its value may affect, for example, the degree of $H_2S$ emission (cf. Figure 2.7). Furthermore, there may also exist a need for calculation of the amount of acid or alkaline chemicals for its control. At wastewater treatment plants, the pH and changes in this value may highly affect the biological processes.

Two parameters quantify such impacts onto the pH: the alkalinity named A or TAL (Total ALkalinity) and the buffer capacity, $\beta$.

Alkalinity is defined as the capacity of an aqueous solution to neutralize acids:

$$A = \sum (\text{acid neutralizing bases}) - C_{H^+} \tag{2.54}$$

In other words, the alkalinity indicates the ability of the water to absorb hydrogen ions.

Equation 2.54 shows that the alkalinity of a water sample can be experimentally determined by titration, i.e., addition of a strong acid to a pH value where all bases have been neutralized by the acid.

The titration curve in terms of pH versus the amount of added acid is therefore of importance for quantifying the buffering characteristics of the wastewater. The buffering characteristics are thereby expressed by the ability to change the pH value by adding acids (or bases).

A corresponding parameter, the buffer capacity, is defined as follows:

$$\beta = -\frac{dC_{H^+}}{dpH} \tag{2.55}$$

where $\beta$ is the buffer capacity (mol L$^{-1}$ or M).

For an acid–base pair as exemplified by Equations 2.43 and 2.45, the buffering characteristics are closely related to the equilibrium between the acid and the corresponding base (cf. Table 2.1):

$$\text{acid} \Leftrightarrow \text{base} + H^+ \tag{2.56}$$

The equilibrium for this system is as follows:

$$\frac{C_{H^+} C_{base}}{C_{acid}} = K_a \tag{2.57}$$

Or expressed in logarithmic form (the so-called Henderson–Hasselbalch equation):

$$pH = pK_a - \log \frac{C_{acid}}{C_{base}} \tag{2.58}$$

Equation 2.58 shows that when adding acids or bases, the change in pH has its minimum value when $C_{acid}$ is equal to $C_{base}$. In other words, when pH = $pK_a$, the buffer capacity, $\beta$, has its maximum value.

## Example 2.10: Titration of a Wastewater Sample

According to Equation 2.55, the buffer capacity, $\beta$, is equal to the reciprocal slope of the titration curve. The titration curve shown in Figure 2.14 is performed by adding the strong acid, hydrochloric acid (HCl), to a municipal wastewater sample originating from a gravity sewer. The titration starts at pH = 7.6, the initial pH value of the sample, and ends at pH = 3.0. The curve shows that the minimum value of the slope occurs at approximately pH = 6.3, corresponding to a maximum value of the buffer capacity, $\beta$. The wastewater sample in question has therefore its maximum "resistance" to a change in pH at this value. At this pH value, the wastewater is characterized as well buffered, i.e., to a minor degree influenced by addition of acids or bases.

Figure 2.14 also shows that the bases in the wastewater sample have been neutralized at about pH = 4.3—where the change in pH is at its maximum value. This pH value is reached by adding 7.7 mmol $L^{-1}$ (mM) of HCl. According to Equation 2.54, the alkalinity is:

$$A = 7.7 \text{ mM}$$

More correctly, the unit for $A$ is expressed in equivalents, where 1 equivalent of mass, 1 eq, equals 1 mol of hydrogen ions. In this example, the alkalinity ($A$) is therefore equal to 7.7 meq $L^{-1}$.

Further details on the experimental determination of alkalinity will be shown in Example 2.11.

It is interesting to compare the characteristics of the titration curve for the wastewater sample with those of the carbonate system. As noted earlier, the maximum value of the buffer capacity, $\beta$, was found at about pH = 6.3. According to Equations 2.43 and 2.44, within the pH range of approximately 5–8, the carbonate system has a maximum value of $\beta$ at:

$$pH = pK_{S1} = -\log(4.47 \ 10^{-7}) = 6.35$$

This pH value for the carbonate system is almost identical with the value pH = 6.3, where the titration curve shows its maximum value of $\beta$. It is therefore

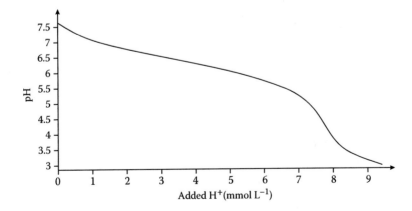

**FIGURE 2.14** Example of a titration curve, i.e., pH versus amount of added strong acid (hydrochloric acid), for a wastewater sample.

possible to conclude that the carbonate system is the dominating buffer system for the wastewater used in this example.

In several regions of the world, the pH value of municipal wastewater is found within the interval 7–8, with the interval 7.5–8 as the most characteristic. The carbonate system for such wastewaters typically plays a central role for the buffering characteristics.

If the carbonate system—with $H^+$ and $OH^-$ from dissociation of water—is the sole source for pH-influencing compounds, the alkalinity is expressed as follows:

$$A = C_{OH^-} + C_{HCO_3^-} + 2C_{CO_3^{2-}} - C_{H^+} \qquad (2.59)$$

Other ions than those included in Equation 2.59 may influence the alkalinity (cf. Section 4.5.1). However, the compounds from the carbonate system are often found to be important acid-neutralizing bases.

It is interesting to note the similarity between the ionic balance (Equation 2.52) and the expression for alkalinity (Equation 2.59). It must, however, be noted that in wastewater where several pH-neutral ions are present (e.g., $Na^+$ and $K^+$), these ions are included in an ionic balance but do not contribute to the alkalinity.

In addition to Example 2.10, Example 2.11 will illustrate and discuss the determination of the alkalinity.

## Example 2.11: Experimental Determination of Alkalinity in Wastewater

A wastewater sample with pH = 8.2 originates from a gravity sewer. It is anticipated that the carbonate system in the wastewater sample dominates the alkalinity. Equation 2.59 is therefore valid for the calculation of $A$.

Based on the measured pH value of the sample, according to Equation 2.51, it can be seen that the concentration of $OH^-$ is negligible:

$$pOH = 14 - pH = 14 - 8.2 = 5.8$$

Corresponding to $C_{OH^-} = 10^{-5.8} = 1.6 \times 10^{-6} \text{ mol L}^{-1}$.
The concentration of $H^+$ is therefore also without importance at pH = 8.2.

Based on Equation 2.46, the pH value of the wastewater sample also shows that the concentration of carbonate, $CO_3^{2-}$, is low and of negligible importance for the alkalinity compared with $HCO_3^-$:

$$\frac{C_{CO_3^{2-}}}{C_{HCO_3^-}} = 4.68 \times 10^{-11} 10^{pH} = 4.68 \times 10^{-11} 10^{8.2} = 7.4 \times 10^{-3}$$

It is therefore concluded that the hydrogen carbonate ion (bicarbonate ion, $HCO_3^-$), is the sole important acid-neutralizing base.

From Equation 2.44, it follows that:

$$pK_{S1} = -\log(4.47 \ 10^{-7}) = 6.35$$

At this pH value, it is furthermore seen that

$$\frac{C_{HCO_3^-}}{C_{CO_2(aq)}} = 1$$

At a pH value 2 units lower, i.e., pH = 4.35, this ratio is $10^{-2}$, and almost all $HCO_3^-$ is therefore transformed to $H_2CO_3$, i.e., practically $CO_2$. When titrating with a strong acid to pH = 4.35, all bases are, according to Equation 2.59, neutralized, and the alkalinity can be determined. The selection of exactly this pH value as the end-point of the titration is somewhat arbitrary. A pH value between 4.0 and 4.5 is equally feasible because no buffer system now exists, and major changes in pH therefore take place when adding acid. Titration is in practice carried out using automatic pH titration equipment. Previously, the pH indicator—methyl orange changing color within a pH range of 4.0–4.5—was used. For this reason, the end-point of the titration is still named the methyl orange endpoint.

A volume of 250 mL filtered wastewater was in the actual case titrated to pH = 4.4 as the end-point using 15.3 mL 0.1 M HCl. The amount of neutralizing acid used with the titration is therefore $0.1 \times 15.3 \times 10^{-3} = 1.53 \times 10^{-3}$ mol of $H^+$ corresponding to an equal initial amount of $HCO_3^-$ in 250 mL of the sample. The concentration of the hydrogen carbonate ion is therefore:

$$C_{HCO_3^-} = 1000/250 \times 1.53 \times 10^{-3} = 6.12 \times 10^{-3} \text{ mol L}^{-1}$$

According to Equation 2.58, the alkalinity of the wastewater is:

$$A = 6.1 \text{ meq L}^{-1}$$

## 2.5   IRON AND OTHER HEAVY METALS IN SEWERS

Heavy metals are common constituents of wastewater, often originating from commercial and industrial sources. Heavy metals constitute a group of substances where most of these are necessary for living organisms. However, the occurrence of these metals in excessive quantities in wastewater exerts potential negative impacts on treatment processes and toxic effects onto the surrounding environment. The heavy metal contents in wastewater are—except for iron—in general regulated by water quality standards.

Compared with other heavy metals, iron is typically found in relatively high concentrations (cf. Table 3.2). Furthermore, in collection systems, iron salts play a specific role as they can be added artificially as agents for sulfide control. A basic understanding of iron chemistry in wastewater is crucial for the optimal use of this method. The chemistry of iron will therefore be our main focus.

### 2.5.1   SPECIATION OF IRON AND SULFIDE

The addition of an iron salt as a method for sulfide control in wastewater collection systems is widely applied. In practice, iron chemistry is therefore closely associated with sulfide. As a consequence, focus on the chemistry of iron is related to the occurrence of sulfide.

Speciation in a chemical system refers to the fact that a substance, depending on external conditions, may appear in different chemical forms (species) with different properties. Speciation is relevant for substances in both soluble and solid forms and concerns, in principle, equilibrium conditions. Among a number of external conditions that affect the speciation of heavy metals, the acid–base characteristics (alkalinity or pH), the concentration ($C$ or $pC$), and the redox potential ($E$ or $p\varepsilon$) are the three main important factors (cf. Section 2.1.3).

The importance of the three parameters (pH, concentration, and redox potential) for the speciation of iron is briefly illustrated and exemplified in relation to sulfide:

1. pH

   As depicted in Figures 2.6 and 2.7, the pH value affects the equilibrium between the molecular form ($H_2S$) and the ionic form ($HS^-$) of sulfide:

$$H_2S(aq) \Leftrightarrow HS^- + H^+ \tag{2.60}$$

Only the ionic form can react with the soluble (ionic) form of ferrous iron ($Fe^{2+}$).

2. Concentration

   $Fe^{2+}$ and sulfide ($HS^-$) form a precipitate (FeS) with a very low solubility product constant ($L_{FeS}$):

$$L_{FeS} = C_{Fe+} \, C_{HS-} = 3.7 \times 10^{-19} \text{ g mol}^2 \text{ L}^{-2} \text{ at } 18°C \tag{2.61}$$

3. Redox potential

   The redox potential affects the relative occurrence of ferrous iron, Fe(II), and ferric iron, Fe(III):

$$Fe^{3+} + e^- \Leftrightarrow Fe^{2+} \tag{2.62}$$

Figure 2.5 shows that the equilibrium between $Fe^{2+}$ and $Fe^{3+}$ is established at a relatively high value of the redox potential. It is often omitted that $Fe^{3+}$ in water exists as a hydrated ion, $Fe(H_2O)_6^{3+}$, surrounded by six water molecules, although this fact causes the acidity of ferric iron.

Further details on the chemistry of iron and sulfide caused by the speciation related parameters are shown in the following.

## 2.5.2 Sulfide Control by Addition of Iron Salts

In brief and from a traditional point of understanding, sulfide control in wastewater collection systems relies on the addition of iron salts forming a precipitate of insoluble iron(II) sulfide, FeS. Emission of gaseous hydrogen sulfide ($H_2S(gas)$) causing toxic impacts, odor problems and concrete corrosion is thereby prevented. It should be noted that odor problems originating from volatile organic compounds are generally not affected by this control methodology. The use of iron salt addition in wastewater collection systems for sulfide control is discussed in Section 7.3.2. In this section, the basic properties of iron are the subject.

Sulfide control by addition of iron salts have been known and applied for decades (Pomeroy and Bowlus 1946). Manuals dealing with this subject have been published by both governmental agencies and societies (e.g., USEPA 1985; ASCE 1989; Melbourne and Metropolitan Boards of Works 1989). More recent studies on sulfide control by addition of iron salts have been conducted by Nielsen et al. (2005a, 2008).

The combined use of ferric iron ($Fe^{3+}$) and ferrous iron ($Fe^{2+}$) affects the efficiency of sulfide control (cf. Equation 2.62). The following two reactions are central for sulfide removal:

$$Fe^{2+} + HS^- \rightarrow FeS + H^+ \tag{2.63}$$

$$2\,Fe^{3+} + HS^- \rightarrow 2\,Fe^{2+} + S^0 + H^+ \tag{2.64}$$

Sulfide in anaerobic wastewater is, according to Equation 2.63, removed by precipitation of insoluble FeS, whereas sulfide, according to Equation 2.64, is oxidized to elemental sulfur, simultaneously reducing Fe(III) to Fe(II) (cf. Equation 2.62). In theory, it is therefore concluded that $Fe^{3+}$ has a higher capacity for control of sulfide than $Fe^{2+}$, because it results in the removal of sulfide by oxidation of $HS^-$ to elemental sulfur followed by precipitation with $Fe^{2+}$. Although ferric iron is in favor of ferrous iron in terms of its theoretical capacity, it is important that the reaction rate of Equation 2.64 is relatively slow. Equation 2.63 is, in terms of kinetics, very fast although it is not an immediate occurring reaction.

The chemistry of iron is related to the biological systems in wastewater. Although Equation 2.64 is a reaction that can proceed chemically, it is also a biochemical redox reaction initiated by a specific group of bacteria—the chemoautotrophic bacteria (cf. Section 3.1.3). Moreover, other groups of bacteria, the chemoheterotrophic bacteria, may reduce $Fe^{3+}$ to $Fe^{2+}$ (cf. Section 3.1.3). If the latter is the case, sulfide is not a part of this process, and ferric iron has therefore—compared with Equation 2.64—no additional capacity for sulfide control.

The kinetics of the two reactions, Equations 2.63 and 2.64, and the importance of the microbial system on the chemistry of iron are complex. The literature gives no conclusive answer on how to optimize sulfide control by addition of iron salts. Investigations by Nielsen et al. (2005a) show that the reaction rate of Equation 2.63 is slow under more or less steady-state anaerobic conditions (redox potential of less than about –200 mV). The reaction rate increases when the concentration of $Fe^{3+}$ is increased and addition of the salt occurs directly to sulfide-containing anaerobic wastewater. Although the solubility product constant of FeS is very low (cf. Equation 2.61), the precipitation pattern of FeS is not correspondingly fully efficient, leaving excess of $HS^-$ in the solution when $Fe^{2+}$ is added. It is likely that the formation of complexes of $Fe^{2+}$ with organic and inorganic complexing agents in wastewater such as EDTA (ethylenediaminetetraacetate) and linear alkylbenzene sulfonate may reduce the amount of Fe(II) available for precipitation. Furthermore, investigations show that FeS and not—as often claimed—other forms of sulfide precipitates (e.g., $Fe_3S_4$) are formed (Nielsen et al. 2005a, 2008).

According to Equation 2.60 and Figure 2.7, lowering the pH value relatively increases the molecular form of sulfide, $H_2S$, which does not react with iron salts. In practice, the pH value should be larger than 7.0, and iron salts are added in excess to obtain an efficient precipitation of sulfide (Nielsen et al. 2008). Not until a pH value of about 8 is reached, does iron salt addition becomes (almost) fully efficient.

The answer on how to manage and operate sulfide control in practice using iron salts will be further discussed in Section 7.3.2. Here, it is briefly summarized that the following conditions may favor optimal sulfide removal:

- pH and iron concentration
    At pH 8, the efficiency of iron salts added is almost 100%, but is only about 40% at pH 7. The pH value can be adjusted by addition of a base, e.g., $Ca(OH)_2$, to improve the efficiency of iron salt addition.
- Redox potential and concentration
    When applying iron salts for sulfide control at upstream locations in the sewer network, a combination of a ferric salt and a ferrous salt can be added, e.g., using a 1:1 mixture. When applied at downstream locations, only the fast reacting ferrous salts are recommended.

### 2.5.3 METALS IN SEWER BIOFILMS

The occurrence of sulfide in anaerobic biofilms of sewers favors conditions for precipitation and accumulation of metal sulfides caused by diffusion of metals from the anaerobic water phase into the biofilm. As an example, zinc sulfide and copper sulfide can accumulate under such conditions (Nielsen et al. 2005b). The concentration level of metals in the biofilm is typically in the order of a few milligrams per gram of total solids, and is typically highest for zinc (Gutekunst 1988; Nielsen et al. 2005b). Furthermore, iron accumulates in biofilms, and particularly seen when added for sulfide control. It is mainly the aerobic biofilms that accumulate metals from the wastewater phase in sewers.

Metals that accumulate in biofilms are slowly released, probably mainly by detachment of the biofilm, and typically within a period of approximately 1 to 2 weeks. Analysis of the metal content in the biofilm can therefore be used for identification of, e.g., illegal abrupt metal discharges to the sewer network (Gutekunst 1988).

### REFERENCES

ASCE (American Society of Civil Engineers) (1989), *Sulfide in Wastewater Collection and Treatment Systems, ASCE Manuals and Reports on Engineering Practice No. 69*, American Society of Civil Engineers, New York.
Atkins, P. and J. de Paula (2002), *Physical Chemistry*, Oxford University Press, UK, p. 1150.
Boon, A.G. (1995), Septicity in sewers: causes, consequences and containment, *Water Sci. Technol.*, 31(7), 237–253.

Characklis, W.G. and K.C. Marshall (eds.) (1990), *Biofilms*, John Wiley & Sons, New York, USA, p. 796.

Christ, O., P.A. Wilderer, R. Angerhöfer, and M. Faulstich (2000), Mathematical modeling of the hydrolysis of anaerobic processes, *Water Sci. Technol.*, 41(3), 61–65.

Dold, P.L., G.A. Ekama, and G.v.R. Marais (1980), A general model for the activated sludge process, *Prog. Water Technol.*, 12(6), 47–77.

Gutekunst, B. (1988), Sielhautuntersuchungen zur Einkreisung Schwermetallhaltiger Einleitungen [Investigations of biofilms for the identification of heavy metal discharges], Band 49, Institut für Siedlungswasserwirtschaft, Universität Karlsruhe, p. 141.

Harremoës, P. (1978), Biofilm kinetics. In: R. Mitchell (ed.), *Water Pollution Microbiology*, Vol. 2, Wiley Interscience, New York, pp. 71–109.

Henze, M., P. Harremoës, J. la Cour Jansen, and E. Arvin (1995), *Wastewater Treatment— Biological and Chemical Processes*, Springer-Verlag, New York, p. 383.

Henze, M., C.P.L. Grady Jr., W. Gujer, G.v.R. Marais, and T. Matsuo (1987), *Activated Sludge Model No. 1, Scientific and Technical Report No. 1*, IAWPRC (International Association on Water Pollution Research and Control), London.

Hvitved-Jacobsen, T., J. Vollertsen, and A.H. Nielsen (2010), *Urban and Highway Stormwater Pollution: Concepts and Engineering*, CRC Press, Boca Raton, FL, p. 347.

Lide, D.R. (ed.) (2009), *Handbook of Chemistry and Physics*, 89th ed., CRC Press, Boca Raton, FL, USA.

Madigan, M.T., J.M. Martinko, and J. Parker (1997), *Biology of Microorganisms*, 8th edn., Prentice-Hall International, London, p. 986.

Melbourne and Metropolitan Boards of Works (1989), *Hydrogen Sulfide Control Manual— Septicity, Corrosion and Odour Control in Sewerage Systems*, Volumes 1 and 2, Technological Standing Committee on Hydrogen Sulphide Corrosion in Sewerage Works, Melbourne, Australia.

Monod, J. (1949), The growth of bacterial cultures, *Annu. Rev. Microbiol.*, 3, 371–394.

Nazaroff, W.W. and L. Alvarez-Cohen (2001), *Environmental Engineering Science*, John Wiley & Sons, New York, p. 690.

Nielsen, A.H., P. Lens, J. Vollertsen, and T. Hvitved-Jacobsen (2005a), Sulfide–iron interactions in domestic wastewater from a gravity sewer, *Water Res.*, 39, 2747–2755.

Nielsen, A.H., P. Lens, T. Hvitved-Jacobsen, and J. Vollertsen (2005b), Effects of aerobic– anaerobic transient conditions on sulfur and metal cycles in sewer biofilms, *Biofilms*, 2, 1–11.

Nielsen, A.H., T. Hvitved-Jacobsen, and J. Vollertsen (2008), Effects of pH and iron concentrations on sulfide precipitation in wastewater collection systems, *Water Environ. Res.*, 80(4), 380–384.

Nielsen, P.H. and T. Hvitved-Jacobsen (1988), Effect of sulfate and organic matter on the hydrogen sulfide formation in biofilms of filled sanitary sewers, *J. Water Pollut. Control Fed.*, 60(5), 627–634.

Pomeroy, R. and F. Bowlus (1946), Progress report on sulfide control research, *Sewage Works J.*, 13, 597–640.

Pourbaix, M. (1974), *Atlas of Electrochemical Equilibria in Aqueous Solutions*, 2nd English ed., National Association of Corrosion Engineers, Houston, TX, USA.

Snoeyink, V.L. and D. Jenkins (1980), *Water Chemistry*, John Wiley & Sons, New York, p. 463.

Speece, R.E. (2008), *Anaerobic Biotechnology and Odor/Corrosion Control*, Archae Press, Nashville, TN, USA, p. 586.

Stumm, W. and J.J. Morgan (1996) *Aquatic Chemistry—Chemical Equilibria and Rates in Natural Waters*, 3rd edn, John Wiley & Sons, New York, p. 1022.

USEPA (US Environmental Protection Agency) (1985), *Odor and Corrosion Control in Sanitary Sewerage Systems and Treatment Plants*, EPA-625/1-85-018, US Environmental Protection Agency, Washington, D.C.

Williamson, K. and P.L. McCarty (1976a), A model of substrate utilization by bacterial films, *J. Water Pollut. Control Fed.*, 48(1), 9–24.

Williamson, K. and P.L. McCarty (1976b), Verification studies of the biofilm model for bacterial substrate utilization, *J. Water Pollut. Control Fed.*, 48(2), 281–296.

# 3 Microbiology in Sewer Networks

Wastewater in sewers is, in general, a high-rate system for microbiological transformations. Wastewater is a very complex microbiological system including a great number of microorganisms divided into specific classes. These microorganisms occur in the water phase, in biofilms (slimes), in sewer deposits, and at walls exposed to the sewer atmosphere. Several fractions of organic matter that are substrates for these microorganisms are found in wastewater. Furthermore, inorganic compounds can microbiologically be transformed.

A detailed understanding of the processes in such a complex system is difficult to achieve because of the great number of possible processes, interactions, and competitions between the microorganisms and the substrates. On the other hand, if one realizes the constraints that the microbial system—including both microorganisms and substrates—is subject to, it will be possible to understand the behavior of the system and apply this understanding. The understanding of wastewater as a microbiological system in a sewer is one of the key approaches to sewer process engineering.

A detailed classification of microorganisms and the microbiological details of their metabolism—the different steps of reactions for maintaining activity—are not the major subject in this book. The formulation in quantitative terms and impacts of sewer processes does not have its starting point in such detailed descriptions. It is, for practical purposes, not feasible. What is important, however, is to determine under which external conditions, e.g., temperature, pH, redox potential, and substrate conditions, a group of microorganisms are active. Microorganisms, which are divided into groups of active biomass combined with their consumption of substrates, constitute under such external conditions, a central and pragmatic platform for description of in-sewer processes.

The microbial system in a sewer system is complex and subject to tremendous variability in time and place. It is, in this respect, central—but also difficult—to balance the understanding of the microbial system between details that cannot be applied for quantification and prediction, and a crude description that—on the other hand—is useless. It is a challenge for this chapter to walk the edge between these two extremes.

## 3.1 WASTEWATER: SOURCES, FLOWS, AND CONSTITUENTS

Sewer processes, in particular those initiated by microorganisms, are the focus in this book. Wastewater characteristics are therefore central, and these aspects will be dealt with in detail. As a starting point, the quality and quantity characteristics of wastewater from a traditional viewpoint will be briefly outlined in this section.

Wastewater is a by-product of human activities. It is a result of our basic need for water (drinking water) as a "process medium" in our bodies, and the way we have "designed" our daily way of living in terms of activities such as toilet flushing and industrial production. The quantity and quality of wastewater therefore depend on culture and vary accordingly in both time and place. These variations in time and between countries are significant. Because wastewater is a complex mixture of potentially harmful substances for both humans and the environment, it is necessary—particularly in urban settlements—to collect and treat it before discharge.

The fact that both quality and quantity of wastewater varies considerably calls for its characterization before any system for its collection, conveyance, and treatment is established. The specific data for wastewater given in this book are therefore just examples, although it is also our aim to impart information at some general level.

### 3.1.1 Sources and Flows of Wastewater

Traditionally, wastewater from urban settlements is termed municipal wastewater. Municipal wastewater can be subdivided into

- Domestic wastewater, i.e., wastewater from residences (households), institutions, and commercial areas. Domestic wastewater is also called sanitary wastewater.
- Industrial wastewater

In addition, stormwater runoff from urban areas and roads is, during runoff events, mixed with municipal wastewater in the combined sewer networks. Furthermore, sewer networks with leaks may interact with the surrounding groundwater and soil system by infiltration or exfiltration, e.g., depending on the groundwater level. Type and performance of the collection system is therefore important for the level of wastewater flow rates, e.g., in combined sewer systems varying with the precipitation pattern. The flow rates also vary considerably over time because of the daily varying activity, and local lifestyle results in variability between countries (cf. Table 3.1).

It should be noted that the flow rates shown in Table 3.1 are typical examples and not values recommended for design of sewer networks.

**TABLE 3.1**

**Examples of Wastewater Flow Rates from Residential Areas in Selected Countries**

| Country | High Flow Rate ($m^3$ capta$^{-1}$ day$^{-1}$) | Low Flow Rate ($m^3$ capta$^{-1}$ day$^{-1}$) |
|---|---|---|
| United States | 0.40 | 0.15 |
| UK | 0.16 | 0.11 |
| Denmark | 0.14 | 0.08 |
| France | 0.11 | 0.07 |

The flow rates in Table 3.1 are shown as person-related numbers. A "unit" expressed in terms of PE (population equivalent) is used worldwide. In general, the following unit is in use:

$$1 \text{ PE} = 0.2 \text{ m}^3 \text{ capta}^{-1} \text{ day}^{-1}$$

The unit of PE should be understood as a "constant" regardless of the fact that the actual flow rate from a single average person is different or whether the wastewater may have originated from industry or commercial sources. The unit 1 PE = $0.2 \text{ m}^3 \text{ day}^{-1}$ is a convenient measure of capacity when comparing and designing both sewer networks and wastewater treatment facilities.

### 3.1.2 Wastewater Quality

The contents and extent of constituents in wastewater reflect the biochemical function of our bodies as well as our general way of life. Globally, therefore, there are similarities as well as considerable variability in which constituents are found in wastewater and in which amount they occur.

The following five main groups of organic and inorganic constituents are typically dealt with because they normally occur in wastewater flows and, in different ways, affect sewer processes and wastewater treatment. Furthermore, they potentially result in adverse effects when discharged into the environment. The grouping is a pragmatic approach that might be differently defined:

- Organic matter of different degree of biodegradability
- Nutrients
- Heavy metals
- Organic micropollutants
- Microorganisms, pathogenic and nonpathogenic

The first four groups in this list concern a number of organic and inorganic substances. The last group includes microorganisms that play a central role in biochemical processes and thereby transformation of the constituents in a sewer. The pathogenic microorganisms potentially affect human health. These five groups of constituents only represent a very rough classification of wastewater components. A number of details are typically needed.

It is important to note that the five groups of constituents can be subdivided in both specific compounds and in which form these constituents occur, e.g., soluble or particulate. In addition to this grouping, a number of specific constituents and parameters are often needed to give a complete characterization of wastewater. The following list includes the commonly selected parameters:

- Solids (suspended solids)—this group deals basically with the solubility characteristics of the constituents in water and their association with a solid phase (particles).

- Dissolved oxygen (DO), nitrate, and sulfate—are substances potentially related to the redox conditions and thereby potential electron acceptors for redox processes.
- Temperature
- Redox potential—a measure of both equilibrium and process-related conditions.
- Conductivity—an overall measure of ionic substances
- pH
- Alkalinity—this is particularly associated with the carbonate system
- Specific constituents and parameters, e.g., chloride, hydrogen sulfide, and odorous substances.

Because of the solute–solid interactions and exchange of substances across interfaces, it is important that the occurrence of the constituents is of interest not just in a water phase (suspension) but also in sediments and biofilms.

The overall variability in quality and quantity of constituents in wastewater of sewers means that no "typical" list of concentrations for wastewater compounds basically exists. The numerical variability of the parameters shown in Table 3.2 should, however, cover what might be realistic for wastewater in the United States, Europe, and Australia, with U.S. figures occupying a place in the lower end of the scale. It should be noted that Table 3.2 is only shown to give an understanding of the possible concentration levels. It is also clear that the level of flow rate per person in a given catchment affects the level of concentration, i.e., infiltration into a sewer system and wet weather flows in combined sewers may typically result in decreased concentrations compared with the data shown in Table 3.2.

Several other values of constituents and parameters than those shown in Table 3.2 are of interest. Such information is often needed when dealing with details of sewer processes.

Quantification of the load of a constituent is, in general, based on flow and concentration measurements. Corresponding to the PE concept shown in Section 3.1.1 for flow (1 PE = 0.2 m$^3$ day$^{-1}$), a corresponding indication is also possible for selected constituents. The most commonly used is:

$$1 \text{ PE} = 60 \text{ g BOD capta}^{-1} \text{ day}^{-1}$$

As mentioned for the flow-related PE, the unit should be understood as a "constant" regardless of whether the actual load of BOD originates from a single average person or from industry or commercial wastewater sources.

### 3.1.3   AN OVERVIEW OF THE MICROBIAL SYSTEM IN WASTEWATER OF SEWERS

The traditional way of characterizing wastewater quality focuses on its chemical composition based on traditional analytical techniques (cf. Section 3.1.2). When dealing with sewer processes, further details will be needed. Such details are, e.g., information on the contents of active (living) biomass in terms of microorganisms and classification of organic matter as "substrates" including their biodegradability.

## TABLE 3.2

## Examples of Wastewater Composition "Typical" for Sewer Networks in the United States, Europe, and Australia

| Constituent | Symbol | Unit | Low Strength | Medium Strength | High Strength |
|---|---|---|---|---|---|
| Total suspended solids (TSS) | $X_{TSS}$ | g TSS m$^{-3}$ | 150 | 300 | 450 |
| Total chemical oxygen demand (COD$_{tot}$) | $C_{COD}$ | g O$_2$ m$^{-3}$ | 200 | 500 | 800 |
| Dissolved chemical oxygen demand (COD$_{sol}$) | $S_{COD}$ | g O$_2$ m$^{-3}$ | 100 | 200 | 300 |
| Biochemical oxygen demand, BOD$_5$ (20°C) | $C_{BOD}$ | g O$_2$ m$^{-3}$ | 100 | 250 | 350 |
| Total organic carbon (TOC) | $C_{TOC}$ | g C m$^{-3}$ | 100 | 150 | 250 |
| Total nitrogen (N) | $C_{TN}$ | g N m$^{-3}$ | 20 | 40 | 80 |
| Total ammonium nitrogen (NH$_4^+$ + NH$_3$) | $S_{NH3}$ | g N m$^{-3}$ | 10 | 30 | 50 |
| Nitrate nitrogen (NO$_3^-$) | $S_{NO3}$ | g N m$^{-3}$ | 0 | 0.5 | 1 |
| Total phosphorous (P) | $C_{TP}$ | g P m$^{-3}$ | 5 | 8 | 15 |
| Orthophosphate | $S_{PO4}$ | g P m$^{-3}$ | 3 | 7 | 10 |
| Copper (Cu) | $C_{Cu}$ | mg Cu m$^{-3}$ | 30 | 60 | 90 |
| Zink (Zn) | $C_{Zn}$ | mg Zn m$^{-3}$ | 100 | 200 | 300 |
| Lead (Pb) | $C_{Pb}$ | mg Pb m$^{-3}$ | 20 | 60 | 80 |
| Cadmium (Cd) | $C_{Cd}$ | mg Cd m$^{-3}$ | 1 | 2 | 3 |
| Iron (Fe) | $C_{Fe}$ | mg Fe m$^{-3}$ | 500 | 1000 | 1500 |
| Polycyclic aromatic hydrocarbons (PAH) | $C_{PAH}$ | mg PAH m$^{-3}$ | 0 | 0.5 | 2 |
| Total coliform bacteria[a] | TC | Number per 100 ml | $10^7$ | $10^8$ | $10^9$ |
| Fecal coliform bacteria[a] | FC | Number per 100 ml | $10^5$ | $10^6$ | $10^7$ |

*Note:* X, concentration of the constituent given for the particulate form; C, total concentration for the constituent; S, concentration value for the soluble form of the constituent.

[a] TC, FC, and the TC/FC ratio vary considerably from one type of wastewater to another.

Furthermore, a sewer is not just a transport system; it is a "reactor" for microbial, chemical, and physicochemical processes.

The sewer system thereby plays a role not just for the sewer network itself but also for the processes occurring in the wastewater treatment plant located downstream and for the impacts onto the surrounding atmosphere and water environment. Concerning wastewater treatment, the microbial processes in the sewers and in the treatment plant should be commonly viewed, i.e., using comparable concepts of process descriptions. In this way, the "treatment starts at the sink" and not just at the inlet to what is traditionally named as a "treatment plant."

Several phenomena contribute to the complexity of microbial processes in waste-water of a sewer system:

- Wastewater is a matrix with a great variety of microorganisms and with different substrates, all varying in time and place.
- The microbial processes may take place under changing aerobic, anoxic, and anaerobic conditions, i.e., at different levels of redox potential.
- The microbial processes proceed in different subsystems: the suspended wastewater phase, the biofilms, the sediments, and the solid surfaces of the sewer in contact with the atmosphere.
- The microbial processes in the sewer interact across the boundaries of the subsystems in terms of exchange of substrates, electron acceptors, biomass, and volatile substances.

The redox conditions established in a sewer network are of major importance for the development of the microbial community, the course of the microbial processes, and the entire impact on the sewer, the treatment processes, and the environment.

The following examples will illustrate why the quality of the wastewater in a sewer is of importance:

- Removal of oxygen-consuming substances of wastewater in sewers, e.g., in terms of biochemical oxygen demand (BOD) or chemical oxygen demand (COD), is important when considering mechanical treatment. The impact on the environment of soluble species is thereby reduced, and production of biomass in the sewer may result in improved removal of particulate COD in the mechanical treatment steps.
- Preservation of readily biodegradable substrates in sewers may be crucial for denitrification (nitrogen removal) and phosphorus removal in advanced treatment processes.
- In-sewer processes, e.g., hydrogen sulfide formation and fermentation, may cause corrosion, toxicity, and odor problems.
- Precipitation of trace elements, caused by sulfide, may have an impact on the performance of biological systems in the treatment plant.

These examples show that when excluding substances such as heavy metals and organic micropollutants, the wastewater quality is closely related to the microbial biodegradability of organic matter. This characteristic feature is not sufficiently reflected by bulk parameters such as COD and BOD. The four generally expressed examples show that it is important to manage the microbial transformations of the wastewater in sewer networks as a basis for its "design" according to given problems and beneficial uses.

Biomass and substrate must be separately described in terms of parameters to establish a concept for classification of wastewater directed toward a description of the microbial processes. For several reasons, e.g., to allow widespread application and to observe a basic mass balance, organic matter expressed in terms of different fractions of COD is acceptable for wastewater quality description. BOD is, in this respect, not a feasible parameter because it does not comply with the mass balance requirement. According to the concepts used in the activated sludge models for wastewater transformations in treatment plants, the fractionation of COD in wastewater of a sewer network can be subdivided as outlined in Figure 3.1 (Henze et al. 1987, 1995, 2000). A description of the interactions between sewer processes and treatment plant processes is thereby within reach.

From the very beginning, it is important to note that activated sludge in treatment plants and wastewater, as it occurs in sewer networks, are two different matrices for microbial activity. The process and model concept developed for activated sludge in wastewater treatment systems is therefore not directly applicable for wastewater in sewers (cf. Chapters 5 and 8).

The concept for COD fractionation as described in Figure 3.1 is simple compared with the complexity of the microbial systems. However, it reflects the basic aspects of what is important related to the processes in a microbial system: a distinction between active biomass and substrate in both soluble and particulate forms. At the same time, an important engineering aspect in terms of a mass balance for COD is a fundamental characteristic that can be observed.

As already outlined in Table 1.1 and further described in Section 2.1, the redox potential determines which microorganisms are potentially effective and which external electron acceptor (typically, $O_2$, $NO_3^-$, and $SO_4^{2-}$) is active. These conditions are the basis for characterization of which type of process (aerobic, anoxic, and anaerobic) is initiated by the microorganisms. In addition, the nutritional requirements of the microorganisms, i.e., their energy source and their carbon source classify the microorganisms:

- *Energy source:* In general, three main types of energy source for microorganisms exist: organic compounds, inorganic compounds, and light (sunlight). It is clear that the latter source of energy is not relevant in sewer networks, and the corresponding organisms called *phototrophic organisms*

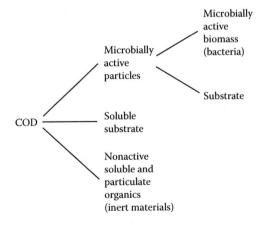

**FIGURE 3.1** Concept for fractionation of COD in wastewater of sewers.

are basically not present in an active state. Microorganisms that use either organic or inorganic compounds as their energy source are named *chemotropic organisms*. If organic matter (as electron donor) is the energy source, the microorganisms are further classified as *chemoorganotrophic organisms*. If an inorganic substance is the energy source, the microorganisms are named *chemoautotrophic organisms* or briefly, just *autotrophic organisms*. These latter types of microorganisms use typically rather simple inorganic (reduced) substances such as ammonia ($NH_3$ and $NH_4^+$) and hydrogen sulfide ($H_2S$ and $HS^-$) as electron donor.

- *Carbon source:* Carbon constitutes the main structural element of any type of organic matter and is therefore the most central and characteristic element of all living organisms. The carbon source for the organisms is either already produced organic matter—basically originating from plants, but in sewers typically as wastes from animals and humans—or inorganic carbon, i.e., carbon dioxide and carbonates. From a gas phase or a low pH water phase, the inorganic carbon occurs as carbon dioxide ($CO_2$). At pH values typical for wastewater, the inorganic carbon occurs as carbonates (normally $HCO_3^-$). If organic matter is the carbon source, the microorganisms are named *heterotrophic organisms*. If the carbon source is inorganic, the microorganisms are classified as *autotrophic organisms*.

The electron acceptor, the energy source, and the carbon source are the three main characteristics that determine the classification of microorganisms in terms of process characteristics. In wastewater of a sewer network, the heterotrophic bacteria dominate the microbial community, i.e., organic compounds are the carbon source. Furthermore, the energy source (the electron donor) for these heterotrophic bacteria is primarily also organic compounds, i.e., the heterotrophs that dominate wastewater in sewers are heterotrophic, chemoorganotrophic bacteria. If DO is the electron acceptor, the bacteria are therefore aerobic, heterotrophic, and chemoorganotrophic. These organisms are named aerobic chemoheterotrophic bacteria or simply aerobic heterotrophic microorganisms.

Furthermore, chemoautotrophic—briefly named autotrophic—microorganisms are found in sewers although they do not dominate in terms of their biomass. However, such microorganisms can play an important role because of the specific processes they cause. Examples of very important autotrophic organisms are the aerobic, chemoautotrophic, bacteria that oxidize hydrogen sulfide to elemental sulfur or sulfuric acid. This oxidation process proceeds in the water phase as well at the sewer walls (cf. Sections 5.5.1). At the sewer walls exposed to the sewer atmosphere, specific species of microorganisms may cause concrete corrosion. Sulfur is in the inorganic compound hydrogen sulfide, $H_2S$, highly reduced and when oxidized to sulfur in either elemental sulfur or sulfuric acid, energy for the bacteria is provided. The carbon source for the bacteria is hydrogen carbonate $HCO_3^-$ or, at the very low pH values at the sewer walls, also $CO_2$. It is important to note that organic matter plays no role in the activity of these aerobic, chemoautotrophic microorganisms.

The classification of the microorganisms is simple and, from a microbiology point of view, rather crude (Stanier et al. 1986). However, this provides an easy overview of the conditions under which the microbial sewer processes proceed. For practical uses, when dealing with sewer processes, the described classification is very central.

## 3.2  MICROBIAL REACTIONS AND QUALITY OF SUBSTRATE

The general overview and fundamental characteristics of wastewater in sewers as a matrix for microbial processes are the subjects of Section 3.1. In this section, the aspects related to the activity of the microbial system will be further described.

### 3.2.1  AEROBIC AND ANOXIC MICROBIAL PROCESSES

An essential part of the microorganisms in wastewater of sewers are potentially active under aerobic conditions. Furthermore, several groups of these microorganisms are also active under anoxic conditions. This means that such microorganisms may use both DO and nitrate as the terminal electron acceptor. This is important because changes in the availability of these substances may not result in dramatic impacts on the outcome of the sewer processes. In contrast to anaerobic degradation, the aerobic and anoxic pathways of organic matter degradation are, in general, identical.

Under aerobic conditions, the complex organic molecules—the electron donors—are broken down (oxidized) in the respiration process by passing electrons to molecular oxygen that is thereby reduced. The organic carbon is simultaneously transformed into inorganic carbon and released as $CO_2$ or $HCO_3^-$. The nitrogen and phosphorus contents of the organic molecules are then released as inorganic substances, ammonia $\left(NH_3/NH_4^+\right)$ and phosphates. The energy produced by the degradation of organic matter is used for both growth processes and maintenance, i.e., nongrowth purposes, by the microorganisms (cf. Figure 2.2).

An example of an aerobic, heterotrophic microbial reaction is the degradation of glucose (cf. Example 2.4):

$$\frac{1}{24}C_6H_{12}O_6 + \frac{1}{4}O_2 \rightarrow \frac{1}{4}CO_2 + \frac{1}{4}H_2O \tag{3.1}$$

The corresponding anoxic process, using nitrate as electron acceptor and ending the process by formation of $N_2$, is (cf. Example 2.5):

$$\frac{1}{24}C_6H_{12}O_6 + \frac{1}{5}NO_3^- + \frac{1}{5}H^+ \rightarrow \frac{1}{10}N_2 + \frac{1}{4}CO_2 + \frac{7}{20}H_2O \tag{3.2}$$

Aerobic respiration with DO as the terminal electron acceptor is an efficient process for energy metabolism. Under DO nonlimiting conditions, the oxygen uptake rate (OUR) may vary considerably for wastewater depending on the density (concentration) and the activity of the bacteria and the amount and biodegradability of the organic substrate. Typically, OUR values have been measured in the range of 2–25 g $O_2$ m$^{-3}$ h$^{-1}$ (Boon and Lister 1975; Matos and de Sousa 1996; Hvitved-Jacobsen and Vollertsen 1998; Vollertsen and Hvitved-Jacobsen 2002). The most biodegradable molecules—those with high biodegradation rates—are often the most volatile, e.g., volatile fatty acids (VFAs). Biodegradable, volatile substances discharged into or produced in an aerobic sewer network are typically efficiently removed.

Anoxic conditions require the absence of DO and the presence of nitrates. Such conditions are typically only found when artificially implemented. The addition of nitrate to wastewater is widely used as a control measure to avoid anaerobic conditions and sulfide formation in sewers (cf. Section 7.3.1.3).

The redox potential—the availability of an electron acceptor—is a central measure for which microorganisms are potentially active. The competition for substrates—the electron donor—between the different species of microorganisms is another dominating factor for which processes might proceed. A third important factor for which species will dominate is the growth rate of the microorganisms. The most fast growing aerobic heterotrophic bacteria can outcompete those species having lower growth rates. As a group, the aerobic, heterotrophic bacteria have relatively high growth rates and are therefore dominating microorganisms. The aerobic, chemoautotrophic and ammonia oxidizing bacteria (nitrifying bacteria), which are well known for nitrification in low-loaded wastewater treatment plants, are outcompeted and washed out in sewers. Nitrification, i.e., oxidation of ammonia to nitrite and nitrate, will therefore not occur under prevailing conditions in sewer networks.

A specific group of aerobic, autotrophic bacteria, *Thiobacilli*, is found at the sewer walls exposed to the atmosphere where it is oxidizing hydrogen sulfide to elemental sulfur and sulfuric acid.

### 3.2.2 ANAEROBIC MICROBIAL PROCESSES

Three main types of anaerobic microbial processes are important in sewer systems:

1. Respiration
2. Fermentation
3. Methanogenesis (methane formation)

Respiration and fermentation proceed to support anaerobic microorganisms with energy (cf. Figure 2.2). The energy source as well as the carbon source is biodegradable organic matter and these microorganisms are therefore anaerobic, heterotrophic organisms. In contrast to respiration, fermentation does not require the participation of an external electron acceptor, for example sulfate. Organic substrates undergo in fermentation processes a balanced series of oxidative and reductive reactions, i.e., organic matter reduced in one step of the process is oxidized in another.

Fermentation results in a partial breakdown of organic matter and yields organic by-products with a low molecular weight (e.g., VFAs, along with $CO_2$). Compared with the aerobic respiration, the fermentation is rather inefficient in terms of energy production. However, the fermentation by-products can, to some extent—and in addition to organic substrate that is fermentable—be consumed by the anaerobic sulfate-reducing bacteria, which thereby make use of sulfate as the terminal electron acceptor (Nielsen and Hvitved-Jacobsen 1988).

A third group of anaerobic microorganisms are the methanogens, also named methane-producing microorganisms. The methanogens can use the low molecular weight fermentation products as their energy source, producing methane ($CH_4$) as a by-product. In general, the redox potential for the methane formation processes is less than about $-0.2$ V, i.e., slightly below the redox potential for sulfate reduction (cf. Figure 2.5 and Section 2.1.3). The sulfate-reducing bacteria (the hydrogen sulfide–producing bacteria) have their activity level at a higher redox potential (typically between $-0.15$ and $-0.10$ V) compared with that of methanogens. As a first estimate, it is therefore assumed that methanogens confine to areas where sulfate is depleted and the redox potential is stabilized at values of $-0.25$ to $-0.3$ V (Cappenberg 1974). However, and as correspondingly described for the aerobic, heterotrophic microorganisms (cf. Section 3.2.1), the competition for substrate will also play a role. The occurrence of low molecular substrates in wastewater of sewers that are needed for both sulfate-reducing bacteria and methanogenic microorganisms allows for hydrogen sulfide formation and methane production to simultaneously take place.

Several species of methanogenic microorganisms exist. Some of these (the chemoautotrophic or hydrogenotrophic species) use $CO_2$ as their carbon source. Furthermore, $CO_2$ acts as the electron acceptor and hydrogen, $H_2$, as the electron donor. For both methanogens and sulfate reducers, it is required that electron acceptors [DO, nitrate, nitrite, ferric iron ($Fe^{3+}$)] resulting in a redox potential higher than about $-0.2$ V be depleted.

The anaerobic decomposition of organic matter by fermentation, methanogenesis (methane formation), and sulfate respiration is exemplified in Table 3.3. In this table, it is readily seen that volatile organic compounds (VOCs) are typical degradation products from these types of sewer processes. The examples in Table 3.3 also show that the anaerobic environment in a sewer can support rather varying types of processes. Furthermore, and under changing aerobic/anaerobic conditions in the sewer, low-molecular organics produced under anaerobic conditions may be degraded aerobically.

In general, the anaerobic microbial processes are physically concentrated in specific segments of a sewer determined by their requirements and characteristics. Fermentation may take place in the three major microbial subsystems: the suspended wastewater phase, the biofilms, and the sediments (Figure 3.2). Because sulfate-reducing bacteria and methanogens are slow-growing microorganisms, they are primarily present in the stationary parts (biofilms and sediments) of the sewer. In the suspended water phase, these microorganisms will normally be washed out. The sulfate-reducing bacteria are present in the biofilm and in the sediments where sulfate and low molecular organics from the wastewater may penetrate (Nielsen and Hvitved-Jacobsen 1988; Hvitved-Jacobsen et al. 1998; Bjerre et al. 1998). However, as a result of biofilm detachment, sulfate reduction may, to some extent, also take place in the suspended wastewater phase (Rudelle et al. 2011). Although the methanogenic microorganisms, as previously mentioned, may coexist with sulfate-reducing bacteria, the methanogenic microbial activity is mainly expected to take place in the deeper parts of the sewer deposits and not in the biofilm. A sewer biofilm is typically

## TABLE 3.3
## Examples of Anaerobic Processes for Decomposition of Low-Molecular Organic Matter

### Fermentation

$$\underset{\text{glucose}}{C_6H_{12}O_6} \rightarrow 2\underset{\text{ethanol}}{CH_3CH_2OH} + 2CO_2$$

$$\underset{\text{glucose}}{C_6H_{12}O_6} + 2H_2O \rightarrow 2\underset{\text{acetic acid}}{CH_3COOH} + 2CO_2 + 4H^+$$

$$\underset{\text{glucose}}{C_6H_{12}O_6} + 2H_2 \rightarrow 2\underset{\text{propionic acid}}{CH_3CH_2COOH} + 2H_2O$$

$$\underset{\text{propionic acid}}{CH_3CH_2COOH} + 2CO_2 \rightarrow \underset{\text{acetic acid}}{CH_3COOH} + CO_2 + 3H_2$$

$$\underset{\text{glucose}}{C_6H_{12}O_6} \rightarrow 2\underset{\text{lactic acid}}{CH_3CHOHCOOH}$$

### Methanogenesis

$$\underset{\text{acetic acid}}{CH_3COOH} \rightarrow CH_4 + CO_2$$

$$\underset{\text{glucose}}{C_6H_{12}O_6} \rightarrow 3CO_2 + 3CH_4$$

$$CO_2 + 4H_2 \rightarrow CH_4 + 2H_2O$$

### Sulfate Respiration

$$2\underset{\text{lactic acid}}{CH_3CHOHCOOH} + SO_4^{2-} + H^+ \rightarrow 2\underset{\text{acetic acid}}{CH_3COOH} + 2CO_2 + 2H_2O + HS^-$$

$$2\underset{\text{lactic acid}}{CH_3CHOHCOOH} + 3SO_4^{2-} + 3H^+ \rightarrow 6CO_2 + 6H_2O + 3HS^-$$

fully penetrated by sulfate, as sulfate in most types of wastewaters occurs in relatively high concentrations.

Although the anaerobic microorganisms concentrate in specific segments of the sewer, it is important that they may interact. As an example, the fermenting bacteria break down complex organic molecules to low molecular organics that can be consumed by sulfate-reducing bacteria and methanogens. Another example of interaction is related to the aerobic *Thiobacilli* bacteria at the sewer walls where they use hydrogen sulfide produced by the sulfate reducers in the submerged biofilm, from where it—via the water phase—is emitted to the sewer atmosphere.

In sewer networks without considerable amounts of sediments, the anaerobic processes are dominated by the acidogenic production of VFAs and $CO_2$ and by sulfate reduction (hydrogen sulfide production). The production of methane can, under such conditions, often be excluded as being of minor importance (Tanaka and Hvitved-Jacobsen 1999).

Depending on the type of substrate and the microorganisms present, fermentation pathways and the compounds produced vary considerably. Figure 3.3 is an example

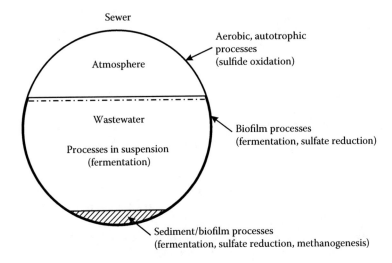

**FIGURE 3.2** Outline illustrating subsystems and occurrence of major microbial processes in a gravity sewer under anaerobic conditions in wastewater phase. Note that aerobic processes may occur at sewer walls exposed to atmosphere.

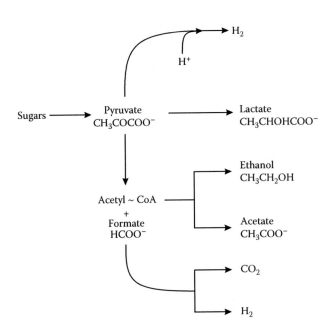

**FIGURE 3.3** Illustration of some major pathways and by-products of bacterial fermentation of sugars from pyruvic acid. CoA (coenzyme A) is an enzyme that is active in biotransformation process.

of the fermentation of sugars, only containing the elements C, H, and O, illustrating that a broad range of VOCs may be formed. It should also be noted that molecular hydrogen ($H_2$) is potentially produced.

### 3.2.3 Microbial Uptake of Substrate and Hydrolysis

The biochemical reactions of living cells are controlled by enzymes, which are defined as temperature-sensitive, organic catalysts. Chemically, the enzymes are proteins that are specific, in that they will act as catalysts to accelerate specific reactions. Some of these enzymes are associated with the material—the protoplasm—of the microorganism cell surrounded by a membrane. Such enzymes perform their function within the cell and are named intracellular enzymes. Other enzymes are excreted by the cell into the surrounding medium and are known as extracellular enzymes, also called exoenzymes.

In terms of the biochemical reactions that proceed, it is important as to whether the processes take place within the cell or outside the cell. Only truly soluble substrates can penetrate the cell membrane and be used within the cell. The particulate substrate in terms of complex organic matter found outside the cell must first undergo an enzymatic breakdown process initiated by the extracellular enzymes. This breakdown process is referred to as hydrolysis. During hydrolysis, complex compounds are converted into smaller, less complex, and more soluble molecules that can penetrate the cell membrane and become readily available substrates for the biomass. The principle of hydrolysis is outlined in Figure 3.4, which illustrates how the $H_2O$ molecule takes part in the process.

As shown in Figure 3.4, the covalent bond, i.e., two common shared electrons, between two carbon atoms in the complex molecule is cleaved when initiated by the exoenzymes. The highly reactive intermediates formed react momentarily and

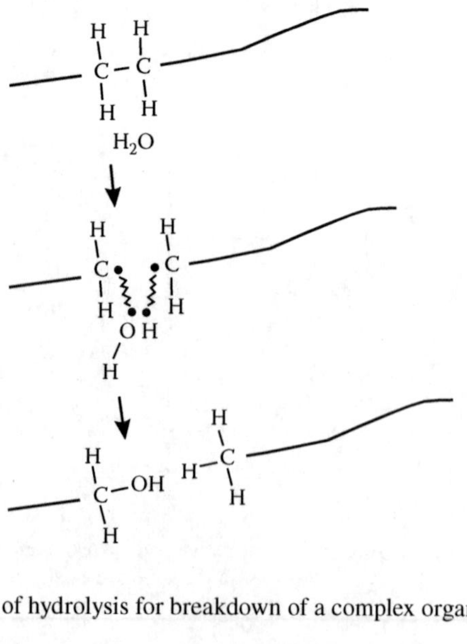

**FIGURE 3.4** Outline of hydrolysis for breakdown of a complex organic molecule.

produce new and stable bonds resulting in two new molecules that may undergo further hydrolysis. Hydrolysis is, thus, an important initial step in the transformation of complex organic matter present in a form that cannot be directly used as a substrate. It is therefore the hydrolysis rate that mainly controls the conversion of slowly biodegradable organic matter into readily biodegradable organic substrate (Morgenroth et al. 2002). Hydrolysis is a process that—with varying reaction rates—proceeds under aerobic, anoxic, and anaerobic conditions.

It is important to note that hydrolysis is a breakdown process that takes place without use of an electron acceptor. Hydrolysis will therefore not change the oxidation level of the substance.

As previously mentioned, the quality of the wastewater as substrate for the microorganisms highly depends on whether the substrate is soluble or particulate. The nomenclature used for substrate in the activated sludge processes of wastewater treatment is also feasible for wastewater in sewers. The nomenclature is further explained in Appendix A.

- $S_S$: soluble substrate (readily biodegradable substrate) directly available for the microorganism in the cell
- $X_S$: particulate substrate (hydrolyzable substrate) available as substrate for the microorganisms after hydrolysis by extracellular enzymes followed by diffusion of the products into the cell

Each of the fractions of substrates ($S_S$ and $X_S$) covers a substantial number of different organic compounds, which—in terms of their quality characteristics—are different. However, from an application point of view, a rather simple description of biodegradability is needed, i.e., a distinction between substrate that is readily biodegradable and substrate that requires an initial hydrolysis. Typically, this simple approach is sufficient when considering the processes and quality changes of wastewater in sewers.

### 3.2.4 Particulate and Soluble Substrate

As discussed in Section 3.2.3, from a microbiological viewpoint, it is important to distinguish between soluble substrates that can penetrate the cell wall and particulate substrates that must undergo hydrolysis before utilization. Methods for the characterization of wastewater organic matter in terms of particle size distribution have been described by a number of researchers (cf. Figure 3.5) (Levine et al. 1985; Logan and Jiang 1990; Levine et al. 1991; Gregory 2006).

Traditionally, the parameter suspended solids in wastewater is defined according to the method used in the analysis of total suspended solids (TSS) (APHA–AWWA–WEF 2005). This definition of suspended solids is based on a filtration procedure—sometimes also centrifugation—and is, therefore, appropriate for particles having a diameter $>0.5–1$ μm (cf. Figure 3.5).

The classification of wastewater in terms of size distribution is normally done from a practical point of view. Typically, a distinction is made between soluble, colloidal, and suspended components (Figure 3.6). This definition for determining what

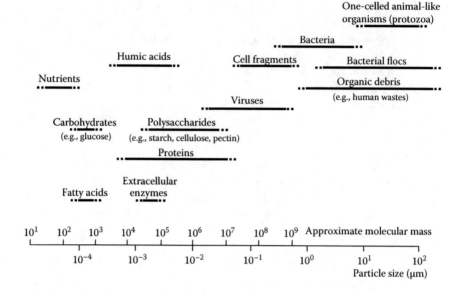

**FIGURE 3.5** Typical levels of particle size for organic substances in settled municipal wastewater.

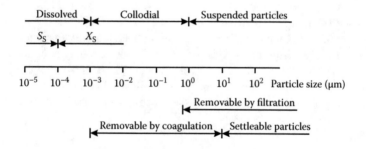

**FIGURE 3.6** Size classification of wastewater components.

"solids" are is rational as far as physical transport processes in sewers are concerned. However, when dealing with the microbial processes in sewers, an extension of the definition of "solids" is required. Particles larger than about $10^{-4}$ μm cannot be transported through a cell wall and are, therefore, from a microbiological point of view, considered particles.

In order to avoid confusion on what is "solids" and what is not, a clear distinction from either a physical or a physiological point of view must be stated.

### 3.2.5 ORGANIC CONSTITUENTS IN WASTEWATER OF SEWER NETWORKS

Wastewater is a mixture of rather complex organic and inorganic compounds. The traditional way of characterizing wastewater components can be found in any textbook dealing with treatment of wastewater (e.g., Tchobanoglous et al. 2003; Henze et al. 2002), and in books on analytical methods (e.g., APHA–AWWA–WEF, 2005). A brief introduction to the traditional approach to quality characterization of wastewater is given in Section 3.1.2. Organic fractions and inorganic compounds (e.g., DO, nitrate, ammonia, sulfate, sulfide, and soluble iron) are of specific importance for the microbial processes in sewers. In addition, settleable solids, heavy metals, and organic micropollutants include substances that potentially result in specific impacts.

The traditional way of characterizing organic matter in domestic wastewater is given in terms of bulk parameters such as BOD, COD, and total organic carbon (TOC). Wastewater characterization by direct measurement of organic constituents has been performed in relatively few studies (Nielsen et al. 1992). In these studies, the main compounds, relevant from a biochemical point of view, have been identified as substances originating from foodstuff:

- Carbohydrates
- Proteins
- Lipids (fats)

In addition to these three main groups of organic components, volatile fatty acids (VFAs), amino acids, detergents, humic substances, organic fibers, etc., have been identified.

Raunkjaer et al. (1994) have modified analytical methods for the determination of carbohydrates, proteins, and lipids to be applied for wastewater with specific relevance for sewer processes. Henze et al. (2002) estimate the average composition of these three organic fractions in wastewater and propose a corresponding conversion to COD units (Table 3.4). The table also includes stoichiometric formulas for carbohydrate and protein, proposed by Kalyuzhnyi et al. (2000).

It should be noted that the formulas shown in Table 3.4 give different compositions of the organic fractions. As an example, the nitrogen content of protein according to the formulas proposed by Henze et al. (2002) and Kalyuzhnyi et al. (2000) is 8.8% and 16.7%, respectively.

Based on the conversion factors given in Table 3.4, the composition of wastewater samples taken at the inlets of four wastewater treatment plants in Denmark is

**TABLE 3.4**

**Average Composition and Corresponding Conversion Factors to COD Units for Carbohydrates, Proteins, and Lipids**

| Organic Fraction | Composition | | Conversion Factor (mg COD mg DM$^{-1}$) |
|---|---|---|---|
| | Henze et al. | Kalyuzhnyi et al. | |
| Carbohydrates | $C_{10}H_{18}O_9$ | $C_6H_{12}O_6$ | 1.13 |
| Proteins | $C_{14}H_{12}O_7N_2$ | $C_4H_6ON$ | 1.20 |
| Lipids | $C_8H_{16}O_2$ | | 2.03 |

*Source:* From Henze, M. et al., *Wastewater Treatment: Biological and Chemical Processes*, 3rd edn, Springer-Verlag, Berlin, p. 430, 2002; Kalyuzhnyi, S. et al., *Water Sci. Technol.*, 41(3), 43–50, 2000. With permission.

shown in Figure 3.7. The wastewater in the corresponding sewer catchments mainly originates from domestic sources, and the network mainly consists of gravity sewer sections. The results show that the three groups—carbohydrates, proteins, and lipids—make up a significant part of the organic matter. Furthermore, the organic residue may include intermediate products from the degradation of these components.

Examples of average dissolved fractions of carbohydrates and proteins are shown in Table 3.5. Proteins occur basically as dissolved substances after transport in a

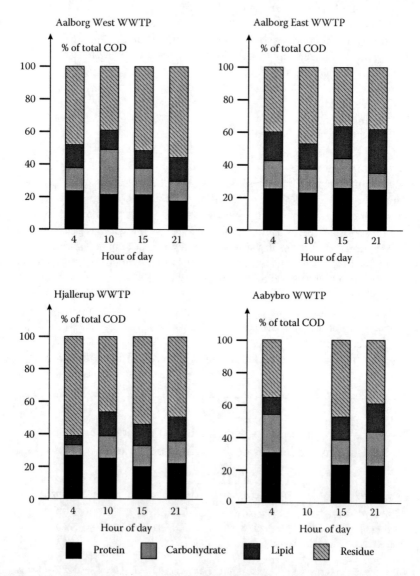

**FIGURE 3.7** Carbohydrates, proteins, and lipids in percentage of total COD in the inlet wastewater of four wastewater treatment plants. (From Raunkjaer, K. et al., *Water Res.*, 28(2), 251–262, 1994. With permission.)

**TABLE 3.5**

**Total and Dissolved Carbohydrates, Proteins, and Lipids as a Percentage of Total and Dissolved COD, Respectively**

| | Carbohydrates (%) | | Proteins (%) | | Lipids (%) | |
|---|---|---|---|---|---|---|
| | Total | Dissolved | Total | Dissolved | Total | Total (%) |
| Average ($n = 13$–16) | 18 | 10 | 28 | 28 | 31 | 78 |
| Standard deviation | 6 | 7 | 4 | 6 | 10 | 11 |

*Source:* Data from the investigations shown in Figure 3.7.

sewer network, whereas lipids—because of their low solubility in water per definition—exist as nondissolved substances.

The biochemical processes are closely related to the specific nature of the different organic substrates. Although each of the fractions of organic matter—carbohydrates, proteins, and lipids—includes a substantial number of specific molecules, there

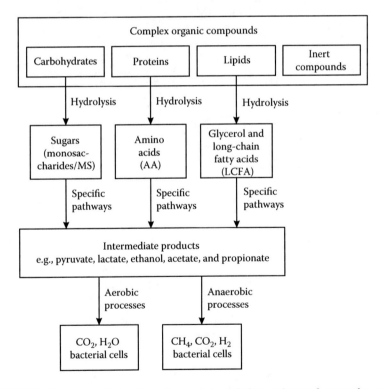

**FIGURE 3.8** Overview of transformation and degradation pathways for complex organic compounds in wastewater (cf. Figure 3.3). (Data from WEF, *Wastewater Biology: The Life Processes*, Water Environment Federation, 184, 1994; Christ, O. et al., *Water Sci. Technol.*, 41(3), 61–65, 2000. With permission.)

are in the group where they belong chemically, common characteristics as substrates for heterotrophic organisms. The overview given in Figure 3.8 is absolutely simplified, and the specific pathways of degradation depend on, for example, the availability of electron acceptor. However, it links together and shows major potential pathways of the substrates. A different behavior of the three groups of organic matter is therefore expected when they are subjected to transformations in sewer networks.

It is evident that Figure 3.8 only roughly shows the overall transformation pathways. As an example of further details, the contents of biodegradable proteins—and particularly the sulfur containing amino acids—may give rise to production of odorous compounds under anaerobic conditions, particularly volatile sulfur compounds (Higgins et al. 2008). These details of transformation are discussed in Chapters 4 through 6. As an example, the formation and occurrence of odorous compounds are discussed in Section 4.3.2.

The transformation of carbohydrates, proteins, and lipids was investigated during transport in an intercepting gravity sewer under aerobic conditions (cf. Figure 3.9).

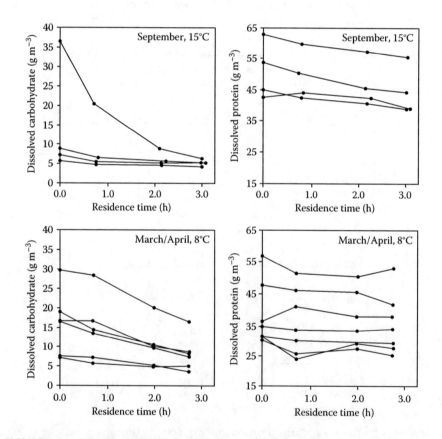

**FIGURE 3.9** Concentration profiles of dissolved carbohydrate and protein in wastewater during transport in a gravity sewer under aerobic conditions. (Data from Raunkjaer, K. et al., *Water Environ. Res.*, 67(2), 181–188, 1995. With permission.)

The dissolved carbohydrates and, to some extent, proteins were—depending on the temperature level—removed, whereas lipids were almost unchanged.

### 3.2.6 WASTEWATER COMPOUNDS AS MODEL PARAMETERS

When wastewater compounds are described in terms of model parameters for simulation of the transformations in a sewer, the requirements for their expression depend on model objectives and level of details. A basic requirement—following the concept depicted in Figure 2.2—is that two types of compounds be included when focusing on the heterotrophic microbial transformations:

1. Active, heterotrophic biomass including cell biomass and associated extracellular polymer substances (EPSs)
2. Substrates that are either readily biodegradable or potentially subject to hydrolysis

If these two types of organic compounds are included in the description, it is basically possible to simulate which part of the organic matter causes the transformations and which part is used for either growth of new biomass or energy purposes. These two compounds can be expressed in units of COD, i.e., g $O_2$ produced or degraded as biomass and g $O_2$ produced or consumed as substrate. This type of description of the microbial transformations is well known from modeling activated sludge processes in wastewater treatment plants (Henze et al. 1987, 2000, 2002). Taking the Activated Sludge Model No. 2 (ASM2) as a basis, the compounds shown in Table 3.6 are included in the formulation of the activated sludge processes (Henze et al. 1995).

In wastewater, as it occurs at the inlet to wastewater treatment plants, the autotrophic biomass, and the phosphorous-accumulating biomass, including the stored polyphosphate, are expected to be close to zero (Henze et al. 2002). This is based on the fact that the conditions for growth of nitrifying biomass and phosphorous-accumulating organisms in sewer systems are far from optimal. Therefore, the

---

**TABLE 3.6**

**Fractions of Organic Matter as Applied in Activated Sludge Modeling**

$S_F$ = readily (fermentable) biodegradable substrate

$S_A$ = volatile acids/fermentation products

$S_S$ = readily biodegradable substrate ($S_S = S_F + S_A$)

$X_S$ = slowly biodegradable substrate

$X_{AUT}$ = autotrophic, nitrifying biomass

$X_{PHA}$ = stored polyhydroxyalkanoate

$X_{PAO}$ = phosphorous-accumulating organisms

$X_I$ = inert, nonbiodegradable, particulate organics

$S_I$ = inert, nonbiodegradable, soluble organics

$X_{Hw}$ = heterotrophic biomass

following relation is approximately considered correct when focusing on wastewater in sewer networks:

$$X_{AUT} + X_{PHA} + X_{PAO} = 0 \qquad (3.3)$$

The sum of the following compounds, considering that $S_S = S_F + S_A$, approximately equals the total COD:

$$S_F + S_A + X_S + X_I + S_I + X_{Hw} = COD_{tot} \qquad (3.4)$$

The organic matter fractions referred to in Equation 3.4 are described according to the activated sludge concept, relevant at the inlet to wastewater treatment plants. Basically, these compounds are also relevant in wastewater of sewer systems. However, when considering sewer processes, a slightly different approach for description of organic matter fractions compared with the activated sludge concept is needed. Further details are given in Chapters 5 and 6. In brief, the cause for this is as follows:

- The biodegradability within a specific period corresponding to the residence time in a sewer network (typically between few hours and a couple of days) must be reasonably detailed described. Therefore, the fast biodegradable fractions considered as $S_S$ and relatively fast hydrolyzable substrate must be included in the description as separate fractions. On the contrary, what is not biodegradable within 1 to 2 days is of minor interest. As a consequence, there is no need to distinguish between a rather slowly biodegradable, particulate fraction of a substrate and a fraction that is microbiologically inert, regardless of whether it is soluble or particulate.
- The number of fractions must be minimized, determined by the details desirable and required, e.g., for modeling purposes. Typically, two fractions of the hydrolyzable substrate are sufficient. However, three fractions may be needed in larger sewer networks with several inlets.
- It is important that the different COD fractions in wastewater can be quantified and determined by direct measurement methods.

Based on these criteria, the organic matter fractions shown in Table 3.7 are assessed, and considered central for the description of heterotrophic processes in wastewater of sewer systems (Hvitved-Jacobsen et al. 1998, 1999):

The mass balance according to the concept for fractionation of wastewater compounds is, therefore, as follows:

$$S_F + S_A + \sum X_{Sn} + X_{Hw} = COD_{tot} \qquad (3.5)$$

The number of fractions of hydrolyzable substrate is determined by the quality of the wastewater, i.e., when fractions can be identified in terms of their different rates of hydrolysis. Typically, two or three—often, just two—fractions are needed. Two

## TABLE 3.7

**Fractions of Organic Matter as Applied in Modeling Sewer Processes in Suspended Water Phase**

$S_F$ = readily (fermentable) biodegradable substrate

$S_A$ = volatile acids/fermentation products

$S_S$ = readily biodegradable substrate ($S_S = S_F + S_A$)

$X_{Sn}$ = hydrolysable substrate, fraction # $n$; $\Sigma n$ typically equals 2 or 3: $n = 1$ (fast degradable), $n = 2$ (slowly degradable); $n = 1$(fast degradable), $n = 2$ (medium degradable), $n = 3$ (slowly degradable)

$X_{Hw}$ = heterotrophic biomass in the water phase

---

fractions are generally sufficient in case of a small number of wastewater samples. When several inlets (with different types of wastewater sources) into a sewer network exist, the variability in the hydrolysis rate may require three fractions of $X_{Sn}$.

The COD fractions are given for the wastewater phase. However, corresponding parameters may be needed when including processes in the biofilm and the sediments. For a rather simple modeling of heterotrophic processes in the biofilm, the heterotrophic active biomass, $X_{Hf}$, expressed in units of g COD m$^{-2}$, may be convenient. Depending on the details in the description of the anaerobic processes in the biofilm and the sediments, a sulfate-reducing biomass and a methane-producing biomass may be needed.

Figure 3.10 shows a "typical" composition of wastewater in terms of COD fractions when applied for process description in the activated sludge model No. 2 (ASM2). The figure also shows the corresponding values, exemplified with two fractions of hydrolyzable substrate, for the characterization of wastewater in sewer networks. It should be noted that the word "typical" here is used for wastewater compositions with a rather uncertain degree of variability.

It is interesting to compare the two sets of COD fractions although they refer to wastewater in two different systems and under quite different conditions. It is logical that there are differences in the COD fractions as they refer to a composition of activated sludge and a composition in the water phase in a sewer, respectively. To some extent, different methods have also been applied for characterization. In addition to what is already mentioned concerning the different interpretation of $X_S$, it is interesting to note that $S_A$ (VFAs)—determined by the same analytical method—are estimated differently. A basis for the determination of the different COD fractions in wastewater of sewer networks is briefly described in the following discussion. Details are, in this respect—in particular related to calibration and verification of sewer process models—major subjects of Chapter 10.

The different COD fractions of wastewater in sewers as shown in Table 3.7 reflect the degree of biodegradability of the organic matter and the corresponding activity of the heterotrophic biomass that causes the transformation. The OUR is therefore by nature a natural basis for the determination of the COD fractions (Bjerre et al. 1995; Vollertsen and Hvitved-Jacobsen 2000b; Vollertsen and Hvitved-Jacobsen 2002). The detailed description of the methodology for this analysis is found in

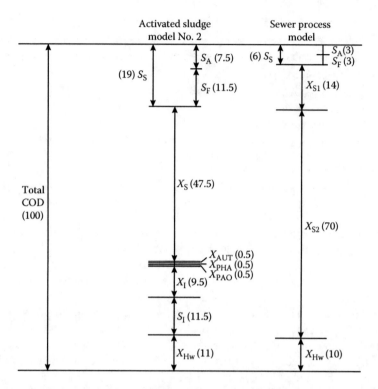

**FIGURE 3.10** Comparison between COD fractions in wastewater, in general determined by different methods and with different objectives: simulation of processes in activated sludge of wastewater treatment plants and simulation of wastewater processes in sewers, respectively. Typical percentages of the different COD fractions are shown in brackets. (Data from Henze, M. et al., *Activated Sludge Models ASM1, ASM2, ASM2d and ASM3, Scientific and Technical Report No. 9*, IWA Publishing, London, p. 121, 2000; Hvitved-Jacobsen, T. et al., *Water Sci. Technol.*, 39(2), 242–249, 1999. With permission.)

Section 10.1.3. However, a brief understanding of the concept in terms of an example will be given with reference to Figure 3.11.

From the shape of the OUR curves shown in Figure 3.11, it is readily obvious that the biodegradability of the two wastewater samples, a and b, is fundamentally different and—particularly for sample a—subject to changes over time. Figure 3.11a shows that the activity of the heterotrophic biomass, $X_{Hw}$, increases considerably within the first 4 h caused by its growth and corresponding consumption of readily biodegradable substrate, $S_S$. When this fraction is more or less used up, the OUR value drops almost immediately to a level that can be considered determined by the hydrolysis rate of the two fractions of particulate substrate, $X_{S1}$ and $X_{S2}$. In contrast, Figure 3.11b shows a wastewater sample with no initial amount of $S_S$ and a more or less constant and low biodegradability determined by the hydrolysis rate of $X_{S1}$ and $X_{S2}$.

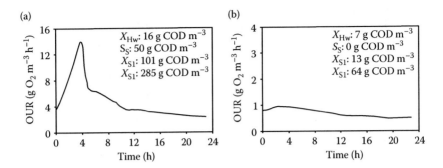

**FIGURE 3.11**   OUR versus time curves measured at 20°C on wastewater samples originating from two gravity sewers and showing the initial water phase composition. OUR versus time experiments are central for assessment of the biodegradability of wastewater samples and determination of COD fractions (cf. Section 10.2.1). Notice the different scales of the OUR unit in (a) and (b).

In Chapter 10, it will be shown in detail how the COD fractions in wastewater are determined. The initial values of the COD fractions for the two rather different wastewater samples are, however, shown in Figure 3.11. These values are the result of model simulation of the measured OUR curves. It is interesting to compare the two sets of COD fractions and how the different composition affects the aerobic biodegradability of wastewater in terms of OUR.

### 3.2.7   BIOFILM CHARACTERISTICS AND INTERACTIONS WITH THE BULK WATER PHASE

The basic characteristics including the kinetics of biofilms are dealt with in Section 2.2.2.1. In the following, specific characteristics related to the microbial activity and processes in biofilms will be described.

Biofilms will potentially occur at all surfaces exposed to a water phase in a sewer and also, to some extent, at the sewer air surfaces where aerosols are present and the humidity is high. Biofilms in sewers are often referred to as "slimes" and consist of the following three major fractions:

1. Microorganisms
2. Relatively large molecular size extracellular polymers, EPSs, produced by the microorganisms outside their cells
3. Adsorbed organic and inorganic compounds from the wastewater

Aerobic biofilms in gravity sewers usually have a thickness of a few millimeters, depending on the flow regime. The anaerobic biofilm thickness in pressure sewers is typically thinner (maximum of a few hundred microns) because of a typically higher water velocity and, therefore, a higher shear stress during the pumping periods, and a lower bacterial growth rate. In each case, however, examples of very thick biofilms may be found where a significant change in friction has led to different hydraulic conditions for the biofilm (Characklis and Marshall 1990; USEPA 1985). The thin biofilms in pressure sewers typically have a rather smooth surface, whereas the biofilms in gravity sewers are fluffy. The real biofilm surface in a gravity sewer is therefore typically much larger than what may be calculated based on the inner surface of the sewer pipe. As already dealt with in Section 2.2.2.1, the exchange of substances between a biofilm and the surrounding water phase is crucial for the processes in both phases.

The microorganisms in sewer biofilms are embedded in a matrix of EPS that mostly consists of polysaccharides produced by the bacteria (Characklis and Marshall 1990). The EPS fraction is typically the largest organic fraction in the biofilm, i.e., up to about 90% of the total organic content. Only limited studies of the total composition of sewer biofilms in terms of carbohydrates, proteins, and humic substances have been undertaken so far (cf. Figure 3.12). Details of the composition of anaerobic biofilms in pressure sewers are not known.

Significant biomass production can take place in a gravity sewer biofilm. The biomass generated in the biofilm detaches and is, together with the biomass produced in the water phase, transported to the treatment plant or via overflow structures into receiving waters. A simple method for estimating the amount of biomass produced in a biofilm is based on the consumption of the electron acceptor, $O_2$, $NO_3^-$, or $SO_4^{2-}$ and the yield constant of the process as shown in Example 3.1.

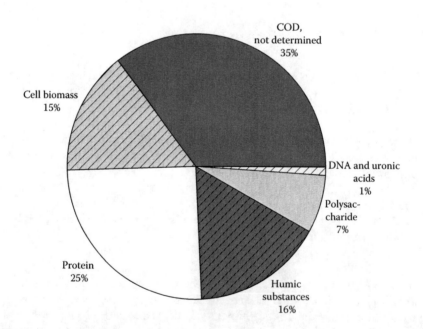

**FIGURE 3.12** Typical composition of a gravity sewer biofilm. (Data from Jahn, A., Nielsen, P.H., *Water Sci. Technol.*, 37(1), 17–24, 1998. With permission.)

### Example 3.1: Biomass Production in the Biofilm of an Intercepting Sewer

A 4-km intercepting sewer with a diameter $D = 0.5$ m is flowing half-full. The DO consumption rate, $r_f$, of the sewer biofilm is measured, and an average value of 0.6 g $O_2$ m$^{-2}$ h$^{-1}$ was estimated. The biofilm yield constant of the heterotrophic biomass was not measured but was estimated as $Y_{Hf} = 0.55$ g COD biomass produced per g COD substrate consumed. Only aerobic heterotrophic transformations in the biofilm are expected to proceed. These data can be used for estimation of the daily production of biomass in the biofilm of the sewer.

The total amount of substrate that undergoes transformation is either used for growth of biomass or it is degraded by respiration (cf. Section 2.1.2). The rate of production of biomass is, therefore:

$$r_B = \frac{Y_{Hf}}{1-Y_{Hf}} 0.6 \text{ g COD m}^{-2} \text{ h}^{-1}$$

$$= \frac{0.55}{1-0.55} 0.6 = 0.73 \text{ g COD m}^{-2} \text{ h}^{-1}$$

The daily production of biomass in the biofilm of the sewer is as follows:

$$M_B = r_B \frac{\pi}{2} D \times 4000 \times 24 \times 10^{-3} \text{ kg COD}$$

$$= 0.73 \frac{\pi}{2} 0.5 \times 4000 \times 24 \times 10^{-3} \text{ kg COD}$$

$$= 55.3 \text{ kg COD}$$

A prerequisite for the validity of this result is that only aerobic microbial processes proceed. If anaerobic processes take place in the inner part of the biofilm, the biomass production is reduced (cf. Section 6.3.1).

The biomass produced in the biofilm is detached to the water phase by erosion and by sloughing. Single bacteria and small parts of the biofilm erode continuously from the surface, whereas bigger parts may slough intermittently. Sloughing may take place when large changes in shear stress or substrate conditions occur, e.g., during wet weather periods. However, the mechanisms are not well understood. In a steady-state situation or over longer periods, the total detached biomass will approximately equal the biomass production.

Small particles can be trapped at the biofilm surface or in voids of the biofilm where organics may be hydrolyzed and further take part in the transformation processes. A number of factors influence adsorption and desorption of particles (e.g., particle size, surface charge, pH) as well as biofilm surface properties and the bulk water flow regime. Studies of model biofilms have shown that water flows into the biofilm in small channels, making the prediction of transport of particles as well as soluble compounds complex (Norsker et al. 1995).

An exchange of heavy metals between the bulk water phase and the biofilm takes place (Gutekunst 1988; Nielsen et al. 2005). The concentrations of heavy metals in the biofilm may be considered indicative of a preceding wastewater transport of heavy metals. Heavy metals originating from, e.g., short-term increases in concentration in the bulk water phase, may be trapped in the biofilm and then slowly released (cf. also Section 2.3.3).

### 3.2.8   Sewer Sediment Characteristics and Processes

The occurrence of sewer sediments is primarily determined by the physical characteristics of wastewater solids, the structure of the sewer network, and the hydraulic conditions. Basically, sewers should be designed and operated in a way that does not result in permanent deposits. This ideal performance of a sewer is generally not observed, and sediments may be more or less temporarily accumulated in sewers. In combined sewer networks, sediments may settle under dry weather conditions when the wastewater velocity and shear stress at the bottom are low, and be eroded and transported in the suspended water phase under wet weather conditions. The solids originating from the sewer sediments may, therefore, turn up in the combined sewer overflows (CSOs), potentially resulting in pollution of adjacent receiving waters.

#### 3.2.8.1   Physical Characteristics and Processes

As briefly mentioned, the occurrence of sediments in sewers is closely related to the hydraulics, physical process characteristics, and the properties of the solids. Except for a brief introduction to water and mass transport in sewers (cf. Section 1.5), this book does not include a detailed quantitative description of the physical properties and processes of sewer solids such as sedimentation, deposition, and erosion. A substantial number of publications and textbooks address these subjects. A comprehensive description with a broad range of literature references relevant for sewer networks can be found in the work of Ashley and Verbanck (1998) and Ashley et al. (2004). Further details are also given by Ashley (1996), Hvitved-Jacobsen et al. (1995), and Hvitved-Jacobsen (1998). Perspectives and relations between the physical characteristics of sewer solids and the chemical and biological sewer processes are outlined by Ashley et al. (1999).

The physical characteristics of sewer deposits can be described in terms of individual particle and bulk properties. The hydraulic and structural conditions in the sewer, together with the nature of the inputs, will control the type of material that deposits at a given location. Crabtree (1989) has proposed a sewer sediment taxonomy that is relevant in terms of physical properties but, to some extent, also to chemical and biological processes (cf. Table 3.8). The taxonomy is based on four primary classes (A, C, D, and E), with a fifth class B comprising agglutinated or cemented class A material. It is clear that the taxonomy and data shown in Table 3.8 just provide a rough description of the complex nature of sewer sediments.

As seen in Table 3.8, organic matter constitutes an essential part of sewer sediments, although, in general with a low biodegradability. As also shown, class D (sewer biofilm) is included in the taxonomy (cf. Section 3.2.7). Class A sewer sediment material is most commonly found in combined sewer networks.

**TABLE 3.8**
**Sewer Sediment Taxonomy as Proposed by Crabtree (1989)**

| Sediment Type | Description/Where Found | Wet Density × $10^3$ (kg m$^{-3}$) | % by Granular Particle Size[a] Sediment Particle Size | | | Organic Content (%) |
| | | | a <0.063 (mm) | b 0.063–2.0 (mm) | c 2.0–50 (mm) | |
|---|---|---|---|---|---|---|
| A | Coarse, granular bed material—widespread | 1.72 | 1–6–30 | 3–61–87 | 3–33–90 | 7 |
| C[b] | Mobile, fine-grained, found in slack zones, in isolation and overlying type A | 1.17 | 29–45–73 | 5–55–71 | 0 | 50 |
| D | Organic pipe wall slimes | 1.21 | 17–32–52 | 1–62–83 | 1–6–20 | 61 |
| E | Fine-grained mineral and organic material found in CSO storage tanks | 1.46 | 1–22–80 | 1–69–85 | 4–9–80 | 22 |

[a] The numbers a–b–c refer to minimum–mean–maximum values given in percent. Notice that $\Sigma b = 100\%$ for each sediment type.

[b] Since these results were published, class C materials include larger (organic) particles.

### 3.2.8.2　Chemical Characteristics

The chemical composition of sewer sediments reflects what is discharged to the network and what can either settle or adsorb to the sewer solids. The contents of, for example, heavy metals, will therefore vary considerably depending on the nature and activities in the catchment. The heavy metal contents in sewer deposits, as shown in Table 3.9, are therefore just examples for illustration of possible levels. Table 3.10 shows a selected number of chemical characteristics for class A sewer sediments.

### 3.2.8.3　Microbial Characteristics and Processes

Although numerous investigations indirectly refer to the outcome of microbial processes in sewer sediments, only a few of them have been directly concerned with the details of the biological processes. However, relatively high anaerobic activity in terms of $H_2S$ formation of sediment deposits compared with what is generally observed in sewer biofilms is observed (cf. Section 6.4.3.2.2). This activity may indicate that sulfide formation in the sediment is caused by the production of readily biodegradable organic matter by anaerobic hydrolysis and fermentation and the impact of such organic products and sulfate being transported by diffusion. It has also been shown that a biofilm may rapidly develop at the surface of sewer sediments and influence the cohesion of the sediment surface, thereby changing the resistance

## TABLE 3.9
**Examples of Selected Heavy Metal Contents in Sewer Sediments Based on Results from Different Sources**

| Heavy Metals | Nonfractionated Sample (mg kg⁻¹ DM) | Fraction <0.2 mm (mg kg⁻¹ DM) | Thresholds for Wastewater Sludge (mg kg⁻¹ DM) |
|---|---|---|---|
| Lead (Pb) | 20 | 150 | 120 |
| Cadmium (Cd) | 0.5 | 2 | 0.4 |
| Copper (Cu) | 30 | 200 | 1000 |
| Nickel (Ni) | 10 | 30 | 30 |
| Zinc (Zn) | 300 | 200 | 4000 |

*Note:* Threshold values for wastewater sludge for application on agricultural land in Denmark are shown for comparison.

## TABLE 3.10
**Typical Values of Selected Pollutants in Class A Sewer Sediments**

| Parameter (Unit) | Average | Min.–Max. |
|---|---|---|
| TS, total solids (g kg⁻¹ DM) | 550–800 | 350–820 |
| VS, volatile solids (%) | 4.5–10 | 1–19 |
| COD (g kg⁻¹ DM) | 25–70 | 6–270 |
| BOD₅ (g kg⁻¹ DM) | 4–14 | 1–90 |
| BOD, 4 h (mg kg⁻¹ DM) | 400 | 100–700 |
| Organic nitrogen (mg kg⁻¹ DM) | 800 | 200–1500 |
| Ammonia, NH₄-N (mg kg⁻¹ DM) | 100 | 10–300 |

*Note:* Numerous studies have been collated to demonstrate the variability.

to resuspension (Vollertsen and Hvitved-Jacobsen 2000). In the same study, it was observed that methanogenesis (methane formation) in the sediments caused the formation of gas cavities that decreased the strength of the sediment surface. The presence of a biofilm layer at the sediment surface of a sanitary gravity sewer pipe was also identified by Chen et al. (2003). At 25°C, a water depth of 0.22 m, and a mean flow velocity of 1.5 m s⁻¹—corresponding to a bottom shear flow velocity of 0.055 m s⁻¹—they measured a sediment oxygen uptake flux of 1.3 g $O_2$ m⁻² h⁻¹. Although DO-consuming processes in the bulk phase of the sediment may also occur, this value indicates the presence of a rather active biofilm at the sediment surface.

Crabtree (1986) postulated the existence of an aging process of sewer deposits caused by interactions between sediments and wastewater. Ristenpart (1995) investigated the occurrence of this aging in terms of the variability of specific components in sediments of different ages (cf. Table 3.11).

The difference in the magnitude of the parameters shown in Table 3.11 corresponds well with the course of (anaerobic) microbial processes in the sewer deposits.

**TABLE 3.11**

**Variability of Selected Parameters in Sediments of Different Ages**

| | Residence Time | | |
|---|---|---|---|
| Parameter (unit) | Short (Newly Deposited) | Medium (Disturbed from Time to Time) | Long (Consolidated) |
| Bulk density (kg m$^{-3}$) | 1200 | 1510 | 1840 |
| TS, total solids (g kg$^{-1}$) | 355 | 705 | 812 |
| VS, volatile solids (% of DM) | 27.0 | 8.8 | 2.4 |
| pH (–) | 5.68 | 7.11 | 7.66 |
| BOD$_5$ (g kg$^{-1}$ wet weight) | 31.6 | 12.5 | 2.7 |
| COD (g kg$^{-1}$ wet weight) | 95.6 | 55.3 | 19.0 |

*Source:* Ristenpart, E., *Water Sci. Technol.*, 31(7), 77–83, 1995. With permission.

Newly deposited sediments not only have the highest pollutant potential but also show the lowest critical shear stress for erosion. Such sediment types may exert the highest impacts on receiving waters from CSOs.

## REFERENCES

APHA–AWWA–WEF (2005), *Standard Methods for the Examination of Water and Wastewater*, 21st edn, APHA (American Public Health Association), AWWA (American Water Works Association), WEF (Water Environment Federation), Washington, D.C., USA.

Ashley, R.M. (ed.) (1996), Solids in sewers, *Water Sci. Technol.*, 33(9), 298.

Ashley, R.M. and M.A. Verbanck (1998), Physical processes in sewers, *Proceedings from Congress on Water Management in Conurbations*, Bottrop, Germany, June 19–20, 1997, pp. 26–47.

Ashley, R.M., T. Hvitved-Jacobsen, and J.-L. Bertrand-Krajewski (1999), Quo vadis sewer process modelling? *Water Sci. Technol.*, 39(9), 9–22.

Ashley, R.M., J.-L. Bertrand-Krajewski, T. Hvitved-Jacobsen, and M. Verbanck (eds.) (2004), *Solids in Sewers, IWA Scientific and Technical Report No. 14*, International Water Association Publishing, London, p. 340.

Bjerre, H.L., T. Hvitved-Jacobsen, B. Teichgräber, and D. te Heesen (1995), Experimental procedures characterizing transformations of wastewater organic matter in the Emscher river, Germany, *Water Sci. Technol.*, 31(7), 201–212.

Bjerre, H.L., T. Hvitved-Jacobsen, S. Schlegel, and B. Teichgräber (1998), Biological activity of biofilm and sediment in the Emscher river, Germany, *Water Sci. Technol.*, 37(1), 9–16.

Boon, A.G. and A.R. Lister (1975), Formation of sulphide in rising main sewers and its prevention by injection of oxygen, *Prog. Water Technol.*, 7(2), 289–300.

Cappenberg, T.E. (1974), Interrelations between sulfate-reducing and methane-producing bacteria in bottom deposits of a fresh-water lake: 1. Field observations, *Antonie van Leeuwenhoek, J. Microbiol.*, 40, 285–295.

Characklis, W.G. and K.C. Marshall (eds.) (1990), *Biofilms*, John Wiley & Sons, New York.

Chen, G.-H., D.H.W. Leung, and J.-C. Hung (2003), Biofilm in the sediment phase of a sanitary gravity sewer, *Water Res.*, 37, 2784–2788.

Christ, O., P.A. Wilderer, R. Angerhöfer, and M. Faulstich (2000), Mathematical modeling of the hydrolysis of anaerobic processes, *Water Sci. Technol.*, 41(3), 61–65.

Crabtree, R.W. (1986), The discharge of toxic sulphides from storm sewage overflows— a potential polluting process, WRC (Water Research Centre) report ER 203E.

Crabtree, R.W. (1989), Sediments in sewers, *J. Inst. Water Environ. Manage.*, 3(6), 569–578.

Gregory, J (2006), *Particles in Water: Properties and Processes*, CRC Press/Taylor & Francis, Boca Raton, FL.

Gutekunst, B. (1988), Sielhautuntersuchungen zur Einkreisung schwermetalhaltiger Einleitungen, Institut für Siedlungswasserwirtschaft, Universität Karlsruhe, Band 49.

Henze, M., C.P.L. Grady Jr., W. Gujer, G. v. R. Marais, and T. Matsuo (1987), *Activated Sludge Model No. 1, Scientific and Technical Report No. 1*, IAWPRC (International Association on Water Pollution Research and Control), London.

Henze, M., W. Gujer, T. Mino, T. Matsuo, M.C. Wentzel, and G.v.R. Marais (1995), *Activated Sludge Model No. 2, Scientific and Technical Report No. 3*, IAWQ (International Association on Water Quality), London, p. 32.

Henze, M., W. Gujer, T. Mino, and M.v. Loosdrecht (eds.) (2000), *Activated Sludge Models ASM1, ASM2, ASM2d and ASM3, Scientific and Technical Report No. 9*, IWA (International Water Association) Publishing, London, p. 121.

Henze, M., P. Harremoës, J. la Cour Jansen, and E. Arvin (2002), *Wastewater Treatment: Biological and Chemical Processes*, 3rd edn, Springer-Verlag, Berlin, p. 430.

Higgins, M.J., G. Adams, Y.-C. Chen, Z. Erdal, R.H. Forbes, D. Glindemann, J.R. Hargreaves, D. McEwen, S.N. Murthy, J.T. Novak, and J. Witherspoon (2008), Role of protein, amino acids, and enzyme activity on odor production from anaerobically digested and dewatered biosolids, *Water Environ. Res.*, 80(2), 127–135.

Hvitved-Jacobsen, T. (ed.) (1998), The sewer as a physical, chemical and biological reactor II, *Water Sci. Technol.*, 37(1), 357.

Hvitved-Jacobsen, T. and J. Vollertsen (1998), An intercepting sewer from Dortmund to Dinslaken, Germany—prediction of wastewater transformations during transport, Report submitted to the Emschergenossenschaft, p. 35.

Hvitved-Jacobsen, T., J. Vollertsen, and P.H. Nielsen (1998), A process and model concept for microbial wastewater transformations in gravity sewers, *Water Sci. Technol.*, 37(1), 233–241.

Hvitved-Jacobsen, T., J. Vollertsen, and N. Tanaka (1999), Wastewater quality changes during transport in sewers—an integrated aerobic and anaerobic model concept for carbon and sulfur microbial transformations, *Water Sci. Technol.*, 39(2), 242–249.

Hvitved-Jacobsen, T., P.H. Nielsen, T. Larsen, and N.A. Jensen (eds.) (1995), The sewer as a physical, chemical and biological reactor I, *Water Sci. Technol.*, 31(7), 330.

Jahn, A. and P.H. Nielsen (1998), Cell biomass and exopolymer composition in sewer biofilms, *Water Sci. Technol.*, 37(1), 17–24.

Kalyuzhnyi, S., A. Veeken, and B. Hamelers (2000), Two-particle model of anaerobic solid state fermentation, *Water Sci. Technol.*, 41(3), 43–50.

Levine, A.D., G. Tchobanoglous, and T. Asano (1985), Characterization of size distribution of contaminants in wastewater; treatment and reuse implications, *J. Water Pollut. Control Fed.*, 57, 805.

Levine, A.D., G. Tchobanoglous, and T. Asano (1991), Size distributions of particulate contaminants in wastewater and their impact on treatability. *Water Res.*, 25, 911.

Logan, B.E. and Q. Jiang (1990), Molecular size distributions of dissolved organic matter, *J. Environ. Eng.*, 116, 1046.

Matos, J.S. and E.R. de Sousa (1996), Prediction of dissolved oxygen concentration along sanitary sewers, *Water Sci. Technol.*, 34(5–6), 525–532.

Morgenroth, E., R. Kommedal, and P. Harremoës (2002), Processes and modeling of hydrolysis of particulate organic matter in aerobic wastewater treatment—A review, *Water Sci. Technol.*, 45(6), 25–40.

Nielsen, A.H., P. Lens, T. Hvitved-Jacobsen, and J. Vollertsen (2005), Effects of aerobic–anaerobic transient conditions on sulfur and metal cycles in sewer biofilms, *Biofilms*, 2, 1–11.

Nielsen, P.H. and T. Hvitved-Jacobsen (1988), Effect of sulfate and organic matter on the hydrogen sulfide formation in biofilms of filled sanitary sewers, *J. Water Pollut. Control Fed.*, 60(5), 627–634.

Nielsen, P.H., K. Raunkjaer, N.H. Norsker, N.Aa. Jensen, and T. Hvitved-Jacobsen (1992), Transformation of wastewater in sewer systems—A review, *Water Sci. Technol.*, 25(6), 17–31.

Norsker, N.-H., P.H. Nielsen, and T. Hvitved-Jacobsen (1995), Influence of oxygen on biofilm growth and potential sulfate reduction in gravity sewer biofilm, *Water Sci. Technol.*, 31(7), 159–167.

Raunkjaer, K., T. Hvitved-Jacobsen, and P.H. Nielsen (1994), Measurement of pools of protein, carbohydrate and lipid in domestic wastewater, *Water Res.*, 28(2), 251–262.

Raunkjaer, K., T. Hvitved-Jacobsen, and P.H. Nielsen (1995), Transformation of organic matter in a gravity sewer, *Water Environ. Res.*, 67(2), 181–188.

Ristenpart, E. (1995), Sediment properties and their changes in a sewer, *Water Sci. Technol.*, 31(7), 77–83.

Rudelle E., J. Vollertsen, T. Hvitved-Jacobsen, and A.H. Nielsen (2011), Anaerobic transformations of organic matter in collection systems, *Water Environ. Res.*, 83(6), 532–540.

Stanier, R.Y., J.L. Ingraham, M.L. Wheels, and P.R. Painter (1986), *The Microbial World*, Prentice Hall, Englewood Cliffs, NJ.

Tanaka, N. and T. Hvitved-Jacobsen (1999), Anaerobic transformations of wastewater organic matter under sewer conditions, in: I.B. Joliffe and J.E. Ball (eds.), *Proceedings of the 8th International Conference on Urban Storm Drainage*, Sydney, Australia, August 30–September 3, 1999, pp. 288–296.

Tchobanoglous, G., F.L. Burton, and H.D. Stensel (2003), *Wastewater Engineering—Treatment and Reuse*, 4th edn, McGraw-Hill, New York, p. 1819.

USEPA (1985), *Odor and Corrosion Control in Sanitary Sewerage Systems and Treatment Plants*, U.S. Environmental Protection Agency, EPA 625/1-85/018, Washington, D.C.

Vollertsen, J. and T. Hvitved-Jacobsen (2000), Resuspension and oxygen uptake of sediments in combined sewers, *Urban Water*, 2(1), 21–27.

Vollertsen, J. and T. Hvitved-Jacobsen (2000b), Sewer quality modeling—a dry weather approach, *Urban Water*, 2(2), 295–303.

Vollertsen, J. and T. Hvitved-Jacobsen (2002), Biodegradability of wastewater—a method for COD-fractionation, *Water Sci. Technol.*, 45(3), 25–34.

WEF (1994), *Wastewater Biology: The Life Processes*, a special publication prepared by Task Force on Wastewater Biology: The Life Processes, chaired by M.H. Gerardi, Water Environment Federation (WEF), p. 184.

# 4 Sewer Atmosphere
## *Odor and Air–Water Equilibrium and Dynamics*

From a chemical and biological process-engineering point of view, it is crucial to realize that a sewer network—in addition to being designed and operated as a collector and transport system for wastewater—also includes an overlying sewer atmosphere. This gas phase is particularly abundant in gravity sewer pipes, manholes, and pumping stations, and plays a central role in such phenomena as microbial and chemical processes (redox conditions) of the wastewater, odor nuisance, sewer corrosion, and health impacts. The sewer atmosphere must therefore be focused on in terms of transfer of substances in both directions across the air–water interface as well as for its interactions with processes at the sewer walls and with the surrounding atmosphere.

It is therefore important in terms of the conditions for both equilibrium and mass transfer to describe the physicochemical phenomena that affect the exchange of substances across the air–water interface. The reactions related to corrosion at the sewer walls and the factors affecting the ventilation and movement of the sewer atmosphere are also aspects of importance (cf. Figure 4.1). The particularly interesting phenomenon of hydrogen sulfide-induced corrosion at the sewer walls is discussed in more details in Chapter 6.

The following phenomena related to the occurrence of volatile substances in the sewer atmosphere are—in terms of equilibrium, mass transfer, reactions, and impacts—important for process engineering of sewers:

- Reaeration
  The occurrence of dissolved oxygen (DO) in wastewater of sewer networks is the dominating factor for maintaining a high redox potential, and the air–water oxygen transfer (reaeration) is, in reality, the sole pathway for the supply of oxygen to the water phase. The extent of reaeration compared with the microbial removal of DO in the water phase determines the potential for aerobic and anaerobic microbial processes and thereby transformation and removal of wastewater compounds.
- Odor nuisance
  Formation of odorous substances in sewer networks primarily takes place under anaerobic conditions by the production of volatile organic compounds (VOCs) and hydrogen sulfide ($H_2S$) (cf. Chapter 3). The formation of odorous substances therefore particularly occurs in rising mains (force mains or pressurized sewers) and (almost) full flowing gravity sewers in areas with

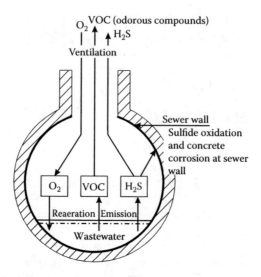

**FIGURE 4.1** Outline of central interfacial related exchange and reactions of volatile compounds in the sewer atmosphere.

relatively high temperatures. In downstream-located partly filled pipes and sewer structures such as drops, pumping stations, and manholes, a free water surface may give rise to emission of upstream produced malodors.

- Concrete corrosion

  Concrete corrosion is associated with $H_2S$ produced under anaerobic conditions in mainly sewer biofilms followed by a buildup in the wastewater phase. When emitted to the sewer atmosphere, $H_2S$ can be oxidized to sulfuric acid ($H_2SO_4$) at the sewer walls by aerobic microbial reactions (cf. Section 3.2.2). $H_2SO_4$ generated at the moist surfaces may react with the alkaline cement in the concrete material and thereby leave a material of loosely bound compounds (gypsum, sand, and gravel). A detailed discussion of concrete corrosion is given in Chapter 6.

- Health impacts

  In addition to being an odorous substance, $H_2S$ is a compound with potential health effects in terms of, e.g., headache symptoms and respiratory injury at rather low concentrations. In high (but still, in practice, occurring) concentrations, it becomes life-threatening. The potential health impacts of $H_2S$ should particularly be considered when working in sewer systems.

It is the general objective of this chapter to concentrate on aspects that relate to the transfer of substances at the air–water interface in sewers, often referred to as partitioning. The basis for the quantification of this phenomenon is, in particular, found within the area of physical chemistry. It is, however, important to stress that this chapter is not a text in physical chemistry. Aspects of physicochemical character that are not crucial for sewer processes will therefore not be addressed, and further details in this regard should be found in the general literature on physical chemistry. In addition to the air–water exchange phenomena, the movement of the sewer atmosphere and ventilation will also be addressed in this chapter.

## 4.1 AIR–WATER EQUILIBRIUM

Equilibrium across a gas–liquid interface is defined as a state where the rates of transfer for a (volatile) compound in each of the two directions are equal. Therefore, the equilibrium is, in principle, dynamic regardless of the fact that no net transfer of the compound takes place.

Although equilibrium for a volatile compound across an air–water interface in a sewer network hardly occurs, it is important to describe it in quantitative terms for the following reasons:

1. It is a rather simple way and a first approach to determine the potential occurrence of a compound at each side of the air–water interface.
2. It is the ultimate occurring state, and the transfer of the volatile compounds proceeds in a direction that approaches this equilibrium state. Furthermore, description of the mass transfer across the air–water interface requires that the equilibrium state be quantified.

### 4.1.1 BASIC CHARACTERISTICS OF THE AIR–WATER EQUILIBRIUM

#### 4.1.1.1 Descriptors for Volatile Substances at the Air–Water Interface

The starting point for quantification of both equilibrium and transfer of volatile substances at an air–water interface is based on their relative occurrence in each of the two phases—the water phase and the gas phase, respectively. The following two descriptors for a compound, A, are therefore basically relevant:

1. In the water phase: the concentration of a compound, $c_A$, in units of, e.g., g m$^{-3}$ or mol L$^{-1}$
2. In the gas phase: the partial pressure of a compound, $p_A$, in units of, e.g., atmosphere (atm) or volumetric expressed parts per million (ppm)

The partial pressure in the gas phase is defined as the pressure that a compound exerts if it, at a constant temperature, is present in the same volume of gas without any other gases. The concept of applying the partial pressure thereby follows Dalton's law stating that the sum of the partial pressures of all compounds of a gas mixture equals the total pressure. This definition is valid for an ideal gas and therefore approximately correct when dealing with the dilute and low-pressure phenomena in the sewer atmosphere.

In this respect, it is important to recall the ideal gas law:

$$pV = nRT \qquad (4.1)$$

where:

$p =$ gas pressure (atm)
$V =$ gas volume (L)
$n =$ equivalent number of molecules (mol)
$R =$ 0.082 L atm K$^{-1}$ mol$^{-1}$
$T =$ absolute temperature (K)

In addition to applying the concentration and partial pressure, the mole fraction of a compound is a convenient measure of concentration in the water phase and the gas phase of both environmental and technical systems. The basics of mole fraction can be illustrated for a binary system—a two-compound system—consisting of compounds A and B. Considering a gas phase, the mole fraction of A is defined as follows:

$$y_A = \frac{N_A}{(N_A + N_B)} \tag{4.2}$$

where:

$y_A =$ mole fraction of compound A in the gas phase (mol mol$^{-1}$)
$N_A =$ moles of compound A (mol)
$N_B =$ moles of compound B (mol)

Applying the concept of mole fraction, the mass balance (in the actual case for a binary gas phase) is simple:

$$y_A + y_B = 1 \tag{4.3}$$

where $y_B$ is the mole fraction of compound B in the gas phase (mol mol$^{-1}$).

As a consequence of Dalton's law, a simple expression exists between the mole fraction and the partial pressure of a compound, A, in the gas phase:

$$y_A = \frac{p_A}{P} \tag{4.4}$$

where:

$p_A =$ partial pressure of compound A (atm)
$P =$ total pressure of the gas phase (atm)

For trace quantities of compound A in the air, $y_A$ can be expressed in terms of the molar concentration of A under conditions where 1 mol of air (a mixture of mainly $N_2$, $O_2$, Ar, and $CO_2$) corresponds to a "molar volume" of approximately 22.4 L mol$^{-1}$ at 0°C (273 K) and 1 atm:

$$y_A = \frac{c_{GA}}{\left(\dfrac{1}{22.4}\right)} = 22.4 c_{GA} = \frac{c_{GA}}{0.0446} \tag{4.5}$$

where $c_{GA}$ is the molar concentration of compound A in the gas phase (mol L$^{-1}$).

In this context it should be noted that the "molar weight" of air under normal atmospheric conditions is approximately 29 g mol$^{-1}$.

For practical purposes, the mole fraction of a compound A in the gas phase, $y_A$, is often by multiplication with $10^6$ converted to parts per million (ppm). The unit in terms of ppm thereby refers to a volumetric fraction.

It should be noted that when dealing with a volumetric descriptor, the temperature plays an important role. As an example, the molar volume at standard conditions of air, i.e., at 25°C (298 K) and 1 atm, is:

$$\frac{22.4 \times 298}{273} = 24.5 \text{ L mol}^{-1}$$

Correspondingly, and for dilute aqueous solutions, the mole fraction in the water phase, $x_A$, can be expressed as follows, as the density of water is approximately equal to 1000 g L$^{-1}$:

$$x_A = \frac{c_{LA}}{\left(\dfrac{1000}{M_{H_2O}}\right)} = \frac{c_{LA}}{\left(\dfrac{1000}{18}\right)} = \frac{c_{LA}}{55.56} \tag{4.6}$$

where:

$x_A$ = mole fraction of a compound A in the water phase (mol mol$^{-1}$)
$c_{LA}$ = molar concentration of compound A in water (mol L$^{-1}$)
$M_{H_2O}$ = molar weight of water = 18 (g mol$^{-1}$)

When dealing with air–water equilibrium and corresponding exchange of substances, the advantage of applying the mole fraction is mainly because these phenomena can be quantitatively expressed in a simple way. Phenomena referring to sewer relevant compound such as $H_2S$, VOCs, and $O_2$ are important examples.

### 4.1.1.2  Partitioning Coefficient

The partitioning coefficient, $K_{LG}$, also named distribution coefficient, is a first approach for the description of the liquid–gas equilibrium for a volatile compound, A. The partitioning coefficient states that the ratio of the concentrations of a volatile compound in the gas phase and the water phase, respectively, is constant at equilibrium:

$$K_{LG} = \frac{y_A}{x_A} \tag{4.7}$$

where:

$K_{LG}$ = partitioning coefficient or distribution coefficient (–)
$y_A$ = mole fraction of compound A in the gas phase (mol mol$^{-1}$)
$x_A$ = mole fraction of compound A in the liquid (water) phase (mol mol$^{-1}$)

In Equation 4.7, the partitioning coefficient is defined based on the mole fractions for A. However, other units for concentration, e.g., mol L$^{-1}$, g m$^{-3}$, or ppm, can also be applied. $K_{LG}$ is temperature dependent but is independent of the quantity of A as long as dilute solutions are dealt with.

The partitioning coefficient will be further discussed in relation to Henry's law (cf. Section 4.1.2.1).

### 4.1.1.3 Relative Volatility

The relative volatility, $\alpha_A$, is a temperature-dependent constant that, under equilibrium conditions, can be used to express the distribution of a volatile compound, A, between a gas phase with both A and water vapor, and a water phase in which A is dissolved. The relative volatility is thereby defined as follows:

$$\alpha_A = \frac{\left(\dfrac{y_A}{y_{water}}\right)}{\left(\dfrac{x_A}{x_{water}}\right)} \tag{4.8}$$

where:

$\alpha_A$ = relative volatility (–)
$y_{water}$ = mole fraction of water vapor in the gas phase (mol mol$^{-1}$)
$x_{water}$ = mole fraction of water in the water phase (mol mol$^{-1}$)

For a dilute solution of A in water, which is a reasonable approximation for typical wastewater, $x_{water}$ is approximately equal to 1, and Equation 4.8 is therefore expressed as follows:

$$\alpha_A = \frac{\left(\dfrac{y_A}{y_{water}}\right)}{x_A} \tag{4.9}$$

As can be seen from Equations 4.8 and 4.9, $\alpha_A$ expresses the occurrence of compound A in the gas phase relative to H$_2$O. It thereby represents a relative tendency for A to be emitted from the water phase.

### 4.1.2 HENRY'S LAW

#### 4.1.2.1 Formulation of Henry's Law

The most widely used and yet simple theoretical approach for describing a gas–liquid equilibrium for a volatile compound, A, is expressed by Henry's law:

$$p_A = y_A P = H_A x_A \tag{4.10}$$

where:

$p_A$ = partial pressure of a compound A in the gas phase (atm)
$P$ = total pressure in the gas phase (atm)
$H_A$ = Henry's law constant for A (atm)

Under equilibrium conditions and at constant temperature, Henry's law expresses the relative amount of a volatile compound in the gas phase as a function of its relative occurrence in the water phase. Henry's law thereby quantifies the tendency of a volatile compound to escape from the liquid phase. The law in its simple formulation applies to dilute solutions of A, i.e., for solutions where $x_A$ is close to 0, and provided that A occurs as a molecular compound and does not dissociate or undergoes reaction in the water phase (cf. Section 4.1.3).

It is crucial to note that a sewer system is highly dynamic in terms of occurrence of substances, and that equilibrium for a volatile compound as required by Henry's law hardly exists. An important reason for this fact is that volatile substances are subject to transport from one location in the sewer to another and therefore occur in concentrations that are both time-varying and varying from location to location in both the water phase and the sewer atmosphere (cf. Section 4.3). In spite of this, Henry's law constant plays a central role when we formulate the rate of the air–water transport processes for volatile compounds (cf. Section 4.2).

Table 4.1 gives some examples of Henry's law constants and boiling points for selected odorous and nonodorous compounds potentially produced in wastewater of sewer systems. Furthermore, the list includes values for hydrocarbons that frequently appear in sewer networks from, e.g., industrial sources and street runoff.

### Example 4.1: Solubility of Oxygen in Water

The partial pressure of molecular oxygen, $O_2$, is measured with a gas detector in a sewer atmosphere and found to be 0.18 atm. This value is obtained because of the oxygen consumption of the wastewater and a limited degree of ventilation, which is typically slightly lower than that in the ambient atmosphere, where it is approximately 0.21 atm.

Determine at 25°C the solubility of $O_2$—i.e., at equilibrium conditions—in the wastewater phase (approximately considered as pure water).

From Equation 4.10 and Table 4.1:

$$x_{O_2} = \frac{p_{O_2}}{H_{O_2}} = \frac{0.18}{43,800} = 4.1 \times 10^{-6}$$

Equation 4.6 converts mole fraction to units of mol $L^{-1}$:

$$c_{O_2} = 55.56 \times x_{O_2} = 55.56 \times 4.1 \times 10^{-6} = 2.28 \times 10^{-4} \text{ mol } L^{-1}$$

And further conversion into units of g $m^{-3}$, knowing that the molar weight of $O_2$ is 32 g $mol^{-1}$, gives:

$$c_{O_2} = 2.28 \times 10^{-4} \cdot 32 \times 10^3 = 7.3 \text{ mg } O_2 \ L^{-1} = 7.3 \text{ g } O_2 \ m^{-3}$$

Whereas a partial pressure of 0.18 atm in a sewer atmosphere is realistic, an equilibrium DO concentration equal to 7.3 $gO_2$ $m^{-3}$ is for wastewater in a sewer rather high and not typically observed because of the microbial consumption of DO. This fact exemplifies that equilibrium conditions for $O_2$ is not common.

## TABLE 4.1
## Boiling Points and Henry's Law Constants for Selected Volatile Compounds

| Compound | Boiling Point at 1 atm Pressure (°C) | Henry's Law Constant at 25°C (atm) |
|---|---|---|
| **Volatile Sulfur Compounds (VSCs)** | | |
| Methyl mercaptan, $CH_3SH$ | 6 | 200 |
| Ethyl mercaptan, $C_2H_5SH$ | 36 | 200 |
| Allyl mercaptan, $C_3H_5SH$ | 69 | |
| Thiophenol, $C_6H_5SH$ | 169 | 19 |
| Dimethyl sulfide, $(CH_3)_2S$ | 37 | 99 |
| Dimethyl disulfide, $(CH_3)_2S_2$ | 109 | 63 |
| **Nitrogenous Compounds** | | |
| Methylamine, $CH_3NH_2$ | −6 | 0.55 |
| Ethylamine, $C_2H_5NH_2$ | 17 | 0.55 |
| Dimethylamine, $(CH_3)_2NH$ | 7 | 1.0 |
| Pyridine, $C_5H_5N$ | 115 | 0.5 |
| Indole, $C_8H_7N$ | 254 | 0.029 |
| Scatole, $C_8H_6(CH_3)N$ | 265 | |
| **Acids (VFAs)** | | |
| Formic acid, $HCOOH$ | 100 | 0.010 |
| Acetic acid, $CH_3COOH$ | 118 | 0.007 |
| Propionic acid, $C_2H_5COOH$ | 141 | 0.009 |
| Butyric acid, $C_3H_7COOH$ | 162 | 0.012 |
| Valeric acid, $C_4H_9COOH$ | 185 | 0.026 |
| **Alcohols and Phenols** | | |
| Methanol, $CH_3OH$ | 65 | 0.25 |
| Ethanol, $C_2H_5OH$ | 78 | 0.28 |
| Phenol, $C_6H_5OH$ | 182 | 0.022 |
| **Aldehydes and Ketones** | | |
| Acetaldehyde, $CH_3CHO$ | 21 | 4.9 |
| Butyraldehyde, $C_3H_7CHO$ | 75 | 6.4 |
| Acetone, $(CH_3)_2CO$ | 56 | 2.0 |
| Butanone, $C_2H_5COCH_3$ | 80 | 2.5 |
| **Inorganic Gases** | | |
| Hydrogen sulfide, $H_2S$ | −60 | 560 |
| Ammonia, $NH_3$ | −33 | 0.9 |
| **Selected Nonodorous Compounds** | | |
| Nitrogen, $N_2$ | −196 | 86,500 |
| Oxygen, $O_2$ | −183 | 43,800 |
| Carbon dioxide, $CO_2$ | −78.5 | 1640 |
| Methane, $CH_4$ | −164 | 40,200 |

**TABLE 4.1 (Continued)**
**Boiling Points and Henry's Law Constants for Selected Volatile Compounds**

| Compound | Boiling Point at 1 atm Pressure (°C) | Henry's Law Constant at 25°C (atm) |
|---|---|---|
| | Hydrocarbons | |
| Pentane, $C_5H_{12}$ | 36 | 70,400 |
| Hexane, $C_6H_{14}$ | 69 | 80,500 |
| Heptane, $C_7H_{16}$ | 98 | 46,900 |
| Octane, $C_8H_{18}$ | 125 | 19,400 |

*Source:* Data mainly from Betterton (1992), Vincent and Hobson (1998), and Sander (2000).

In Equation 4.10, Henry's constant, $H$, is expressed based on the mole fractions of a compound at the air–water interface. The unit used for $H$ is therefore atmosphere. However, and referring to Section 4.1.1.1, there are other possibilities for selecting descriptors for a compound that will affect the unit of Henry's law constant. In addition to the units of mole fraction, two examples are outlined in Table 4.2.

In addition to what is already shown in Table 4.2, it should be noted that numerous other units for Henry's law constant are also in use, e.g., (kg bar mol$^{-1}$), (Pa m$^3$ mol$^{-1}$), and (mol m$^{-3}$) (mg L$^{-1}$)$^{-1}$, with the latter unit corresponding to the use of mg L$^{-1}$ in the water phase and mole m$^{-3}$ in the vapor phase for the volatile compound. Betterton (1992) lists Henry's law constants as $K_H$ values in reciprocal units of L atm mol$^{-1}$ equivalent to M atm$^{-1}$ for a wide variety of substances. These different units for Henry's law constant may cause confusion if not carefully considered.

Referring to Table 4.2, it is important to note that the dimensionless Henry's law constant is compound-specific, whereas the ratio between the first and the second values of $H$ shown in Table 4.2 is constant and equal to 55.56 irrespective of

**TABLE 4.2**
**Water Phase and Gas Phase Descriptors used in Different Forms of Henry's Law**

| Water Phase Descriptor (unit) | Gas Phase Descriptor (unit) | Units of Henry's Law Constant, $H$ |
|---|---|---|
| Mole fraction (–) | Mole fraction (–) | Atm |
| Concentration (mol L$^{-1}$) | Partial pressure (atm) | L atm mol$^{-1}$[a] |
| Concentration (mol L$^{-1}$) | Partial pressure (bar) | L bar mol$^{-1}$[a] |
| Concentration (g m$^{-3}$ or mol L$^{-1}$) | Concentration (g m$^{-3}$ or mol L$^{-1}$) | Dimensionless[b] |

[a] The reciprocal value of H, often denoted as either $k_H$ or $K_H$, is also in use.
[b] The dimensionless expressed Henry's law constant is also named partitioning coefficient (cf. Section 4.1.1.2).

which compound is considered (cf. Equation 4.6). In the literature, the dimensionless expressed constant is often referred to as Henry's law constant; however, it is more correct to consider this term a partitioning coefficient (cf. Section 4.1.1.2).

### 4.1.2.2 Temperature Dependency of Henry's Law Constant

Physical, chemical, and biological "constants" used in the description of sewer processes are in general temperature dependent. Such constants are, for example, rate constants for chemical and biological processes and diffusion-related transport coefficients. Different approaches are used to determine a temperature dependency, typically a simplified version of the Arrhenius equation (cf. Section 2.2.3).

The van't Hoff equation used in the following is, in its thermodynamic basis, enthalpy related and thereby, in this respect, formulated identically with the Arrhenius equation. It is the thermodynamic properties of the solvent (water) and the volatile compound in question that determine the temperature dependency of Henry's law constant. For aqueous solutions of a volatile compound, the temperature dependency of Henry's law constant is, in terms of the van't Hoff equation, expressed as follows:

$$H_T = H_{298} \exp\left(-C\left(\frac{1}{T} - \frac{1}{298}\right)\right) \qquad (4.11)$$

where:
$H_T$ = Henry's law constant at temperature $T$ (unit, cf. Table 4.2)
$H_{298}$ = Henry's law constant at 298 K (unit, cf. Table 4.2)
$C$ = temperature coefficient (K)
$T$ = absolute temperature (K)

Corresponding to the values shown in Table 4.1, Equation 4.11 is formulated with 298 K (25°C) as the standard temperature. As can be seen in Equation 4.11, Henry's law constant is temperature dependent with a positive $dH_T/dT$ value.

The constant $C$ in Equation 4.11 is, in addition to the influence of the solvent (water), compound-specific. As an example for aqueous solutions of DO and hydrogen sulfide ($H_2S$) at temperature intervals relevant for sewer systems, $C$ is 1700 and 2200 K, respectively.

### Example 4.2: Temperature Dependency of Henry's Law Constant for $H_2S$

Referring to Table 4.1, Henry's law constant for $H_2S$, $H_{H_2S,298}$, is 560 atm at standard temperature, 298 K (25°C). Using a temperature coefficient of $C = 2200$ K and applying the van't Hoff equation (Equation 4.11), the value of Henry's law constant at 15°C is:

$$H_{H_2S,288} = 560\exp\left(-2200\left(\frac{1}{288} - \frac{1}{298}\right)\right) = 560\exp(-0.256) = 433 \text{ atm}$$

As can be seen from this example, there is a considerable reduction (23%) in the value of Henry's law constant for $H_2S$ from 25°C to 15°C.

Alternatively, a modified form of the van't Hoff equation can be used (Tchobanoglous et al. 2003):

$$\log H_T = -\frac{A}{T} + B$$

Applying the empirical constants $A = 885$ K and $B = 5.70$ for $H_2S$ proposed by Tchobanoglous et al. (2003), Henry's law constant for $H_2S$ at 15°C is:

$$H_{H_2S,288} = 427 \text{ atm}$$

Taking into account the uncertainties in the determination of the temperature coefficients, $A$, $B$, and $C$, the two calculated values of $H_{H_2S,288}$, 433 and 427 atm, are considered to be within an acceptable range.

### 4.1.3 Water–Air Equilibrium for Dissociated Substances

As previously mentioned, the expressions for air–water equilibrium of volatile compounds—including Henry's law—require that these substances exist in a molecular and nondissociated form in the water phase. Furthermore, no chemical or biological reaction with neither water nor other compounds must take place. For several substances, such requirements are not fulfilled, and the expressions for air–water equilibrium must be modified.

Hydrogen sulfide is an important example of a compound that does not observe the requirements of being "inert" in this respect:

- Hydrogen sulfide is a weak acid that undergoes dissociation dependent on the pH value.
- As a weak acid, hydrogen sulfide reacts with, e.g., heavy metals forming precipitates.

Because of odor, corrosion, and health impacts, hydrogen sulfide is a highly essential compound in sewers. Sulfide is therefore selected as the example of a compound that requires a modification of the simple air–water equilibrium expressions.

The central point is the dissociation of $H_2S$ expressed by the following equilibrium reactions:

$$
\begin{aligned}
&\text{Air-Water} \\
&\text{transfer} \quad pK_{a1} = 7.0 \quad pK_{a2} = 14.0 \\
&H_2S(g) \Leftrightarrow H_2S(aq) \Leftrightarrow HS^- + H^+ \Leftrightarrow S^{2-} + 2H^+ \\
&\text{(gas)} \qquad \text{(aqueous)}
\end{aligned}
\tag{4.12}
$$

The $pK = -\log K$ values shown in Equation 4.12 are given at 20°C and defined by the following equilibrium equations:

$$K_{a1} = \frac{C_{H^+} C_{HS^-}}{C_{H_2S,(aq)}} \tag{4.13}$$

$$K_{a2} = \frac{C_{H^+} C_{S^{2-}}}{C_{HS^-}} \tag{4.14}$$

These equilibrium constants are—similar to all equilibrium constants—temperature dependent. The temperature dependency is typically determined by the Arrhenius equation (Equation 2.37) or a simplified version of it (cf. Equation 2.39). Alternatively, the van't Hoff equation (Equation 4.11) can be used.

The release of $H_2S$ into the atmosphere as shown by Equation 4.12 is strongly dependent on the actual pH value because only the molecular form—and not the dissociated forms—can be emitted. As an example, at pH = 7.0, an equal amount of $H_2S$ and $HS^-$ is found in the water phase and only 50% of the total amount of sulfide is therefore subject to be emitted. At equilibrium conditions and at a constant total concentration of sulfide in the water phase, increase in pH value will reduce the partial pressure of $H_2S$ in the atmosphere above the water surface. Therefore, when applying Henry's law (Equation 4.10), only the actual occurring molecular form, $H_2S$, should be taken into account.

As a consequence, it becomes crucial to determine the amount of the different forms of sulfide, i.e., $H_2S$, $HS^-$, and $S^{2-}$, and Equations 4.12 through 4.14 are therefore central. However, as shown by the value of $pK_{a2}$, the sulfide ion, $S^{2-}$, only exists in measurable amounts above a pH of about 12. Only the equilibrium between $H_2S$ (aq) and $HS^-$ as shown in Equation 4.12 is therefore generally relevant for wastewater systems. Equation 4.13 can therefore be reformulated to Equation 4.15, which determines the ratio between the concentrations of the two relevant aqueous forms of sulfide, $C_{HS^-}$ and $C_{H_2S(aq)}$, at the actual pH:

$$pH = pK_{a1} + \log \frac{C_{HS^-}}{C_{H_2S,(aq)}} \tag{4.15}$$

A graphical representation of Equation 4.15 is shown in Figure 4.2. As shown in this figure, the pH-dependent distribution between $H_2S$ and $HS^-$ is temperature dependent.

Equation 4.15 and its representation in Figure 4.2 is the basis for modifying air–water equilibrium expressions for sulfide, e.g., Henry's law (cf. Equation 4.10). As an example, the result of combining Equations 4.10 and 4.15 for sulfide at 20°C is shown in Figure 4.3 at a total sulfide concentration in the water phase of 1 g S m$^{-3}$.

As already discussed in Section 4.1.2.1, air–water equilibrium for a volatile compound—and therefore also for sulfide—typically does not occur in sewer networks. Although the curves shown in Figures 4.2 and 4.3 represent equilibrium, they are essential when assessing the potential risk for odor and health impacts, in particular under turbulent conditions in the water phase. Under such conditions, the emission of volatile substances is increased and semiequilibrium conditions being

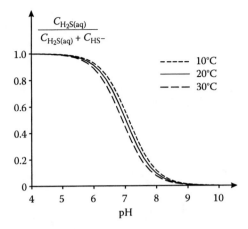

**FIGURE 4.2** Distribution between $H_2S$ and $HS^-$ at equilibrium in aqueous solution (cf. Equation 4.15).

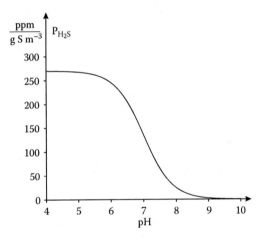

**FIGURE 4.3** Partial pressure of $H_2S$ (g) in ppm versus pH at equilibrium, 20°C and $C_{H_2S(aq)} + C_{HS^-} = 1$ g S m$^{-3}$.

approached. Release of $H_2S$ into the sewer atmosphere at sewer drops, pumping stations and hydraulic jumps may be rather intensive. Further details on $H_2S$ emission are given in Section 4.3.2.

### Example 4.3: Air–Water Equilibrium for Sulfide in a Sewer

This example concerns the calculation of the partial pressure of $H_2S$ in units of ppm in a sewer atmosphere at 15°C in equilibrium with wastewater at pH = 7.5 and a dissolved sulfide concentration of 1 g S m$^{-3}$.

Although $pK_{a1} = 7.0$ is given at 20°C (cf. Equation 4.12), it is considered approximately correct to use this value at 15°C, whereas the temperature dependency of Henry's law constant, $H_{H_2S}$, must be taken into account. According to Table 4.1, $H_{H_2S} = 560$ atm at 25°C, a value that is reduced to 433 atm at 15°C (cf. Example 4.2).

According to Equation 4.15, the concentration of the molecular fraction of the total sulfide concentration (with atomic weight of $S$ equal to 32 g S mol$^{-1}$) can be calculated as:

$$\log\left(\frac{C_{H_2S}}{C_{HS^-}}\right) = pK_{a1} - pH = 7.0 - 7.5 = -0.5$$

$$\frac{C_{H_2S}}{1 - C_{H_2S}} = 0.316$$

$$C_{H_2S} = 0.24 \text{ g S m}^{-3} = \frac{0.24}{32} = 0.0075 \text{ mol m}^{-3}$$

According to Henry's law, Equation 4.10, and Equation 4.6, the equilibrium partial pressure of H$_2$S is:

$$p_{H_2S} = H_{H_2S} \cdot x_{H_2S} = 433 \frac{0.0075 \times 10^{-3}}{55.56} = 5.85 \times 10^{-5} \text{ atm}$$

$$p_{H_2S} = 58.5 \text{ ppm}$$

The numerical value of $p_{H_2S}$ is in agreement with Figure 4.3. This figure also shows that there is a rather steep decline in the equilibrium value of $p_{H_2S}$ within the pH range 7–8, an interval that is typical for municipal wastewater. From a practical point of view, it is important that the air–water equilibrium for H$_2$S normally does not occur because of, e.g., ventilation and reaction with concrete materials.

The potential effects in terms of odor nuisance and human health impacts related to the concentration of H$_2$S in the sewer atmosphere are dealt with in Section 4.3.4. Concrete corrosion caused by H$_2$S in the sewer atmosphere is a central subject of Chapter 6.

In addition to what has been discussed in this section (air–water equilibrium of dissociated compounds), it is important to note that other phenomena also have an impact. As an example, sulfide (HS$^-$) may form rather insoluble precipitates with several heavy metals whereby the concentration of the molecular form, H$_2$S, is considerably reduced. This phenomenon is, by addition of iron salts to wastewater in sewers, widely used for sulfide control (cf. Chapter 7).

Sulfide has been focused on in this section; however, it must be noted that sulfide was used only as an example. Other volatile compounds may exert similar impacts, and a modification of the simple air–water equilibrium must be done. An example is ammonia that exists as NH$_3$ and NH$_4^+$ influenced by pH. Other examples are mercaptans (thiols) and volatile fatty acids (VFAs) that may undergo microbial degradation in wastewater in addition to the fact that VFAs also exist in molecular and ionic forms.

## 4.2 AIR–WATER TRANSPORT PROCESSES

The equilibrium air–water phenomena were the focal points of Section 4.1. In this section, the transport processes across the air–water interface are dealt with, i.e., phenomena that refer to the dynamics and kinetics, and thereby the importance of a net transfer of compounds from one phase into another. An important and illustrative example is the reaeration process, i.e., the transfer of oxygen ($O_2$) from the atmosphere into the wastewater phase.

### 4.2.1 OVERVIEW OF THEORETICAL APPROACHES

Several approaches to the air–water mass transport exist. In particular, and also relevant to transport processes in sewer networks, the main developments have been directed toward the reaeration process. Briefly, the theoretical descriptions of interfacial transport processes include the following:

- Two-film theory
  The two-film theory is based on the molecular diffusion of volatile substances through two stagnant films at the interface, a liquid film and a gas film (Lewis and Whitman 1924). Fick's first law of molecular diffusion is consequently valid (cf. Section 4.2.2).
- Penetration theories
  According to the penetration theories, the diffusion of volatile molecules across the interface takes place at the elements of water that is transported by turbulence to the surface (Higbie 1935; Danckwerts 1951; Dobbins 1956). The contact of these water elements to the gas phase is considered short and constant depending on the level of turbulence.
- Surface renewal theory
  This theory describes the replacement and movement of a surface liquid film by the action of eddies (Danckwerts 1951; King 1966). The surface renewal theory is, in principle, similar to the penetration theories except for the fact that the contact time at the interface varies.

In this book, the principal theory for mass transport across the air–water interface in a sewer network is based on the two-film theory. Further basic details on the air–waster mass transfer phenomena can be found in the reports of Thibodeaux (1996) and Stumm and Morgan (1996).

### 4.2.2 TWO-FILM THEORY

As briefly mentioned, the two-film theory is based on the molecular diffusion of volatile components through stagnant liquid and gas films. This theory is the classical way of understanding mass transfer across the air–water boundary. To some extent, in contrast to the other theories mentioned in Section 4.2.1, the two-film theory gives a fundamental understanding that leads to rather simple expressions for gas–liquid mass exchange phenomena that are useful in practice.

The two-film theory describes the transfer of volatile compounds in "both directions," i.e., as absorption from gas phase to liquid phase and as desorption, emission, or release from a liquid phase into a gas phase.

### 4.2.2.1  Expressions for Mass Transfer across Air–Water Interface

According to the two-film theory, the principle of the transfer mechanisms of a volatile compound across the air–water interface is depicted in Figure 4.4. The figure illustrates that the bulk phases are completely mixed and that the concentration gradients in the two films express the driving forces for the mass transport and thereby determine the direction and the magnitude of the flux. According to this understanding, the overall resistance to the mass transport is the sum of the resistances in each of the two films:

$$r_O = r_L + r_G \tag{4.16}$$

where:

$r_O$ = overall resistance to transport of a compound (m$^2$ s)
$r_L$ = resistance to transport of a compound in the liquid film (m$^2$ s)
$r_G$ = resistance to transport of a compound in the gas film (m$^2$ s)

The two-film theory states that resistance to the transport of mass resides within thin water and gas layers located at the interface, i.e., the two films where the molecular diffusion results in concentration gradients (cf. Figure 4.4). The resistance to the mass transfer at the interface itself is assumed to be negligible resulting, in no accumulation of mass at the interface. Accordingly, equilibrium conditions therefore exist at the interface itself and Henry's law (Equation 4.10) can be applied:

$$y_{Ai} = \frac{H_A}{P} x_{Ai} \tag{4.17}$$

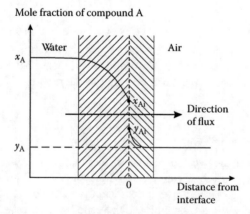

**FIGURE 4.4**  Principle of mechanisms governing the transfer of a volatile compound across the air–water interface according to two-film theory.

where:

$y_{Ai}$ = mole fraction of compound A at the gas (air) side of the interface (mol mol$^{-1}$)

$x_{Ai}$ = mole fraction of compound A at the liquid (water) side of the interface (mol mol$^{-1}$)

Because of this conceptual understanding of the transport mechanisms, the theory is referred to as the two-film theory (Lewis and Whitman 1924). According to the two-film theory, it is appropriate to consider the transport of volatile compounds between the water phase and gas phase in two steps: from the bulk water phase to the interface and from the interface to the air, or vice versa.

The mass transfer rates of a compound, A, per unit surface area from the water phase to the interface and from the interface to the air phase is determined from the difference between the actual mole fractions, $x_A$ and $y_A$, and the corresponding equilibrium values, $x_{Ai}$ and $y_{Ai}$ at the interface:

$$J_A = -k_{L,A}(x_{Ai} - x_A) \tag{4.18}$$

$$J_A = -k_{G,A}(y_{Ai} - y_A) \tag{4.19}$$

where:

$J_A$ = flux rate of compound A (mol mol$^{-1}$ m$^{-2}$ s$^{-1}$)

$k_{L,A}$ = water phase mass transfer coefficient (m$^{-2}$ s$^{-1}$)

$k_{G,A}$ = gas phase mass transfer coefficient (m$^{-2}$ s$^{-1}$)

Which of Equations 4.18 and 4.19 is the governing equation depends on which part of the boundary, the liquid film or the gas film, has the major resistance to the mass transport. As an example, if the major resistance exists in the water film of the boundary, the consequence is that $k_{L,A} \ll k_{G,A}$. Although both equations are equally valid, Equation 4.18 is, in this case, the relevant equation for determining the flux rate.

Equations 4.18 and 4.19 are theoretically correct; however, no means exist to determine $x_{Ai}$ and $y_{Ai}$. Two fictitious (nonexisting) equilibrium values of the mole fractions, $x_A^*$ and $y_A^*$, referring to $y_A$ and $x_A$, respectively, are therefore introduced. The point is that although $x_A^*$ and $y_A^*$ are fictitious, these values observe Henry's law (Equation 4.10):

$$y_A^* = \frac{H_A}{P} x_A \tag{4.20}$$

$$y_A = \frac{H_A}{P} x_A^* \tag{4.21}$$

Equations 4.17 through 4.21 constitute the basis for development of expressions that, for practical purposes, can be used to determine mass transport across the air–water interface. An important point in this respect is in the continuation of Equations 4.18

and 4.19 to define "overall mass transfer coefficients" that replace the "phase mass transfer coefficients." Equations 4.18 and 4.19 are therefore transformed as follows:

$$J_A = -K_L \left( x_A^* - x_A \right) \tag{4.22}$$

$$J_A = K_G \left( y_A^* - y_A \right) \tag{4.23}$$

where:
$K_L$ = overall mass transfer coefficient referring to the liquid phase ($m^{-2} s^{-1}$)
$K_G$ = overall mass transfer coefficient referring to the gas phase ($m^{-2} s^{-1}$)

Equations 4.20 and 4.21 can be substituted into Equation 4.23 and Equation 4.22, respectively, resulting in:

$$J_A = K_L \left( x_A - \frac{y_A}{\frac{H_A}{P}} \right) \tag{4.24}$$

$$J_A = -K_G \left( y_A - \frac{H_A}{P} x_A \right) \tag{4.25}$$

The two equations, Equations 4.24 and 4.25, are equally valid; however, Equation 4.24 with a mass transfer coefficient, $K_L$, that refers to the water phase is typically applied, and therefore will be discussed further.

As an overall mass transfer coefficient, $K_L$ will include contributions of resistance to transport across both the liquid film and the gas film. Equation 4.16 that expresses the overall resistance to mass transport across the two films is therefore the basis for the formulation of the overall mass transfer coefficient, $K_L$:

$$\frac{1}{K_L} = \frac{1}{k_{L,A}} + \frac{P}{H_A k_{G,A}} \tag{4.26}$$

Equations 4.24 and 4.26 are, for practical purposes, the two central expressions for the determination of air–water mass transfer rates. Further details in this respect are dealt with in Sections 4.2.2.3 and 4.4.2.

### 4.2.2.2 Molecular Diffusion at the Air–Water Interface

The flux rate for a compound, A, as expressed by Equation 4.24 is basically a result of molecular diffusion and therefore relates to Fick's first law of molecular diffusion. For the water phase, i.e., in the water film, Fick's first law is, for practical purposes, expressed as follows:

$$J_A = -D_{L,A} \frac{\partial C_A}{\partial z} \tag{4.27}$$

where:

$J_A$ = flux rate of compound A (mol m$^{-2}$ s$^{-1}$)

$D_{L,A}$ = molecular diffusion coefficient (diffusivity) for compound A in the water phase (m$^2$ s$^{-1}$)

$C_A$ = concentration of component A (mol m$^{-3}$)

$z$ = direction (m)

Note that $J_A$ in Equation 4.27 is expressed in different units compared with Equation 4.18.

As a consequence of the two-film theory, there is a relation between a mass transfer coefficient and the diffusion coefficient. It is important that the mass transfer coefficients $k_{L,A}$ and $k_{G,A}$ can be interpreted in terms of the corresponding molecular diffusion coefficients. These mass transfer coefficients, expressed in units of m$^{-2}$ s$^{-1}$—and thereby referring to a flux rate—are often replaced by a so-called exchange constant, $k$, in units of m s$^{-1}$, i.e., with a velocity dimension. Further details in this respect are given in Section 4.2.2.3. For both water and gas phases, the relation between the exchange constant and the diffusion coefficient is as follows:

$$k = \frac{D}{z_f} \quad (4.28)$$

where:

$k$ = exchange constant (m s$^{-1}$)

$z_f$ = thickness of a liquid or a gas film (m)

The exchange constant, $k$, is a measure of the transport of a volatile compound per unit concentration gradient. The magnitude of $k$ depends mainly on the degree of turbulence of both the liquid phase and the gas phase. Although values of the diffusion coefficients are well known, Equation 4.28 is, for practical purposes, not directly applicable because the film thicknesses are unknown. In relative terms, however, knowledge on molecular diffusion coefficients is important for the determination of overall mass transfer coefficients, $K_L$ values, for volatile substances (cf. Section 4.2.2.3).

The diffusion coefficient is temperature dependent according to the so-called Eyring equation that relates a chemical reaction rate to temperature (Eyring 1935). For practical purposes, this equation is equivalent to the empirically formulated version of the Arrhenius equation (cf. Section 2.2.3):

$$D_T = D_{298} \, \beta^{T-298} \quad (4.29)$$

where:

$D_T$ = molecular diffusion coefficient in the water phase at temperature $T$ (m$^2$ s$^{-1}$)

$D_{298}$ = molecular diffusion coefficient in the water phase at temperature 298 K (m$^2$ s$^{-1}$)

$\beta$ = temperature coefficient (–)

$T$ = temperature (K)

The use of Equation 4.29 is further discussed in Example 4.4.

### Example 4.4: Temperature Dependency of the Molecular Diffusion Coefficient

The temperature dependency of a diffusion coefficient for a volatile compound is in the order of 3% per °Celsius temperature change at temperatures relevant for wastewater in sewers. The constant, $\beta$, in Equation 4.24 is therefore approximately equal to 1.03.

$D_{298}$ values for the compounds $O_2$ and $H_2S$, two important examples of volatile compounds of wastewater in sewers, are approximately $2.3 \times 10^{-9}$ and $1.9 \times 10^{-9}$ $m^2$ $s^{-1}$, respectively. Applying Equation 4.29, these values are, at 10°C (283 K), reduced to about $1.5 \times 10^{-9}$ and $1.2 \times 10^{-9}$ $m^2$ $s^{-1}$, respectively.

Volatile compounds with a molecular size larger than oxygen and hydrogen sulfide typically show lower diffusion coefficients, often with $D_{298}$ values reduced by 50% to 70%. Several empirical equations exist for estimating diffusion coefficients for gases dissolved in water (Yongsiri et al. 2004a).

### 4.2.2.3 General Characteristics of Air–Water Mass Transfer Coefficients

Equations 4.16 and 4.26 state that the total resistance to mass transfer across the air–water boundary is equal to the sum of the resistance values across the liquid film and the gas film. This understanding is a central point of the two-film theory, which also implies equilibrium at the interface itself with no accumulation of mass. In Equation 4.26, it is furthermore seen that the magnitude of Henry's law constant, $H_A$, in terms of mass transport plays an important role. For substances with relatively high values of $H_A$, exemplified by $O_2$ and $H_2S$, the resistance mainly exists in the water film, and turbulence of the wastewater flow will, therefore, enhance the transfer of such substances from the water phase into the sewer atmosphere. As seen from the $H_A$ values in Table 4.1, the importance of turbulence in the water phase is relatively reduced for several odorous compounds with rather low $H_A$ values. Consequently, movements in the sewer atmosphere will correspondingly increase the release rate for such compounds. As also shown in Equation 4.26, the release rates for volatile substances also depend on the $k_{L,A}/k_{G,A}$ ratio, which varies according to component as well as system characteristics.

Liss and Slater (1974) have, based on the value of Henry's law constant, assessed which type of mass transfer resistance exists. Pragmatically, they propose the following criteria, which are valid for most cases (cf. Table 4.1):

- The mass transfer is controlled by the liquid film if $H_A > 250$ atm.
- The resistance in both liquid and gas films is of importance if $H_A$ is between 1 and 250 atm.
- The flux rate is controlled by the air film if $H_A < 1$ atm. This situation corresponds not only to compounds with a relatively low volatility but also to compounds that are reactive in the water phase [e.g., ammonia ($NH_3$)].

As can be seen in Table 4.1, all three cases are relevant for volatile compounds. Again, it is important to note that the resistance for the air–water oxygen transfer (reaeration) and for the hydrogen sulfide emission resides in the water film.

A major problem in the quantification of air–water transport phenomena in terms of the rate equation, Equation 4.24, is to find values for $K_L$. Such values are, by nature, compound specific. As far as sewer systems are concerned, the most detailed knowledge concerning air–water mass transfer exists on reaeration, i.e., on oxygen transfer (cf. Section 4.4). An important task based on this knowledge is to estimate the $K_L$ values for other compounds.

In this respect, an approach has been suggested for the determination of a mass transfer coefficient, $K_{L,A}$, for a volatile compound A. The basic idea behind this approach is the fact that the ratio between $K_{L,A}$ and the $K_L$ value of any reference compound, $K_{L,ref}$, is constant:

$$\frac{K_{L,A}}{K_{L,ref}} = \text{constant} \tag{4.30}$$

Oxygen is often selected as the reference compound because the air–water oxygen transfer coefficient, $K_{L,O_2}$, can be determined from empirical expressions that have been developed based on numerous reaeration studies (cf. Section 4.4.2 and Table 4.3). Furthermore, the constant in Equation 4.30 can be determined from the molecular diffusion coefficients, $D$, for the actual compound, A, and oxygen (cf. Section 4.2.2.2).

The relation combining the mass transfer coefficients and the diffusion coefficients is formulated as follows:

$$\frac{K_{L,A}}{K_{L,O_2}} = \left( \frac{D_{L,A}}{D_{L,O_2}} \right)^n \tag{4.31}$$

---

**TABLE 4.3**

**Empirical Expressions Proposed for the Determination of the Overall Oxygen Transfer Coefficient $K_L a(20) = K_{L,O2}(20)$ in Gravity Sewers**

| References | Expressions for $K_L a(20)$ (h$^{-1}$)[a] |
|---|---|
| 1. Krenkel and Orlob (1962) | $0.121(u \cdot s)^{0.408} d_m^{-0.66}$ |
| 2. Owens et al. (1964) | $0.00925\, u^{0.67} d_m^{-1.85}$ |
| 3. Parkhurst and Pomeroy (1972) | $0.96(1 + 0.17\, Fr^2)(s \cdot u)^{3/8} d_m^{-1}$ |
| 4. Tsivoglou and Neal (1976) | $B \cdot u \cdot s$ |
| 5. Taghizadeh-Nasser (1986) | $0.4u \left( \dfrac{d_m}{R} \right)^{0.613} d_m^{-1}$ |
| 6. Jensen (1994) | $0.86(1 + 0.20\, Fr^2)(s \cdot u)^{3/8} d_m^{-1}$ |
| 7. Huisman et al. (1999) | $0.66(1 + 0.20\, Fr^2)(s \cdot u)^{3/8} d_m^{-1}$ |

[a] Where $Fr = u(gd_m)^{-0.5}$ is the Froude number (–); $u$ is the mean velocity of flow (m s$^{-1}$); g is the gravitational acceleration (m s$^{-2}$); $d_m$ is the hydraulic mean depth (m); $s$ is the slope (m m$^{-1}$); $R$ is the hydraulic radius, i.e., the cross-sectional area of the water volume divided by the wetted perimeter (m); and $B$ is the coefficient given as a function of water quality and intensity of mixing (–) (here about 2360).

where:

$K_{L,O_2}$ = overall mass transfer coefficient for oxygen (reaeration constant) refer-
    ring to the water phase ($m^{-2}\ s^{-1}$)

$n$ = number (–)

The value of $n$ in Equation 4.31 needs further explanation. It was, by way of
Equation 4.28, formulated that the exchange constant (mass transfer coefficient), $k$,
according to the two-film theory, is equal to $D/z_f$, where $z_f$ is the thickness of each
of the two films. In contrast to this theory, the penetration theories and the surface
renewal theory implies that

$$k = \frac{D^{0.5}}{z_f} \tag{4.32}$$

The discrepancy between the two-film theory and the other mass transfer theories
has implications for the choice of the value of $n$ in Equation 4.31. For practical purposes, it appears that $n$ is about 1 in a slow-flowing sewer and that $n$ approaches 0.5
in turbulent conditions. It is in general, the case that turbulence in sewers occurs in
a more widespread fashion than slow-flowing conditions, and that $n$ typically varies
within the interval 0.50–0.67.

As previously mentioned, the resistance to oxygen transfer across the air–water
interface occurs mainly in the water film. Therefore, Equation 4.31 can only be
applied to compounds that, in this respect, are comparable to oxygen, which according to Liss and Slater (1974) have $H_A$ values larger than about 250 atm.

The emission of hydrogen sulfide into the sewer atmosphere is an important
example for illustrating odor problems and other negative effects such as corrosion
(cf. Chapter 6). According to Table 4.1, $H_{H_2S} = 560$ atm, and $H_2S$, therefore, observes
the requirement for applying Equation 4.31 to determine the mass transfer coefficient, $K_{L,H_2S}$.

Further details and examples when determining mass transfer coefficients and
diffusion coefficients directly related to sewer systems will be given in Section 4.4.2.

### Example 4.5: Estimation of Overall Mass Transfer Coefficient (Emission Constant) for $H_2S$

This example will illustrate how an overall mass transfer coefficient can be estimated. The example concerns $H_2S$ in terms of calculation of $K_{L,H_2S}$, the emission
constant for $H_2S$. The estimation is based on the ratio of diffusion coefficients for
$H_2S$ and $O_2$, and an empirical expression for calculating the overall oxygen transfer coefficient (reaeration constant), $K_{L,O2}$.

A number of approaches have been suggested for the determination of the
water phase molecular diffusion coefficient, $D$, of a volatile compound. The studies conducted by Othmer and Thakar (1953), Scheibel (1954), Wilke and Chang
(1955), and Hayduk and Laudie (1974) are examples in this respect. Based on these
four studies, the diffusion coefficient ratio $D_{H_2S}/D_{O2}$ was found to vary within the
interval 0.86–0.89 with the following arithmetic mean value:

$$\frac{D_{H_2S}}{D_{O_2}} = 0.87$$

This value can be inserted in Equation 4.31:

$$\frac{K_{L,H_2S}}{K_{L,O_2}} = 0.87^n$$

With a typical variability of $n$ in sewer pipes between 0.50 and 0.67, the $K_{L,H_2S}/K_{L,O_2}$ ratio will correspondingly vary from 0.93 to 0.91.

As will be further discussed in Section 4.4.2, the empirical expressions shown in Table 4.3 can be used to estimate the value of $K_{L,O_2}$ and thereby the emission constant, $K_{L,H_2S}$, for $H_2S$, taking into account the actual flow and system characteristics.

A different approach to this procedure is to derive (simple) equations for a direct determination of an emission rate as exemplified for $H_2S$ by Lahav et al. (2006). It should, however, be stressed that such equations must be carefully calibrated and verified.

## 4.3 SEWER ATMOSPHERE AND ITS SURROUNDINGS

In general, municipal wastewater collection systems are designed to transport sewage flows, and little consideration is given to the sewer atmosphere. The movement of air into, along, and out of the sewer is therefore typically uncontrolled. However, the sewer atmosphere plays a central role when we deal with odor, human health impacts, concrete corrosion, and reaeration of the wastewater. In terms of sewer processes, the sewer atmosphere cannot be overlooked. It is an integral and important element of the entire sewer network (cf. Figure 4.5).

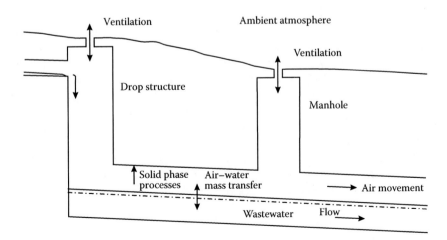

**FIGURE 4.5** Outline of major processes and interactions between sewer atmosphere and its surroundings.

Referring to Figure 4.5, the mass transfer across the air–water interface is central for reaeration of the wastewater in the sewer and for the emission of volatile substances with effects in terms of odor, corrosion, and human health impacts. These basic interfacial aspects with focus on equilibrium and mass transfer are dealt with in Sections 4.1 and 4.2, respectively. In this section, knowledge on these phenomena is expanded, and focus is centered on the occurrence, characteristics, and impacts of emitted volatile substances. The odor problem and impacts related to hydrogen sulfide are, in this respect, central. The reaeration process resulting in aeration of the wastewater phase and thereby central for the redox conditions will be dealt with in Section 4.4, whereas the details of processes at the solid surfaces, e.g., adsorption and concrete corrosion processes at the sewer walls, are the major focus of Section 6.5.2.

### 4.3.1 Odors: Properties and Characteristics

Odors constitute an inhomogeneous group of numerous volatile substances, with the common characteristic of affecting the olfactory receptor cells located in the upper part of the human nose. The detection of odor is complex, and a direct relation between the molecular structure of an odorous compound and its perception has not been identified. However, today, it seems clear that the human detection of an odor is an objective phenomenon, and it is common for most humans to distinguish between a pleasant and an unpleasant odor. The general and common properties and characteristics of odors are discussed in this section, whereas the more specific aspects are dealt with in Sections 4.3.2 and 4.3.5.

In environmental engineering, it is common practice to measure a pollutant in units of concentration and its transport in units of mass per time that passes a given cross-sectional area. When it comes to odors, this type of identification for a "parameter" is no longer appropriate. The main reason is that it is a mix of numerous—typically not easily detectable—gaseous substances (odorants) that affect humans in terms of smell and malodors. The solution to this problem is to use odor parameters that refer to a direct effect on the human sensory system.

Methods and instrumentation applied for odor analysis are discussed in Section 10.1.5. In this section, the parameters used as odor descriptors are dealt with. In brief, the methodology of an odor evaluation—referred to as a dynamic dilution olfactometry technique—is based on the assessment of an odor panel (a number of persons trained to detect odors), which—under controlled and reproducible laboratory conditions—determine which dilution level of a sample makes its odor nondetectable. This level of dilution is defined as the detection threshold value or the odor threshold value.

The following four odor parameters are used as odor descriptors. It is important to note that all parameters rely on well-defined monitoring techniques using the human nose as sensor. These odor parameters are thereby considered quantitative-based descriptors:

- Odor concentration
  The method for odor concentration measurement is based on an odor panel's determination of a dilution-to-threshold ($D/T$) value of a collected gas

sample. For example, $D/T$ = 100 means that 99 volume units of odor-free air must be mixed with 1 volume unit of a sample to dilute it to a concentration where the odor becomes nondetectable. The odor threshold value is thereby defined as the lowest concentration in air that half of the members of the odor panel can detect. The European Committee for Standardization [Comité Européen de Normalisation (CEN)] defines the odor concentration equal to the dilution factor, $D/T$, in units of $ou_E$ m$^{-3}$ (CEN, 2003). At the odor threshold, the odor concentration is thereby defined as equal to 1 $ou_E$ m$^{-3}$. In countries where the CEN standard is not in use, the odor concentration is normally also defined as the $D/T$ value, although it is not typically assigned an odor-specific unit.

- Odor intensity

  Odor intensity measurements refer to a comparison with a referencing scale. Typically, the referencing scale consists of $n$-butanol ($C_4H_9OH$) in the air of varying concentrations, where 1 $ou_E$ m$^{-3}$ is defined as equal to an $n$-butanol concentration of 40 ppb (volume based). The panelists compare an (unknown) odor sample with varying concentrations of $n$-butanol in air to find a best-intensity mach. By this comparison, the panelists must focus on the intensity of the smell and omit all other odor impressions.

  The intensity can be categorized by applying a scale, e.g., ranging from 1 (no odor) to 10 (extremely strong).

- Odor character

  The odor character is a descriptor that refers to other types of (well-known) odors. The odor character is thereby an addition factor to other odor descriptors. Examples of characters are:
  - Fragrant (having a sweet or pleasant smell)
  - Pungent (having a sharp or strong—pleasant or unpleasant—smell)
  - Acrid (having a bitter smell)

  The odor character can also be referred to in terms of the source (e.g., aerobic or anaerobic wastewater) or to a specific chemical (e.g., ammonia).

- Odor persistency

  In brief, odor persistency refers to the psychophysical effect of odor in terms of a dose–response relation. For practical purposes, odor persistency thereby describes how odor intensity decreases when odorants are diluted, e.g., when emitted from a sewer network to the urban atmosphere. Odor concentration and odor intensity are the two main parameters in this respect. These two parameters observe the so-called Stevens' power law (Stevens 1957). Stevens' power law states the relationship between the magnitude of a physical stimulus and its impact on activity. In physiology, the law refers to how external conditions influence an organism or an organ. Examples of stimuli to humans are heat, electric shock, and vibration—and, of course, odor. A different example of a stimulus is light to growth of plants.

  In terms of odor, Stevens' power law is expressed as follows:

$$I = aC^b \tag{4.33}$$

where:

$I$ = odor intensity relative to *n*-butanol (–)
$C$ = odor concentration, $D/T$ (–) or ($ou_E$ m$^{-3}$)
$a, b$ = empirical constants (–)

The constants $a$ and $b$ are odor specific and depend on the odor concentration range. Typically, the constants vary from 0.2 to 0.7 (Duffee et al. 1979).

### Example 4.6: Odor Reduction Expressed in Terms of Odor Persistency

In principle, a target odor level can be expressed as both odor concentration and odor intensity, although odor concentration is normally considered the best measure. In this example, both $a$ and $b$ (in Stevens' power law) are arbitrarily set as equal to 0.5. The relation between concentration and intensity (odor persistency) in terms of Stevens' power law, $I = 0.5C^{0.5}$, is illustrated in Figure 4.6.

Figure 4.6 shows that if an odor concentration of $D/T = 200$ is required to be reduced to a target level of 10, i.e., a 95% reduction, the corresponding reduction in odor intensity from about 7.1 to 1.6 corresponds to a 77% reduction. The example shows that when dealing with odor reduction in percent, it is extremely important to state which units the reduction refers to. The geometric form of the curve also shows that even major changes in odor concentrations at relatively high values may result in minor changes in odor intensity. Compared with concentration, odor intensity is therefore often less advisable as a measure of odor.

In addition to these four descriptors of odor, the hedonic tone is as a more or less subjective parameter used to distinguish between pleasant, neutral, and unpleasant.

In continuation of the description of the odor parameters, it is relevant to mention how they relate to modeling. Odor modeling is a technique used to determine and predict the extent of odor in the (urban) atmosphere caused by an odor-emitting source. An odor model is, in principle, an air dispersion model with typically the odor concentration (i.e., the $D/T$ ratio), as the key impact parameter and not the concentration of a specific compound. Within the area of wastewater systems, the modeling

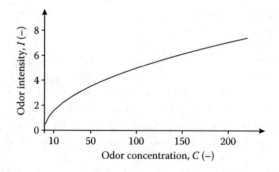

**FIGURE 4.6** Illustration of Stevens' power law. If plotted on a log–log scale, the curve becomes a straight line.

technique has been applied with wastewater treatment facilities as the odor source. However, when it comes to more diffuse sources of a sewer network, modeling is more difficult and therefore often less valuable in practice. One exception, however, is when the use of rather detailed formulated and well-calibrated conceptual models is possible (cf. Chapters 8 and 9). The result of odor models is typically given in statistical terms, e.g., by mapping averaging times of odor impact levels.

It is not a well-defined task to assess when odor becomes a nuisance to the public. Correspondingly, it is difficult to produce guidelines for acceptable ambient odor levels. It is therefore also seen that odor-abatement policies issued by regulatory institutions across the world vary considerably. Most often, guidelines are expressed as acceptable odor concentrations expressed in units of $ou_E$ m$^{-3}$ or $D/T$ numbers. As an example, a rule requiring operators not to exceed a 1-h average $D/T$ value of 5–10 in 99% of the time can be prescribed. In existing catchments, assessment by monitoring is possible. In case of new or expanded systems, dispersion modeling is, in principle, an option. However, sewer networks are rather complex systems when dealing with formation and release of odorous substances. It is therefore not surprising that dispersion models are not reliable tools for prediction of odors originating from sewers.

Odor is in relation to sewer networks an issue of increasing public concern. Municipalities and water companies across the world are aware of this, and policies and control strategies directed to reduce odor problems in urban areas are increasingly being implemented. Further details on odor-specific aspects related to wastewater systems can be found in several publications (e.g., WEF 2004; Stuetz and Frechen 2001).

### 4.3.2 OCCURRENCE OF VOLATILE SUBSTANCES IN SEWER ATMOSPHERE

It is mainly the anaerobic microbial processes in the wastewater of a sewer that give rise to the formation of low molecular, volatile substances. When emitted, these substances are—in the gas phase—identified by a number of problems such as malodors, health risks, and corrosion of the sewer network. The conditions and processes in terms of formation of both VOCs and low molecular inorganic gases (in particular, hydrogen sulfide ($H_2S$) and ammonia ($NH_3$)) are dealt with in Sections 2.1.3 and 3.2. Further information on measurement, modeling and control of these volatile substances is discussed in Chapters 7 through 10 (cf. also Stuetz and Frechen 2001).

As shown in Figure 3.3 and Table 3.3, fermentation processes in wastewater may result in the formation of numerous VOCs. Sulfate reduction result in the formation of hydrogen sulfide and volatile hydrocarbons may originate from industrial sources. In general, several low molecular components are malodors. Common volatile organic and inorganic substances associated with odors are shown in Table 4.4. A great number of these compounds have been identified in domestic wastewater (Raunkjaer et al. 1994; Hvitved-Jacobsen et al. 1995; Hwang et al. 1995). As examples, VFAs are known as anaerobic decomposition products of carbohydrates (e.g., starch), and mercaptans (thiols) are primarily produced from proteins. Several of the compounds shown in Table 4.4 arise from the decomposition of organic matter containing sulfur and nitrogen.

**TABLE 4.4**

**Examples of Volatile Odorous Organic and Inorganic Substances Potentially Occurring in a Sewer Atmosphere (cf. also Dagúe 1972; Vincent and Hobson 1998)**

| Group of Substance | Substance | Formula | Threshold Odor (ppb) |
|---|---|---|---|
| Volatile sulfur compounds (VSCs) | Methyl mercaptan | $CH_3SH$ | 1 |
| | Ethyl mercaptan | $C_2H_5SH$ | 0.2 |
| | Allyl mercaptan | $CH_2=CHCH_2SH$ | 0.05 |
| | Benzyl mercaptan | $C_6H_5CH_2SH$ | 0.2 |
| | Dimethyl sulfide | $(CH_3)_2S$ | 1 |
| | Dimethyl disulfide | $CH_3SSCH_3$ | 0.3–10 |
| | Thiocresol | $CH_3C_6H_4SH$ | 0.1 |
| Nitrogenous compounds | Methylamine | $CH_3NH_2$ | 1–50 |
| | Ethylamine | $C_2H_5NH_2$ | 2400 |
| | Dimethylamine | $(CH_3)_2NH$ | 20–80 |
| | Pyridine | (pyridine ring structure) | 4 |
| | Indole | (indole ring structure) | 1.5 |
| | Scatole | (scatole ring structure, —$CH_3$) | 0.002–1 |
| Acids (VFAs) | Acetic | $CH_3COOH$ | 15 |
| | Butyric | $C_2H_5COOH$ | 0.1–20 |
| | Valeric | $C_3H_7COOH$ | 2–2600 |
| Aldehydes and ketones | Formaldehyde | $HCHO$ | 370 |
| | Acetaldehyde | $CH_3CHO$ | 0.005–2 |
| | Butyraldehyde | $C_2H_5CHO$ | 5 |
| | Acetone | $CH_3COCH_3$ | 50–10,000 |
| | Butanone | $C_2H_5COCH_3$ | 270 |
| Hydrocarbons | Benzene | $C_6H_6$ | 200–20,000 |
| | Phenol | $C_6H_5OH$ | 40 |
| | Toluene | $C_6H_5CH_3$ | 50–2500 |
| Inorganic | Ammonia | $NH_3$ | 300–500 |
| | Hydrogen sulfide | $H_2S$ | 0.5–200 |

Only a few studies have been concerned with the measurements of specific odorous compounds in sewer systems. Hwang et al. (1995) have analyzed the influent wastewater in a study of malodorous substances in wastewater at different steps of sewage treatment (Table 4.5). Although the results in Table 4.5 only represent examples, they are interesting for several reasons. Table 4.5 shows that several of the odorous

**TABLE 4.5**

**Sulfur and Nitrogen-Containing Odorous Compounds in the Influent Wastewater at a Treatment Plant**

| Compound | Average Concentration ($\mu$g L$^{-1}$) | Range of Concentrations ($\mu$g L$^{-1}$) |
|---|---|---|
| Hydrogen sulfide | 24 | 15–38 |
| Carbon disulfide | 1 | 0.2–1.7 |
| Methyl mercaptan | 148 | 11–322 |
| Dimethyl sulfide | 11 | 3–27 |
| Dimethyl disulfide | 53 | 30–79 |
| Dimethylamine | 210 | — |
| Trimethylamine | 78 | — |
| $n$-Propylamine | 33 | — |
| Indole | 570 | — |
| Skatole | 700 | — |

*Source:* Hwang, Y. et al., *Water Res.*, 29(2), 711–718, 1995. With permission.

compounds may appear in the wastewater phase from a sewer system in relatively high concentrations, especially when compared with $H_2S$. In addition to the importance of Henry's law constant, several system and hydraulic conditions in the sewer network affect to what degree the compounds will appear in the sewer atmosphere.

Thistlethwayte and Goleb (1972) report investigations of the composition of sewer air. Majority of the samples were taken in a sewer transporting mixed municipal wastewater with a maximum residence time of about 4 h. The $BOD_5$ of the wastewater varied between 300 and 350 g m$^{-3}$, and the temperature was typically about 24°C. The authors divided the compounds in the sewer air into four groups. These groupings were selected based not only on the chemical nature of the compounds but also according to their respective concentrations (note that ammonia was not included in this scheme):

1. Carbon dioxide ($CO_2$)
2. Hydrocarbons and chlorinated hydrocarbons
3. Hydrogen sulfide, $H_2S$
4. Odorous gases and vapors such as mercaptans, amines, aldehydes, and VFAs

The typical composition of sewer air shown in Table 4.6 is based on the results reported by Thistlethwayte and Goleb (1972) and other similar studies. It is evident that the range of "typical composition" must be evaluated in terms of which conditions prevail in the wastewater, e.g., example redox and flow conditions, and sewer network characteristics related to emission of volatile substances. In Table 4.6, no distinction is made

**TABLE 4.6**

**Typical Composition of Sewer Air Reported in the Literature and Corresponding to Dry Weather Conditions and Anaerobic Conditions in the Sewer (cf. Tables 4.1 and 4.4)**

| Group of Compounds | Typical Concentration Range by Volume |
|---|---|
| 1. Carbon dioxide, $CO_2$ | 0.05–1.0%[a] |
| 2. Hydrocarbons and chlorinated hydrocarbons | |
|    a. Hydrocarbons, mainly aliphatic $C_6$–$C_{14}$ and mostly $C_8$–$C_{12}$ originating from petrol | 0–500 ppm |
|    b. Chlorinated hydrocarbons, mostly trichlorethylene, ethylene dichloride and carbon tetrachloride originating from industrial sources | 0–100 ppm |
| 3. Hydrogen sulfide, $H_2S$ | 0–100 ppm |
| 4. Odorous organic compounds | |
|    a. Sulfides and mercaptans (thiols) | 0–50 ppb |
|    b. Amines | 0–50 ppb |
|    c. Aldehydes | 0–50 ppb |

[a] The $CO_2$ concentration in an urban atmosphere is approximately 0.04%.

between compounds from different inlets (e.g., industrial sources) and compounds produced in the sewer.

The level of concentration of group 1 ($CO_2$) is related to the contents of $CO_2$ in the drinking water but may also indicate that microbial degradation of wastewater organic matter takes place in the sewer. In terms of odor, the other groups (2–4) are relevant. In spite of the fact that Table 4.6, does not refer to the sources of the substances found in the sewer atmosphere, the compounds listed under group 2 are probably a result of inputs to the system. The compounds included in groups 3 and 4, however, are interpreted as resulting from anaerobic processes.

The results reported by Thistlethwayte and Goleb (1972) indicate that the concentrations of the constituents of groups 3 and 4 tend to be related—i.e., the constituents of group 4 (a, b, and c) tend to vary according to the levels of the $H_2S$ concentrations, roughly in the ratio of 1:50 to 1:100. Therefore, the authors conclude that this observation suggests that although the $H_2S$ concentration alone may not be a sufficient measure of potential sewer air odor levels, $H_2S$ concentration measurements play a central role when assessing odor.

Based on case studies, it was shown that, in general, emission rates of VOCs in sewers increase with increased pipe diameter and increased ventilation rates (Melcer et al. 1997).

### 4.3.3  Odorous, Corroding, and Toxic Substances in Sewers

This section will, in addition to the more basic aspects discussed in Sections 4.1 and 4.2, give further details and examples related to the air–water transfer of volatile substances and their occurrence in the sewer. The main pathways of hydrogen sulfide with detrimental effects in terms of human health impacts, odor, and concrete corrosion are included as examples although further details of $H_2S$ relevant processes and phenomena relating mainly to corrosion are focal subjects of Chapter 6.

Release of odorous, corroding, and toxic substances from the wastewater into the sewer atmosphere is a fundamental process when assessing the corresponding effects and impacts. As long as these volatile compounds remain in the water phase, such problems do not exist.

It is furthermore important that volatile substances can undergo transformation and degradation in the wastewater phase under aerobic conditions. Organic volatile sulfur compounds are expected to be faster degradable than nitrogenous compounds (Hwang et al. 1995).

Three major transport phenomena are important for the occurrence of volatile compounds in the sewer atmosphere and in the ambient atmosphere (cf. Figure 4.1):

1. The air–water transfer process describing the emission from the wastewater into the sewer atmosphere
2. The ventilation of the sewer network responsible for transfer of volatile compounds from the sewer atmosphere into the ambient atmosphere
3. Adsorption of gases—in particular, $H_2S$—at the moist sewer walls followed by various transformation processes

The fundamentals of the first mentioned point are dealt with in Sections 4.1 (equilibrium relations) and 4.2 (transport processes). Further aspects and examples of emission processes are included in this section. Ventilation and air movement in the sewer are discussed in Section 4.3.4, and details of the last mentioned item—processes at the sewer wall—are discussed in Section 6.5 with $H_2S$ as the central compound.

A great number of processes and phenomena related to the sulfur cycle in a sewer affect to which extent hydrogen sulfide is a problem. Figure 4.7 outlines the main steps of the sulfur cycle that will be detailed in Chapter 6. Although not all aspects of Figure 4.7 can be easily quantified, they should be included in an assessment of odor problems and other negative effects associated with sulfide in sewer networks.

The transformation and transport rates involved in the sulfur cycle shown in Figure 4.7 determine to what extent the relevant components will occur in the different phases of the sewer system. As already shown—and further focused on when dealing with concrete corrosion in Section 6.5—the impacts of hydrogen sulfide are associated with its occurrence in the gas phase. In this respect, it is interesting that although there is a rather fast release of hydrogen sulfide from the water phase into the sewer atmosphere, the concentrations measured in the atmosphere may be as low as 2% to 20% of the equilibrium concentration (Pomeroy and Bowlus 1946; Matos and de Sousa 1992). Adsorption of $H_2S$ on the sewer walls potentially followed by oxidation to elemental sulfur or sulfuric acid and ventilation are considered the main reasons for this observation. Another important aspect for nonequilibrium is the fact that the flow regime of water and air are different and therefore result in a rather fast separation of the air volume relative to the water volume from where the emission took place. This latter aspect may, of course, also result in a concentration in the

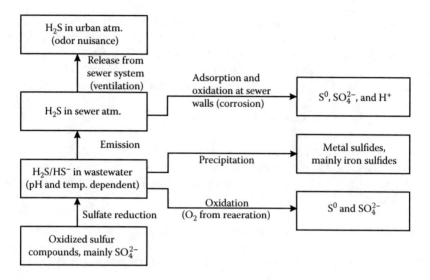

**FIGURE 4.7**  Main pathways in terms of transformation and transport related to sulfur cycle in sewer networks.

sewer air being higher than equilibrium at the point where the air volume is actually located. At any rate, the relative transport phenomena of volatile compounds in the two-phase system of a sewer are complex and therefore cannot be simply quantified.

Equation 4.34 is a simple and, in many cases, a realistic mass balance for sulfide in the sewer atmosphere. Compared with the absorption rate of $H_2S$ at the sewer walls, the amount released by ventilation from a sewer into the urban atmosphere is often, i.e., in well-designed sewers without forced ventilation, relatively low.

$$r_{atm} = r_{emit} - r_{ads} - r_{vent}$$  (4.34)

where:

$r_{atm}$ = rate of $H_2S$ buildup in the sewer atmosphere (e.g., g s$^{-1}$)
$r_{emit}$ = rate of $H_2S$ emission from the wastewater (e.g., g s$^{-1}$)
$r_{ads}$ = rate of $H_2S$ adsorption and potential oxidation at the sewer walls (e.g., g s$^{-1}$)
$r_{vent}$ = rate of $H_2S$ release from the sewer into the urban atmosphere (e.g., g s$^{-1}$)

Matos and de Sousa (1992) and Matos and Aires (1995) have, based on empirical expressions for the emission rate of $H_2S$ and the adsorption rate on the sewer walls, used Equation 4.34 as a basis for forecasting the buildup of $H_2S$ in the sewer atmosphere. The expressions included in this model are detailed by Matos and de Sousa (1992). Equation 4.34 is, in principle, valid for any volatile substance. Typically, the processes at the sewer walls can be neglected for most VOCs.

In addition to the VOCs produced in sewers under anaerobic conditions, such compounds may also originate from external sources (e.g., industries) (Corsi et al. 1995; Olson et al. 1998). Typically, these VOCs are hydrocarbons and other similar products (e.g., chlorinated hydrocarbons). The movement and kinetics of such incoming VOCs to the sewer follow the general approach described in this chapter for odorous substances. The fate of such compounds is thereby influenced by a number of sewer system parameters, hydraulic conditions, wastewater characteristics, and physicochemical properties for the VOC compounds in question.

### 4.3.4  AIR MOVEMENT AND VENTILATION IN SEWERS

The flow of wastewater in a municipal wastewater collection system with its pipes, drop structures, overflow structures, pumping stations, and manholes is focused on when hydraulic engineers design and analyze the performance of a network. The dynamics of the headspace in terms of movement along the system, pressurization, and ventilation is typically either omitted or not dealt with to the same extent. Although this book has its focus on the microbial and chemical (physicochemical) process engineering aspects and not the hydraulics, the fact that the sewer air dynamics is generally overlooked calls for its attention when addressing odor, human health impacts, concrete corrosion, and reaeration.

In continuation of the fact that the headspace in a sewer in general plays an unobtrusive role, it is important to note that computational fluid dynamic techniques and procedures are being developed for the description of the airflow dynamics (Edwini-Bonsu and Steffler 2004, 2006a, 2006b). In particular, the impacts of wastewater drag, sewer headspace pressurization, and ventilation have been focused on. It is, however, important to state that these conceptual and semiempirical descriptions of the air movement at present are at an early stage. The computational time is high, and the models are not applicable to entire catchments.

In the context of this book, a rather simple distinction is made between air movement and ventilation. Air movement will be understood as a transport process of the sewer air from one location in the network to another, whereas ventilation is the exchange of air (in both directions) between the sewer atmosphere and the ambient atmosphere, i.e., in particular the urban atmosphere. As seen in Figure 4.5, air movement mainly occurs in the horizontal direction, whereas the direction of ventilation is vertical. It is clear that air movement and ventilation are highly interacting phenomena, as ventilation results in a corresponding movement of the sewer air. However, it is still considered relevant to make a basic distinction between the two phenomena.

An overall result of air movement and ventilation is that equilibrium for volatile substances at the air–water interface hardly exists. Furthermore, volatile substances (e.g., $H_2S$), produced (emitted) at one location in the sewer may appear and result in effects at quite different points.

#### 4.3.4.1  Ventilation

There are two types of ventilation: natural and forced ventilation. Natural ventilation occurs via small openings in manholes and other types of leaks in the collection system, whereas forced ventilation is designed and controlled by installation of blowers or fans. Forced ventilation is required when managing sewer gases for treatment and dilution (cf. Section 7.3.7).

Natural ventilation is affected by conditions in the ambient atmosphere and flow-related phenomena in the sewer (Pescod and Price 1981, 1982; Olson et al. 1997). Further details observed by several researchers on the phenomena that affect natural ventilation are:

- Wind speed
  Wind speed at the urban surface may cause suction or pressure across openings in the manhole cover and thereby create a net inflow or outflow of air.
- Barometric pressure variations
  Changes in atmospheric pressure may lead to air volume compression or expansion and thereby result in inflow or outflow of air in the sewer.
- Wastewater drag
  Because of shear forces at the air–water interface, a momentum (liquid drag) is transferred from the wastewater flow to the overlying sewer atmosphere.
- Temperature differences
  Temperature differences between the sewer air and the urban atmosphere create differences in density, which may cause a corresponding airflow.
- Variations in wastewater level
  The level of wastewater is time variant, and the rise and fall of the water level will thereby cause a corresponding change in the sewer air volume.

It is readily obvious from these five points that air movement and ventilation are highly interacting phenomena.

It is evident that the details of the above-mentioned five points and their relative importance for ventilation of a sewer become complex when applied for real systems. The structure of a sewer network in terms of, e.g., hydraulic jumps, drop structures, junctions, siphons, and pump stations, and the size and locations of openings to the ambient atmosphere, will highly influence the ventilation rate. Furthermore, the air–water related kinetics and hydraulics make a model description complex. Olson et al. (1997, 1998) have produced valuable mathematical formulations and approaches for the description of "cocurrent ventilation" in sewers. However, it must be realized that when modeling the emission of volatile compounds relative to the flows of the water and gas phases in a real sewer network, both quantity and details of data must be available at a very advanced level. Further details and examples related to modeling and experimental procedures will be dealt with in Chapters 8 and 10, respectively.

As a first approach and for practical purposes, it is therefore crucial that results from both laboratory-scale and full-scale investigations have been produced (Pescod and Price 1981, 1982; Olson et al. 1997; Madsen et al. 2006). Results in terms of data from such studies are, however, typically considerably scattered and difficult to interpret. Although site-specific results to some extent are contradictory, they serve as a starting point for analyses of other systems. As an example, Olson et al. (1997) conclude that ventilation rates caused by either wind speed or temperature differences are higher than those caused by wastewater drag. In contrast, and based on pilot-plant studies, Monteith et al. (1997) conclude that wind speed and wastewater drag are the most important phenomena affecting the gas phase velocity.

A simple means of quantifying ventilation rates is to use turnovers per day (TPD) as a measure of how many times per day the sewer air is replaced with air from the ambient atmosphere. Natural ventilation rates vary considerably with number of openings and size of the sewer. Rates varying from 1 to 25 TPD have been reported (Corsi et al. 1989; Melcer et al. 1997).

It must be realized that the present basic knowledge on the different phenomena that control ventilation and air movement is incomplete. The details are either not sufficiently known, or they are in a complex form formulated in general terms that cannot be applied for practical purposes in terms of modeling. From a modeling point of view, a rather simplistic and pragmatic approach is needed. In this respect, it seems appropriate to use the wastewater drag as the governing phenomenon when predicting natural ventilation and air movement in sewers (cf. Section 4.3.4.2). As an alternative to modeling, it is, in concrete cases, recommended to use experimental techniques to determine the extent of air movement and ventilation (cf. Section 4.3.4.3).

Forced ventilation is, in some cases, an appropriate solution to control sulfide-induced concrete corrosion, extraction of foul air, and treatment of sewer gas, thereby replacing the sewer gas with fresh air. It can involve the use of blowers or large fans (Corsi et al. 1989). Forced ventilation is typically applied for odor and corrosion control and treatment of sewer air in relatively large diameter pipes and deep tunnel systems.

### 4.3.4.2 Air Movement and Wastewater Drag

An important aspect of air movement in a sewer is that volatile substances produced at a specific site may appear and result in effects at a quite different location. An important example is the occurrence of concrete corrosion caused by hydrogen sulfide. Hydrogen sulfide produced in upstream sewer reaches (e.g., force mains) can—by the movement of sewer air to downstream gravity sewer pipes—result in concrete corrosion at such locations.

As discussed in relation to ventilation, a pragmatic approach for prediction of air movement is its relationship to wastewater drag. Two types of driving forces are important in this respect: the shear stress at the air–water interface causing a drag-induced transport of the sewer air and the shear stress of the air at the pipe wall. It is theoretically possible to describe the work generated at an air–water interface and the frictional losses of the air along a pipe reach of an open-ended pipe resulting in a potential air velocity (Olson et al. 1997). However, when it comes to a real sewer, this approach becomes far more complex.

The pragmatic empirical approach for prediction is to relate airflow to wastewater flow, assuming that the wastewater drag is the major phenomenon affecting air movement. In a gravity sewer, the theoretical maximum velocity of the air occurs close to the water surface for gradually being reduced at the pipe wall. It is expected that the largest average air velocity is reached when the water surface exposure to the overlying atmosphere is at a maximum, i.e., corresponding to a half-full sewer. Under such conditions, for relatively small pipe diameters and moderate water flows (less than about 0.6 m s$^{-1}$), Prescod and Price (1981) conclude that an average air velocity is 35–50% of the water flow velocity. Based on other investigations, however, and for larger diameter sewers, these percentages are suggested to be overestimated.

### 4.3.4.3 Experimental Techniques for Monitoring Air Movement and Ventilation

The lack of sufficient conceptual knowledge on sewer air movement and ventilation require experimental means for their quantification. In general, three categories exist:

1. Direct measurements of the air velocity applying velometers
2. Flow visualization by injection of an inert flue gas
3. Tracer gas measurements

Based on numerous studies, the tracer gas method is, for general purposes, considered the most valuable and reliable technique for the assessment of air movement and ventilation. The method is conducted by injecting a tracer gas to the sewer atmosphere followed by monitoring the gas concentration at downstream locations. Relatively inert tracer gasses not present in the sewer atmosphere (e.g., carbon monoxide, CO, and sulfur hexafluoride, $SF_6$) have been applied (Pescod and Price 1982; Melcer et al. 1997; Monteith et al. 1997; Parker and Ryan 2001). However, a drawback in using these tracer gases is that CO is toxic and $SF_6$ is classified as a severe greenhouse gas with restricted use. A nontoxic alternative is to use propane, $C_3H_8$ (Madsen et al. 2007). Although $O_2$ is present in the sewer atmosphere, and reactive, pulse-injection of pure $O_2$ changing the composition of the sewer atmosphere can be used for monitoring the gas phase flow (Madsen et al. 2006).

### 4.3.5 ODOR AND HEALTH PROBLEMS OF VOLATILE COMPOUNDS IN SEWERS

Two different types of odor measurements can be performed: analytical measurements or sensory measurements (cf. Sections 4.3.1 and 10.1.5). Sensory measurements are either performed by the human nose or by electronic detectors and, therefore, relate to the effects of the odor (Sneath and Clarkson 2000; Stuetz et al. 2000). Sensory measurements are useful; however, in terms of modeling, analytically based measurements are often preferred. On the other hand, the large number of potential odorous substances is a major obstacle for an easy and fast determination of a "level of problem." As is known from other areas of environmental engineering, the answer may be found by defining an appropriate and general reliable indicator.

Thistlethwayte and Goleb (1972) made an important, but somewhat dubious statement when they concluded that although $H_2S$ concentration alone may not be a sufficient measure of potential sewer air odor levels, $H_2S$ concentration measurements are probably appropriate for most studies of sewer gases. Their statement was based on rather limited measurements in sewers; however, it corresponds well to the fact that odor formation in sewers is a result of anaerobic microbial processes (cf. Chapter 3). Other authors, who primarily focused on odors related to wastewater treatment, have also observed that $H_2S$ can be used as a relevant indicator for an odor level (Gostelow and Parsons 2000). From an engineering point of view, it is interesting if hydrogen sulfide can be attributed a level of importance in this respect.

In addition to odor, several human health-related problems relevant for sewer networks are potentially associated with the occurrence of hydrogen sulfide. In this respect, it is interesting to compare values from Table 4.7 with Figure 4.3. The levels indicated in Table 4.7 depend on human sensitivity and time of exposure.

It is important to note that $H_2S$ loses its characteristic smell at about 50 ppm, and, therefore, no immediate possibility for its detection exists. As it is typically not detected by its smell at those concentrations, where it is life-threatening, sensors and alarm systems for its detection must be in constant use when working in sewer systems. It is, in this respect, important that $H_2S$ is often seen to accumulate in, e.g., pumping stations and manholes where gas movement and ventilation are reduced.

Although the concentration of hydrogen sulfide in the sewer atmosphere or the surroundings is the direct and more correct indicator of odor problems, a pragmatic approach is, as a first estimate, to use hydrogen sulfide in the wastewater phase of a sewer network as an indicator of the potential risk. It must, however, be stressed that because of non-equilibrium conditions and a number of sinks for hydrogen sulfide in the atmosphere, the ratio between the concentration in the atmosphere and that of the water phase is highly variable (cf. Figure 4.7). It should be noted that an indication of "minor" or "medium" identified problems must not be considered equivalent to "no need for control."

As an example, the atmospheric partial pressure of $H_2S$ on a volumetric basis in equilibrium with a water phase of sulfide ($H_2S + HS^-$) is at a pH of 7 approximately equal to 130 ppm $(gS\ m^{-3})^{-1}$ (cf. Figure 4.3). It is therefore clear that under equilibrium conditions, much lower concentrations than those corresponding to the values shown in Table 4.8 may result in detrimental odor and human health impacts. It is also evident from the fact that Henry's law constant for $H_2S$ is rather high, $H_{H_2S} = 560$ atm at 25°C (cf. Table 4.1). In real sewer networks, however, conditions close to equilibrium rarely exist because of, e.g., ventilation and processes at the sewer walls. Typically, the gas concentration of $H_2S$ found in the sewer atmosphere ranges from 2% to 20% of the theoretical equilibrium value and is normally found to be less than 10% (Melbourne and Metropolitan Board of Works, 1989).

---

**TABLE 4.7**

**Odor and Human Health-Related Effects of Hydrogen Sulfide in the Atmosphere**

| Odor or Human Effect | Concentration in Atmosphere (ppm) |
|---|---|
| Threshold odor limit | 0.0001–0.002 |
| Unpleasant and strong smell | 0.5–30 |
| Headache; nausea; eye, nose, and throat irritation | 10–50 |
| Eye and respiratory injury | 50–300 |
| Life threatening | 300–500 |
| Immediate death | >700 |

*Source:* U.S. National Research Council, Hydrogen sulfide, report by Committee on Medical and Biological Effects of Environmental Pollutants, Subcommittee on $H_2S$, 1979; ASCE, *Sulfide in Wastewater Collection and Treatment Systems, Manuals and Reports on Engineering Practice* 69, ASCE, New York, 1989. With permission.

---

**TABLE 4.8**
**Levels of Total Hydrogen Sulfide Concentration in Wastewater
of a Sewer System and Potential Associated Problems in
Terms of Malodors, Health, and Corrosion**

| H$_2$S Concentration Level (g S m$^{-3}$) | Identified Problems |
|---|---|
| <0.5 | Minor |
| 0.5–2 | Medium |
| >2 | Considerable |

Another reason why a total sulfide concentration of 0.5 g S m$^{-3}$ typically does not give rise to problems is that wastewater includes a small—but not insignificant—amount of heavy metals that can react with free sulfide to form insoluble metal sulfides. In particular, iron is typically present in wastewater in relatively high concentrations (cf. Table 3.2).

### 4.3.6 ODOROUS SUBSTANCES IN THE URBAN ATMOSPHERE

The so-called dispersion models are tools that can simulate the wind transport of vented gases in the urban atmosphere and thereby assess the impacts of gaseous odorants. The interesting aspect is their movement in the city atmosphere from the emission source to the receptor, typically defined as residential and commercial urban areas.

The analytical principle of most dispersion models is based on the Gaussian equation. This equation describes the ground level concentration of a compound in an emitted plume depending on wind speed and dispersion in vertical and horizontal direction (Gostelow et al. 2001; WEF 2004).

Although dispersion models in principle are useful tools for assessing the spreading of polluting gasses, they are also complex. Several phenomena influence the result of computation, in particular the local meteorological conditions (wind speed and direction, temperature, and atmospheric stability) and the topography of the surrounding area including, e.g., buildings. Dispersion models are simplest to use when applied in cases of one major point source (e.g., a wastewater treatment plant). For sewer networks where several small ventilation outlets are typically located in urbanized catchments, the modeling results may be uncertain.

## 4.4 REAERATION IN SEWER NETWORKS AND ITS ROLE IN PREDICTING AIR–WATER MASS TRANSFER

The DO mass balance, and thereby the redox potential, of wastewater in sewer systems is fundamental for the course and extent of the microbial processes (cf. Chapters 2 and 3). The low solubility of oxygen in water, illustrated by a very high Henry's law constant, a relatively high resistance to mass transfer across the air–water interface, and a potential high removal rate of DO in wastewater are the major reasons why DO is often a limiting factor for the aerobic microbial processes. A quantification of oxygen transfer across the air–water interface in a sewer, taking into account both process and system relevant parameters, is highly important and central for a quantification of the microbial processes. In addition to the temperature, the oxygen transfer—i.e., the reaeration—is highly affected by the turbulence level of the wastewater, and the flow regime thereby also becomes an important factor.

Reaeration in flowing waters such as streams, rivers, and channels has been recognized to be of central importance for maintaining aerobic conditions in the water phase. Numerous investigations and approaches for its determination and prediction have been performed and reported. The studies by Streeter (1926), Parkhurst and Pomeroy (1972), and Tsivoglou and Neal (1976) are classic examples. Approaches and methodologies used for determination of reaeration in flowing waters in terms of radiotracer and gas tracer techniques have been adopted for sewer pipe flows.

It is important to note that the mathematical formulations related to the reaeration process are not just dealt with to determine the air–water oxygen transfer. There are further aspects of importance. As already illustrated in Example 4.5, the oxygen transfer coefficient (reaeration constant) plays a central role for determination of the mass transfer coefficient for $H_2S$ as an example of an odorous, toxic, and corroding gas.

### 4.4.1 Solubility of Oxygen

The first and fundamental aspect of quantifying the DO mass balance in a sewer is related to the solubility of oxygen in water and wastewater in equilibrium with an overlaying atmosphere (cf. Henry's law, Equation 4.10, and Example 4.1).

In addition to using Henry's law constant and the concept illustrated in Example 4.1, the following empirical formula can be applied for the determination of the solubility of oxygen in distilled water in equilibrium with the atmosphere:

$$S_{OS} = \frac{P - p_S}{760 - p_S}(14.652 - 0.41022T + 0.00799T^2 - 0.0000777T^3) \qquad (4.35)$$

where:

$S_{OS}$ = dissolved oxygen saturation concentration in bulk water phase (in equilibrium with the atmosphere) (g $O_2$ m$^{-3}$)
$P$ = actual air pressure (mm Hg)
$p_S$ = saturated vapor pressure at temperature $T$ (mm Hg)
$T$ = temperature (°C)

In general, the temperature dependency of $S_{OS}$ is more important than the dependency on pressure. As seen in Equation 4.35, $S_{OS}$ is close to 14.65 g $O_2$ m$^{-3}$ when the temperature is approaching 0°C, and at 15°C, it is 10.04 g $O_2$ m$^{-3}$. A reduced partial pressure of $O_2$ in a sewer atmosphere compared with that of the urban atmosphere must be considered (cf. Example 4.1).

Equation 4.35 refers to clean water. Inorganic and organic soluble components in wastewater have an impact on the solubility:

$$S_{OS, ww} = \beta S_{OS} \qquad (4.36)$$

where:

$S_{OS,ww}$ = dissolved oxygen saturation concentration in wastewater (g $O_2$ m$^{-3}$)
$\beta$   = correction factor equal to the ratio of solubility of $O_2$ in wastewater to that of
       clean water (–)

The value of $\beta$ varies depending on the type of wastewater, typically in the interval of 0.8–0.95.

## 4.4.2 Empirical Models for Air–Water Oxygen Transfer in Sewer Pipes

Emission in terms of mass transfer rates of volatile compounds requires a fundamental understanding as dealt with in the preceding sections, particularly in Section 4.2.2.1. The determination of an overall mass transfer coefficient, $K_L$ or $K_G$, as shown in Equations 4.24 and 4.25, respectively, is central. As previously noted, $K_L$ is typically applied.

The mass transfer rate across the air–water interface expressed in terms of a flux rate, $J$, and in units of mol mol$^{-1}$ m$^{-2}$ s$^{-1}$—or basically just m$^{-2}$ s$^{-1}$—is discussed in Section 4.2.2.1. When addressing sewer pipes in practice, mass transfer is described using a similar approach, but often expressed in different units. The flux rate can thereby be expressed as mass per volume rate (e.g., g m$^{-3}$ s$^{-1}$), and the mole fraction of a compound is thereby typically replaced with its concentration in units of g m$^{-3}$. With reaeration as an example, the rate of oxygen transfer is thereby formulated as follows:

$$F = K_L a (S_{OS} - S_O) = K_{L,O_2}(S_{OS} - S_O) \qquad (4.37)$$

where:

$F$ = rate of oxygen transfer (g $O_2$ m$^{-3}$ s$^{-1}$)
$K_L a = K_{L,O2}$ = overall oxygen transfer coefficient, also named reaeration constant
       (s$^{-1}$)
$S_O$ = dissolved oxygen concentration in bulk water phase (g m$^{-3}$)

As discussed in Section 4.2.2.3 and formulated in Equation 4.31, $K_{L,O2}$ plays a central role for the determination of $K_L$ values for other volatile compounds. In addition to quantification of reaeration, it is therefore an important task—based on data from relevant experiments, typically performed in pipe and channel systems—to develop empirical equations for determination of $K_{L,O2}$.

It should be noted that although Equation 4.37 is formulated for the reaeration process, similar expressions, i.e., according to Equation 4.24, are formulated for other volatile compounds.

The overall oxygen transfer coefficient is related to system characteristics as follows:

$$K_L a = K_L' a = K_L' \frac{A}{V} = K_L' d_m^{-1} \qquad (4.38)$$

where:

$K'_L$ = oxygen transfer velocity (m s$^{-1}$)

$a$ = specific surface area, i.e., water–air surface area, $A$, to volume of water, $V$ (m$^{-1}$)

$d_m$ = hydraulic mean depth of the water phase, i.e., the cross-sectional area of the water volume divided by the water surface width (m)

Equation 4.37 can be reformulated to include temperature dependency and a correction factor dependent on the wastewater matrix compared with clean water:

$$F = \alpha K_{LO_2}(20)(\beta S_{OS} - S_O)\alpha_r^{T-20} \tag{4.39}$$

where:

$\alpha$ = correction factor for wastewater constituents (–)

$\alpha_r$ = temperature coefficient for reaeration (–)

The value of $\alpha$ depends on the amount and quality of wastewater constituents that affect the overall oxygen transfer (e.g., surface-active agents). For household wastewater types, the value of $\alpha$ for oxygen is relatively close to 1 (i.e., about 0.95). For other volatile compounds it may be different. As an example, Yongsiri et al. (2005) found $\alpha = 0.6$ for $H_2S$ independent on the turbulence level.

In general, is accepted that the temperature coefficient, $\alpha_r$, for the reaeration process is 1.024 (Elmore and West 1961). This value may, to some extent, depend on which volatile compound it is concerned with. In a study on the effect of temperature on the air–water transfer of $H_2S$, Yongsiri et al. (2004b) found a temperature coefficient of 1.029 and 1.021 at a pH of 6.5 and 7.0, respectively. The numerical value of the temperature coefficient corresponds well with the fact that the reaeration process it is diffusion limited (cf. Section 2.2.3).

Table 4.3 summarizes a number of empirical expressions proposed for the determination of the reaeration constant, $K_L a$, in gravity sewers.

The formulas for $K_L a$ presented by Parkhurst and Pomeroy (1972), Taghizadeh-Nasser (1986), Jensen (1994), and Huisman et al. (1999), shown in Table 4.3, have all been developed based on sewer pipe studies. Taghizadeh-Nasser (1986) performed the investigation in a pilot sewer, whereas the other three formulas were developed based on measurements in real sewers. Parkhurst and Pomeroy (1972) conducted their investigations based on an oxygen mass balance in sewers that were cleaned for sediments and biofilm. Jensen (1994) based his formula on the one developed by Pomeroy and Parkhurst (1972), but performed measurements of the reaeration by a direct methodology using krypton-85 as the radiotracer. Huisman et al. (1999) also used a tracer method; however, he based his work on the inert gas sulfur hexafluoride, $SF_6$. Further details on experimental techniques are found in Chapter 10.

The importance of the different terms in the equations shown in Table 4.3 can be exemplified by Equation 3. The term $(1 + 0.17Fr^2)$ accounts for an increased interfacial area resulting from disturbance of the water surface, and the term $(su)$ corresponds to the rate of potential energy dissipation of the water flow. The 3/8 power is based on the penetration theory (cf. Section 4.2.1). The impact of molecular diffusion of oxygen is included in the coefficient 0.96.

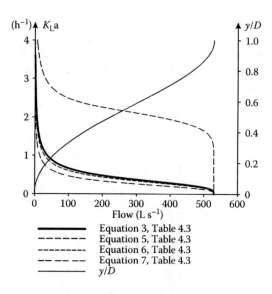

**FIGURE 4.8** Example of varying reaeration constant ($K_L a$) and water depth/diameter ratio ($y/D$) versus the flow in a gravity sewer pipe with diameter $D = 0.7$ m, slope $s = 0.003$, and wall roughness $k = 0.001$ m at a temperature of 20°C.

The formulas shown in Table 4.3 are important because they relate the reaeration constant, $K_{L,O2}$ to the system and flow characteristics of a gravity sewer pipe. In other words, the expressions show that it is the sewer system parameters and flow characteristics that determine the magnitude of the reaeration constant. As an example, Figure 4.8 illustrates how $K_L a = K_{L,O2}$ varies with the flow in a gravity sewer pipe with diameter $D = 0.7$ m, slope $s = 0.003$, and wall roughness $k = 0.001$ m at a temperature of 20°C based on results when applying the Manning formula. The figure also depicts the corresponding water depth/diameter ratio ($y/D$) and a full-flowing pipe capacity of 530 L s$^{-1}$.

Referring to Figure 4.8, it is likely that the model by Taghizadeh-Nasser (1986) depicts an unrealistic high potential of reaeration at conditions of an almost full-flowing pipe. The three other models are more or less identical, as also indicated by the mathematical expressions. It is noted that $K_L a$ shows a rather steep decline when $y/D$ increases from 0 to about 0.2.

### 4.4.3 MASS TRANSFER RATES FOR VOLATILE SUBSTANCES RELATIVE TO REAERATION RATE

Determination of the mass transfer rate for volatile substances across the air–water interface in collecting systems plays a central role when quantifying the impacts of malodors and corroding substances such as $H_2S$. The basic understanding and theories in this respect were discussed in Section 4.2, and it was shown that the key parameter is the mass transfer coefficient, $K_L a$.

For practical purposes, it is important that the ratio between mass transfer coefficients of two volatile substances is constant and related to the corresponding diffusion coefficients (cf. Equations 4.30 and 4.31). In this respect, oxygen plays a central role as a reference substance because of an extensive knowledge in terms of the results from reaeration experiments (cf. Section 4.4.2 and Table 4.3). In Example 4.5, it is furthermore shown how a $K_L a$ value for an arbitrary volatile substance, A, exemplified by $H_2S$, can be determined based on the corresponding information on oxygen.

In accordance with Equations 4.24 and 4.37 and, e.g., with correction factors as shown in Equation 4.39, the mass transfer rate can be determined for an arbitrary volatile substance, A:

$$F = K_L a \, (C_{L,A} - H_A C_{G,A}) \qquad (4.40)$$

where:

$F$ = rate of mass transfer of a volatile substance A (g m$^{-3}$ h$^{-1}$)
$K_L a$ = overall mass transfer coefficient for A (h$^{-1}$)
$C_{L,A}$ = concentration of A in the water phase (g m$_L^{-3}$)
$C_{G,A}$ = concentration of A in the gas phase (g m$_G^{-3}$)
$H_A$ = Henry's law constant for A (–)

As previously shown in this chapter, several different units for the parameters in Equation 4.40 can be selected. What is important to note is that, in addition to Henry's law constant, the molecular diffusion coefficients for the two substances, A and oxygen, and $K_L a$ for oxygen, $K_{L,O2}$, are the key constants required.

Equation 4.40, or any similar expressions describing the exchange of volatile substances between the water phase and the gas phase, is a term that is a central element in any sewer process model including the sewer atmosphere. Examples where the emission plays a central role are hydrogen sulfide-induced concrete corrosion and odor problems related to VOC formation in collection systems.

### 4.4.4 Air–Water Mass Transfer at Sewer Falls and Drops

Sewer structures such as junctions, manholes, bends, weirs, and drops may give rise to increased turbulence compared with the hydraulic conditions that exist under normal flow conditions in a sewer pipe. In particular, at sewer falls and drops, the mass transfer between water and air is significantly increased caused by phenomena such as splashing droplets and entrainment of air in the water phase (cf. Figure 4.9). Such structures are therefore locations in a sewer network where both VOC and $H_2S$ stripping and reaeration are being promoted.

The air–water mass exchange at sewer falls and drops is complex, and each structure has, in principle, its own specific characteristics. For practical purposes, the description of both stripping and uptake in the water phase in terms of expressions and models is therefore rather simple and incompletely described.

In principle, the same fundamental mechanisms are governing both emission (stripping) from and absorption in the water phase of volatile substances at sewer

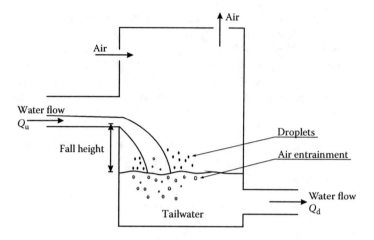

**FIGURE 4.9**  Principle of phenomena related to air–water mass transfer at a sewer fall.

falls and drops. However, more studies have been devoted to reaeration than to emission of VOCs at these structures. The description of reaeration is therefore focused on as a representative phenomenon. In several cases, however, the expressions valid for reaeration may be transformed to describe stripping of VOCs.

### 4.4.4.1  Mass Transfer at Sewer Falls and Drops

Referring to Figure 4.9 and under well-defined and steady-state conditions, a mass balance for a volatile compound, A, can be established. In principle, this mass balance can be set up in both the water phase and the gas phase. For the water phase and under steady-state conditions and assuming no chemical or biological reaction of the volatile compound, the mass balance expresses that the loss of volatile mass from the water flow is only due to stripping. Defining an overall mass transfer coefficient, $K_L a$, for the drop structure, the mass balance is:

$$\frac{Q_L}{V_{tail}}(C_{u,A} - C_{d,A}) - K_L a(C_{d,A} - H_A C_{G,A}) = 0 \qquad (4.41)$$

where:

$Q_L$ = water flow rate through the drop structure (m³ h⁻¹)
$V_{tail}$ = tailwater volume (m³)
$C_{u,A}$ = upstream concentration of A in the water phase (g m⁻³)
$C_{d,A}$ = downstream concentration of A in the water phase (g m⁻³)
$K_L a$ = overall mass transfer coefficient for A (h⁻¹)
$C_{G,A}$ = mass concentration of A in the gas phase (g $m_G^{-3}$)
$H_A$ = Henry's law constant for A (–)

It is noteworthy that Equation 4.41 is generally valid for volatile compounds irrespective of the process being absorption or stripping. Important examples are absorption of oxygen and stripping of hydrogen sulfide.

Referring to Equation 4.41, the hydraulic residence time of the tailwater is:

$$t_w = \frac{V_{tail}}{Q_L}$$

(4.42)

It should be noted that the mass transfer coefficient in Equation 4.41 varies with the physical characteristics and performance of the drop structure. In principle, it is Equation 4.41 that defines its numerical value. As a pragmatic and simple alternative to $K_L a$, the absorption or stripping characteristics of a volatile compound at a drop structure can be related to the fall height, $H$, as exemplified for reaeration by Equation 4.44.

### 4.4.4.2 Reaeration at Sewer Falls and Drops

Because of the turbulence at drop structures that considerably increases the air–water oxygen transfer (reaeration), the formulas in Table 4.3 are no longer valid. Such sewer structures typically have their own site-specific characteristics in terms of geometric configuration, and several factors influence their performance (Almeida et al. 1999). A simple empirical description of the reaeration at sewer drops and falls therefore only includes the most important parameters. The following equations express a simple pragmatic solution to this approach.

Equation 4.43 introduces the DO deficit ratio:

$$r_O = \frac{S_{OS} - S_{u,O}}{S_{OS} - S_{d,O}}$$

(4.43)

where:

$r_O$ = DO deficit ratio for reaeration at a sewer drop (–)
$S_{OS}$ = dissolved oxygen saturation concentration (in equilibrium with the atmosphere) (g $O_2$ m$^{-3}$)
$S_{u,O}$ = DO concentration upstream sewer drop (g $O_2$ m$^{-3}$)
$S_{d,O}$ = DO concentration downstream sewer drop (g $O_2$ m$^{-3}$)

Based on the definition of the DO deficit ratio, $r_O$, Pomeroy and Lofy (1977) introduced a fall or drop reaeration coefficient, $K_H$.

$$\ln r_O = K_H H$$

(4.44)

where:

$K_H$ = fall reaeration coefficient (m$^{-1}$)
$H$ = fall height, i.e., difference between the energy line upstream and the energy line downstream of the fall (m)

Table 4.9 shows three empirical expressions derived for determination of the DO deficit ratio.

Thistlethwayte (1972) proposed that $K_H = 0.21$ m$^{-1}$, and Pomeroy and Lofy (1977) used in their expression $K_H = 0.41$ m$^{-1}$. Because of the complexity of real sewer drops,

## TABLE 4.9
### Empirical Expressions for Determination of the DO Deficit Ratio $r_O$, for Reaeration at a Sewer Fall in a Gravity Sewer

| Reference | Expressions for $r_O$ (–) |
|---|---|
| 1. Thistlethwayte (1972) | $1 + K_H H$ |
| 2. Pomeroy and Lofy (1977) | $e^{K_H H}$ |
| 3. Matos (1992) | $e^{0.45H - 0.125H^2}$ |

it is not surprising that results from different field measurements may differ from what is estimated by the empirical expressions shown in Table 4.9. As an example, Almeida et al. (1999) found that the expression by Thistlethwayte (1972) underestimated the DO deficit ratio, $r_O$, whereas the other equations overestimated $r_O$.

The expression by Thistlethwayte (1972) shown in Table 4.9 is derived from experiments in polluted watercourses. The other two expressions are based on data obtained in sewer systems. The expression proposed by Matos can be applied in small sewers, and is considered appropriate to fall heights lower than about 1.75 m.

The three empirical expressions shown in Table 4.9 are compared in Figure 4.10. The constants included in the expressions are a result of the investigations performed by the authors. These constants may, of course, vary according to a number of site-specific conditions including impacts of temperature and wastewater quality characteristics.

In addition to the DO deficit ratio, the following empirical parameter, $\gamma$, is also used to characterize reaeration at sewer drops:

$$\gamma = 1 - s_0^{-1} \tag{4.45}$$

where $\gamma$ is the efficiency coefficient for reaeration at a sewer drop (–).

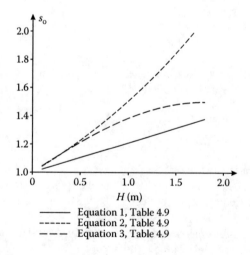

FIGURE 4.10   Comparison between expressions for DO deficit ratio for reaeration, $s_0$, presented in Table 4.9.

Both $s_O$ and $\gamma$ depend on the same basic factors as the $K_L a$ value, i.e., sewer structure characteristics, flow conditions, wastewater quality, and temperature.

As previously noted, the air–water mass transfer at sewer drops is complex. The construction of the drop structure will, for example, affect reaeration and stripping. Furthermore, it has been shown that the mass transfer at drop structures is the largest at low flow rates because of large residence times and significant interfacial area (Chanson 2004).

Further details on air–water mass transfer at sewer falls and drops are given by WERF (1998).

## 4.5  ACID–BASE CHARACTERISTICS OF WASTEWATER IN SEWERS: BUFFERS AND PHASE EXCHANGES

The acid–base chemistry in the water phase of sewers was the subject of Section 2.4. The objective of this section is to expand this subject to include the relation between the classic acid–base chemistry and the interactions with particularly volatile substances emitted to or absorbed from the sewer atmosphere. The acid–base characteristics in the water phase may thereby be affected. The interaction between the water phase and the solid state will furthermore be described. These phase exchanges of specific compounds is, in terms of acid–base relations, important for the understanding of the potential pH changes during transport of wastewater in sewers. The buffering characteristics of wastewater with its different contents of acid–base systems are therefore also an important issue. As already shown in, e.g., Figures 2.7 and 4.3, even relatively small changes in pH may considerably affect the distribution between $H_2S$ and $HS^-$ in the water phase and thereby influence the potential for emission of the molecular form, $H_2S$. A solid basis for the determination of the pH value in sewer process models is therefore fundamental.

The acid–base chemistry discussed in Section 2.4 constitutes the prerequisite for understanding the extended pH-related aspects focused on in this section. Section 2.4 will therefore only in selected cases be referred to in the following discussion.

### 4.5.1  BUFFER SYSTEMS IN WASTEWATER OF SEWERS

The constituents of the carbonate system in groundwater originate in several parts of the world from limestone and dolomite rocks. In such types of groundwater, these substances typically play a dominating role as the buffer system (cf. Section 2.4). Consequently, the carbonate system will also play an important role as the buffer system of the corresponding drinking water and wastewater. For more "soft" water systems, a number of other acid–base systems originating from different types of wastes may relatively, in the carbonate system, contribute to the buffering characteristics of wastewater. Such types of wastewater may therefore be found in areas where the soil contents of limestone or dolomite are low and where the drinking water source is a "soft" type of surface water or urban runoff water.

Potential buffer systems in wastewater supplementary to the carbonate system are schematically shown in Figure 4.11. Although the pH value of wastewater typically varies from 7 to 8, values outside this range may, as an example, occur when adding

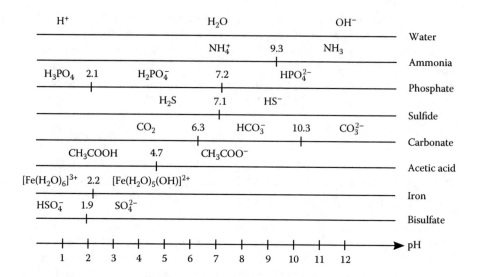

**FIGURE 4.11** Buffer systems potentially relevant for wastewater. $pK_a$ values for acid–base pairs are given at 25°C.

acids or bases for sulfide control. The concentrations of the different compounds shown in Figure 4.11 and the corresponding $pK_a$ values determine to which extent—and at which pH values—the buffer systems are important (cf. Equation 2.55). Furthermore, such wastewater compounds also affect the alkalinity (cf. Equation 2.54).

In addition to the buffer systems shown in Figure 4.11, it is important to mention that amino groups ($R-NH_2$), in addition to ammonia, influence the buffer capacity at pH values >7.5–8.0. These amines [e.g., methylamine ($CH_3-NH_2$) and ethylamine ($C_2H_5-NH_2$)] are degradation products of proteins. They do not appear in Figure 4.11 because of a general lack of details on their occurrence in wastewater. However, as a group they play an important role (cf. Example 4.8).

It is, in addition to Figure 4.11, also important that acetic acid only appears in the list as a representative for VFAs. Several other fatty acids [e.g., propionic acid ($C_2H_5COOH$)] add to the buffer capacity with $pK_a$ values of about 5 (cf. Example 4.8).

Example 4.7 will illustrate the pH-dependent variability of the alkalinity in a wastewater where both the carbonate system and ammonia are potentially important. Furthermore, the dissociation of water must be taken into account when calculating the alkalinity at both high and low pH values.

## Example 4.7: Alkalinity of Wastewater at Varying pH Values

This example concerns the pH-dependent changes in alkalinity for a soft wastewater type, i.e., a wastewater with relatively low contents of calcium and magnesium.

In these types of wastewater, other buffer systems than the carbonate system may—in contrast to hard water—typically play a role. At certain pH values—depending on the concentrations and $pK_a$ values of the buffers—the carbonate system has less influence on the magnitude of the alkalinity.

The ammonia content of the wastewater is, in this example, considered of importance as a buffer and therefore also of potential importance for the alkalinity. The following three reaction equilibria outline the buffer systems in the wastewater:

$$H_2O + CO_2 \Leftrightarrow H_2CO_3 \Leftrightarrow HCO_3^- + H^+ \quad pK_a = 6.3$$
$$HCO_3^- \Leftrightarrow CO_3^{2-} + H^+ \quad pK_a = 10.3$$
$$NH_4^+ \Leftrightarrow NH_3 + H^+ \quad pK_a = 9.3$$

In addition to these equilibria, the dissociation of $H_2O$ to $H^+$ and $OH^-$ contributes to the alkalinity at both low and high pH values (cf. Equations 2.49 through 2.51).

According to Equation 2.54, the alkalinity is:

$$A = C_{OH^-} + C_{HCO_3^-} + 2C_{CO_3^{2-}} + C_{NH_3} - C_{H^+}$$

The chemical analysis of the central compounds for the alkalinity resulted in the following:

- Total inorganic carbon (cf. Equation 2.53) is equal to 15 mg C $L^{-1}$ (1.25 mM).
- Total ammonia concentration is equal to 40 mg N $L^{-1}$ (2.86 mM).

From the expression of the alkalinity, it is readily seen that it varies with pH. The pH variations of the alkalinity can therefore be calculated based on the expression for the alkalinity, $A$; the results of the analysis; and the chemical equilibrium equations for the three reaction equilibria and water dissociation. The result is graphically depicted in Figure 4.12.

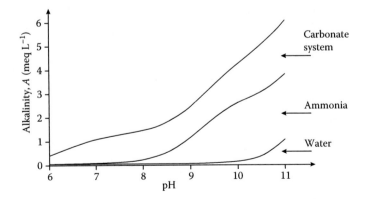

**FIGURE 4.12** Graphic presentation of alkalinity versus pH (6–11) for a soft wastewater type. The figure shows contributions of carbonate system, ammonia, and water.

Figure 4.12 shows that the relative amount of the molecular form of ammonia, $NH_3$, with increasing pH, has an increasing influence on the alkalinity. It is, however, evident that the carbonate system even for this relatively soft type of water is important.

In most types of wastewater, the carbonate system and ammonia are the two sets of components that mostly determine the alkalinity and the buffer characteristics. The first and the third reaction equilibriua having a $pK_a$ value of 6.3 and 9.3, respectively, constitute the two pH buffer systems that result in a rather stable pH value of about 7–8 for most types of wastewaters.

Finally, it should be noted that the alkalinity within the pH range of 7–8 in this example is approximately 1 meq $L^{-1}$ and mainly determined by the carbonate system. The alkalinity for this soft wastewater type is relatively low in contrast to an interval of 3–7 meq $L^{-1}$ that is usually claimed to be typical for the alkalinity of wastewaters.

The buffer capacity, $\beta$, is a measure of the resistance to pH changes and, in addition to the alkalinity, a central parameter for wastewater in a sewer (cf. Equation 2.54). As an example, the magnitude of the buffer capacity influences the amount of bases to be added in order to raise the pH value when using alkaline substances for sulfide control (cf. Section 7.3.4). Furthermore, a high buffer capacity of the wastewater ensures a relatively high stability of the pH value.

### Example 4.8: Buffer Capacity of Wastewater at Varying pH Values

This example illustrates the magnitude of the buffer capacity, $\beta$, at varying pH values. The occurrence of the different acid–base buffer systems is—as depicted in Figure 4.11, with a relatively high-strength wastewater type in this example—illustrated by the composition shown in Table 4.10.

Calculation of the buffer capacity

$$\beta = -\frac{dC_{H^+}}{dpH}$$

is carried out by modeling in terms of a fictitious addition of $H^+$ to the wastewater, i.e., a titration procedure, with the different acid–base systems as shown in Table 4.10. The result of this titration is shown in Figure 4.13.

---

### TABLE 4.10
### Alkalinity and Composition of the Buffer Systems in a High-Strength Wastewater Type

| Composition of Wastewater | Concentration (g m$^{-3}$) | Concentration (mM) |
| --- | --- | --- |
| Alkalinity | | 4 (meq L$^{-1}$) |
| Ammonia | 50 | 3.57 |
| Amino groups | – | 3.5 |
| Phosphates | 10 | 0.32 |
| VFAs | 50 | 0.83 |
| Bisulfate and carboxyl groups | 100 | 3.13 |

---

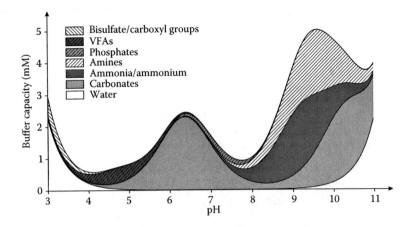

**FIGURE 4.13** Buffer capacity versus pH for wastewater with an acid–base composition as shown in Table 4.10.

Figure 4.13 exemplifies the importance on the buffer capacity of the different potential acid–base compounds in wastewater. The figure shows that the carbonate system dominates within the pH range of 6–7. However, ammonia and amino groups play an increasing role with increasing pH. Within a typical pH range of 7.5–8.0 for wastewater, it is readily seen that these compounds cannot necessarily be excluded when assessing the influence on the buffer capacity.

### 4.5.2 Impacts of Volatile Substances on pH of Wastewater

In the context of sewer processes, the importance of acid–base relations is indirectly connected with the emission of volatile substances such as $CO_2$ and $H_2S$. The general understanding of acid–base relations, as discussed in Section 2.4, forms the fundamental principle. However, specific aspects must be included to understand and quantify the relation between the acid–base chemistry in terms of pH and the release of volatile substances. The characteristics of both the wastewater matrix and the structure of the sewer network are, in this case, important.

What has been described in general terms may seem somewhat cryptic. It is therefore convenient to exemplify what it is all about. The acid–base relations of hydrogen sulfide are, in this respect, an illustrative and important example:

$$H_2S \Leftrightarrow HS^- + H^+ \qquad (4.46)$$

Only the molecular form of hydrogen sulfide, $H_2S$, can be emitted from the water phase into the sewer atmosphere. Figures 4.2 and 4.3 clearly illustrate that the pH value of wastewater plays a central role for the emission of $H_2S$ to the sewer atmosphere.

## Example 4.9: Emission of Wastewater Compounds

In continuation of the discussion on the potential risk for emission of $H_2S$, it is important that the $pK_a$ value of the $H_2S/HS^-$ acid–base system is 7.1 (cf. Table 2.1). A varying and typically essential fraction of sulfide (at constant total concentration of sulfide) is therefore as $H_2S$ potentially emitted to the sewer atmosphere at pH values of 7–8, a range often observed in wastewater.

Ammonia ($NH_3$) and VFAs [e.g., acetic acid ($CH_3COOH$)] are other examples of molecular substances with potential for emission and a corresponding odor and health-related impacts. In this respect, the following acid–base equilibria are important:

$$NH_4^+ \Leftrightarrow NH_3 + H^+ \qquad pK_a = 9.3$$
$$CH_3COOH \Leftrightarrow CH_3COO^- + H^+ \qquad pK_a = 4.7$$

From the $pK_a$ values, it is readily seen that within the pH range of 7–8, the relative major part of the two substances are $NH_4^+$ and $CH_3COO^-$, respectively. It is important to note that ammonia and amines with high $pK_a$ values and VFAs with low $pK_a$ values under normal pH conditions in wastewater are mainly present in dissociated forms. They therefore remain mainly in the water phase in contrast to, for example, sulfide.

In the context of emission of $H_2S$ and odorous substances, it is important that $CO_2$ emission has an impact on the pH value of the wastewater. First, it is interesting that the emission of $CO_2$ does not affect the alkalinity. At approximately neutral pH values, the following reaction proceeds when $CO_2$ is being emitted (cf. Equation 2.43):

$$H^+ + HCO_3^- \rightarrow CO_2 + H_2O \qquad (4.47)$$

From this equation, it is readily seen that emission of $CO_2$ removes the same equivalent number of an acid, $H^+$, and a base, $HCO_3^-$. Consequently, the alkalinity is not affected (cf. Equation 2.54). However, Equation 4.47 also shows that emission of $CO_2$ results in the removal of a strong acid and a weak base in the water phase. The overall result of $CO_2$ emission is therefore an increase in the pH value of the wastewater. The extent of this increase depends on the buffer systems in the wastewater that tend to stabilize the pH value (cf. Sections 2.4.2 and 4.5.1). As shown in Equation 4.47, the pH change caused by $CO_2$ emission is formulated by expressing the equilibrium in mathematical terms and when including the mass balance for $CO_2$.

The degradation of organic matter in wastewater of a sewer network will produce inorganic carbon, and excess $CO_2$ from this degradation—depending on the actual conditions—may therefore be emitted to the sewer atmosphere. Under aerobic conditions, the amount of $CO_2$ produced—or the production rate of it—can be estimated based on the consumption of $O_2$. As an example, Equation 3.1 shows that the molar ratio of $CO_2/O_2$ equals –1 for the aerobic degradation of glucose. The emitted amount or rate of emission for $CO_2$ can thereby be calculated from a mass balance. Depending on air movement and ventilation, the $CO_2$ emitted may accumulate in the sewer atmosphere, and concentrations as high as 0.5% to 1.0% have been found in contrast to about 0.04% $CO_2$ in the urban atmosphere.

Compounds with characteristics similar to $CO_2$ may also affect the pH value and the buffer system in the wastewater. As an example, molecular hydrogen sulfide, $H_2S$, has a corresponding effect: emission of $H_2S$ removes $H^+$ and $HS^-$ from the water phase, resulting in an increased pH value and no change in alkalinity.

The transformation processes of organic matter not only produce $CO_2$, but also, under anaerobic conditions, potentially VFAs such as acetic acid or acetates. The fermentation of organic compounds may therefore, in different ways affect the pH value, the buffer system, and the alkalinity (cf. Table 3.3 and Figure 4.11).

### 4.5.3   WATER–SOLID INTERACTIONS AND IMPACTS ON pH VALUE

As shown in Example 4.9, the air–water exchange (emission) of $CO_2$ has no effect on the alkalinity. In contrast, there is, at the water–solid interface for the carbonate system, a change in alkalinity (cf. Figure 2.13). With limestone as an example, the exchange process is:

$$CaCO_3 \Leftrightarrow Ca^{2+} + CO_3^{2-} \tag{4.48}$$

Equation 4.48 shows that dissolution of limestone will increase the concentration of $CO_3^{2-}$ and thereby the alkalinity by 2 equivalent units. Precipitation of limestone consequently results in the opposite change. This equation also shows that dissolution of limestone increases the pH value.

Typically, the exchange processes at the water–solid interface in a sewer are of minor importance in terms of impact on pH and alkalinity.

### 4.5.4   FINAL COMMENTS

The acid–base characteristics of wastewater have, in several ways, significant impacts on sewer processes. The pH-dependent emission of hydrogen sulfide resulting in corrosion and odor problems is an example of severe effects. In general, it is an important task to obtain reliable estimates of pH and alkalinity changes that occur in the sewer networks.

As described in Section 2.4 and this section, there are several parameters and complex interactions that determine the level of pH and alkalinity of wastewater. It is evident that a formulation in model terms of all changes of these two parameters requires an insight in an actual distribution of the acid–base compounds in wastewater that may not be available. From a pragmatic point of view, it is therefore important to focus on those aspects that, in each specific case, mostly affect such acid–base characteristics. The carbonate system, in general, VFAs at low pH values and ammonia and amines at relatively high pH values, are those substances that may particularly affect pH, alkalinity, and buffer characteristics.

# REFERENCES

Almeida, M.C., D. Butler, and J.S. Matos (1999), Reaeration by sewer drops, in: I.B. Joliffe and J.E. Ball (eds.), *Proceedings of the 8th International Conference on Urban Storm Drainage*, Sydney, Australia, August 30–September 3, 1999, pp. 738–745.

ASCE (1989), *Sulfide in Wastewater Collection and Treatment Systems, Manuals and Reports on Engineering Practice* 69, ASCE, American Society of Civil Engineers, New York, p. 324.

Betterton, E.A. (1992), Henry's Law constants of soluble and moderately soluble organic gases: effects on aqueous phase chemistry, in: J.O. Nriagu (ed.), *Gaseous Pollutants: Characterization and Cycling*, John Wiley & Sons, New York, pp. 1–50.

CEN (2003), EN 13725, Air quality–determination of odor concentration by dynamic olfactometry.

Chanson, H. (2004), Understanding air–water mass transfer in rectangular dropshafts, *J. Environ. Eng. Sci.*, 3, 319–330.

Corsi, R.L., D.P.Y. Chang, and E.D. Schroeder (1989), Assessment of the effect of ventilation rates on VOC emissions from sewers, in: *Proceedings of the 82nd Annual Meeting of the Air and Waste Management Association*, Anaheim, CA, USA, June 25–30, 1989, Air and Waste Management Association, Pittsburgh, PA, USA.

Corsi, R.L., S. Birkett, H. Melcer, and J. Bell (1995), Control of VOC emissions from sewers: a multi-parameter assessment, *Water Sci. Technol.*, 31(7), 147–157.

Dagúe, R.R. (1972), Fundamentals of odor control, *J. Water Pollut. Control Fed.*, 44, 583–595.

Danckwerts, P.V. (1951), Significance of liquid-film coefficients in gas absorption. *Ind. Eng. Chem.*, 43(6), 1460–1467.

Dobbins, W.E. (1956), The nature of the oxygen transfer coefficient in aeration systems, in: B.J. McCabe and W.W. Eckenfelder Jr. (eds.), Section 2.1 of *Biological Treatment of Sewage and Industrial Wastes,* Reinhold Publishing Corp., New York, pp. 141–148.

Duffee, R.A, F.D. Flesh, D.M. Benforado, and W.S. Cain (1979), *Odors from Stationary and Mobile Sources*, National Academy of Sciences, Washington, D.C., USA, p. 491.

Edwini-Bonsu, S. and P.M. Steffler (2004), Air flow in sanitary sewer conduits due to waste-water drag: a CFD approach, *J. Environ. Eng. Sci.*, 3, 331–342.

Edwini-Bonsu, S. and P.M. Steffler (2006a), Modeling ventilation phenomenon in sanitary sewer systems: a system theoretic approach, *J. Hydraulic Eng.*, 132(8), 778–790.

Edwini-Bonsu, S. and P.M. Steffler (2006b), Dynamics of air flow in sewer conduit headspace, *J. Hydraulic Eng.*, 132(8), 791–799.

Elmore, H.L. and W.F. West (1961), Effects of water temperature on stream reaeration, *J. Sanit. Eng. Div.*, 87, 59.

Eyring, H. (1935), The activated complex in chemical reactions, *J. Chem. Phys.*, 3, 107–115.

Gostelow, P. and S.A. Parsons (2000), Sewage treatment works odour measurement, *Water Sci. Technol.*, 41(6), 33–40.

Gostelow, P., S.A. Parsons, and A. McIntyre (2001), Dispersion modelling, in: R. Stuetz and F.-B. Frechen (eds.), *Odours in Wastewater Treatment: Measurement, Modelling and Control*, IWA Publishing, London, 232–249.

Hayduk, W. and H. Laudie (1974), Prediction of diffusion coefficients for non-electrolysis in dilute aqueous solutions, *J. AIChE*, 20, 611–615.

Higbie, R. (1935), The rate of absorption of a pure gas into a still liquid during short periods of exposure, *Am. Inst. Chem. Eng. Trans.*, 31, 365–389.

Huisman, J.L., C. Gienal, M. Kühni, P. Krebs, and W. Gujer (1999), Oxygen mass transfer and biofilm respiration rate measurement in a long sewer, evaluated with a redundant oxygen balance, in: I.B. Joliffe and J.E. Ball (eds.), *Proceedings of the 8th International Conference on Urban Storm Drainage*, Sydney, Australia, August 30–September 3, 1999, pp. 306–314.

Hvitved-Jacobsen, T., K. Raunkjaer, and P.H. Nielsen (1995), Volatile fatty acids and sulfide in pressure mains, *Water Sci. Technol.,* 31(7), 169–179.

Hwang, Y., T. Matsuo, K. Hanaki, and N. Suzuki (1995), Identification and quantification of sulfur and nitrogen containing odorous compounds in wastewater, *Water Res.,* 29(2), 711–718.

Jensen, N.A. (1994), Air–water oxygen transfer in gravity sewers, Ph.D. dissertation, Environmental Engineering Laboratory, Aalborg University, Denmark.

King, C. J. (1966), Turbulent liquid phase mass transfer at a free gas–liquid interface, *Ind. Eng. Chem.,* 5, 7.

Krenkel, P.A. and G.T. Orlob (1962), Turbulent diffusion and the reaeration coefficient, *J. Sanit. Eng. Div.,* 88(SA2), 53.

Lahav, O., A. Sagiv, and E. Friedler (2006), A different approach for predicting $H_2S_{(g)}$ emission rates in gravity sewers, *Water Res.,* 40, 259–266.

Lewis, W.K. and W.G. Whitman (1924), Principles of gas absorption, *Ind. Eng. Chem.,* 16(12), 1215.

Liss, P.S. and P.G. Slater (1974), Flux of gases across the air–sea interface. *Nature,* 247, 181–184.

Madsen, H.I., T. Hvitved-Jacobsen, and J. Vollertsen (2006), Gas phase transport in gravity sewers—a methodology for determination of horizontal gas transport and ventilation, *Water Environ. Res.,* 78(11), 2203–2209.

Madsen, H.I., J. Vollertsen, and T. Hvitved-Jacobsen (2007), Air–water mass transfer and tracer gases in stormwater systems, *Water Sci. Technol.,* 56(1), 267–275.

Matos, J.S. (1992), Aerobiose e septicidade em sistemas de drenagem de águas residuais, Ph.D. thesis, IST, Lisbon, Portugal.

Matos, J.S. and C.M. Aires (1995), Mathematical modelling of sulphides and hydrogen sulphide build-up in the Costa do Estoril sewerage system, *Water Sci. Technol.,* 31(7), 255–261.

Matos, J.S. and E.R. de Sousa (1992), The forecasting of hydrogen sulphide gas build-up in sewerage collection systems, *Water Sci. Technol.,* 26(3–4), 915–922.

Melbourne and Metropolitan Board of Works (1989), *Hydrogen Sulphide Control Manual— Septicity, Corrosion and Odour Control in Sewerage Systems,* Technological Standing Committee on Hydrogen Sulphide Corrosion in Sewerage Works, vols. 1 and 2, Melbourne, Victoria, Australia.

Melcer, H., P. Tam, and R.L. Corsi (1997), Ventilation rate and its impact on estimating VOC emissions in collection systems, in: *Proceedings of the Water Environment Federation (WEF) Specialty Conference on Control of Odors and VOC Emissions,* Houston, TX, USA, April 20–23, 1997, 2.37–2.46, WEF, Alexandria, VA, USA.

Monteith, H., J. Bell, and T. Harvey (1997), Assessment of factors controlling sewer ventilation rates, in: *Proceedings of the Water Environment Federation (WEF) Specialty Conference on Control of Odors and VOC Emissions,* Houston, TX, USA, April 20–23, 1997, WEF, Alexandria, VA, USA.

Olson, D., S. Rajagopalan, and R.L. Corsi (1997), Ventilation and industrial process drains: mechanisms and effects on VOC emissions, *J. Environ. Eng.,* 123(9), 939–947.

Olson, D.A., S. Varma and R. L. Corsi (1998), A new approach for estimating volatile organic compound emissions from sewers: methodology and associated errors, *Water Environ. Res.,* 70(3), 276–282.

Othmer, D.F. and M.S. Thakar (1953), Correlating diffusion coefficients in liquids, *Ind. Eng. Chem.,* 45(3), 589–593.

Owens, M., R.W. Edwards, and J.W. Gibbs (1964), Some reaeration studies in streams, *Ing. J. Air Pollut.,* 8, 469.

Parker, W.J. and H. Ryan (2001), A tracer study of headspace ventilation in a collector sewer, *J. Air Waste Manage. Assoc.,* 51(4), 582–592.

Parkhurst, J.D. and R.D. Pomeroy (1972), Oxygen absorption in streams, *J. Sanit. Eng. Div., ASCE*, 98(SA1), 101–124.

Pescod, M.B. and A.C. Price (1981), Fundamentals of sewer ventilation as applied to the Tynside sewerage scheme, *Water Pollut. Control*, 90(1), 17–33.

Pescod, M.B. and A.C. Price (1982), Major factors in sewer ventilation, *J. Water Pollut. Control Fed.*, 54(4), 385–397.

Pomeroy, R.D. and F.D. Bowlus (1946), Progress report on sulfide control research, *J. Sewage Works*, 18, 597–640.

Pomeroy, R.D. and R.J. Lofy (1977), *Feasibility Study on In-Sewer Treatment Methods*, EPA/600/2-77/992, US Environmental Protection Agency (USEPA), Cincinnati, OH.

Raunkjaer, K., T. Hvitved-Jacobsen, and P.H. Nielsen (1994), Measurement of pools of protein, carbohydrate and lipid in domestic wastewater, *Water Res.*, 28(2), 251–262.

Sander, R. (2000), Henry's law constants, in: W.G. Mallard and P.J. Lindstrom (eds.), *Chemistry WebBook*, NIST (National Institute of Standards and Technology) Standard Reference Database Number 69, USA, http://webbook.nist.gov/chemistry.

Scheibel, E.G. (1954), Liquid diffusivities, *Ind. Eng. Chem.*, 46, 2007–2008.

Sneath, R.W. and C. Clarkson (2000), Odour measurement: a code of practice, *Water Sci. Technol.*, 41(6), 23–31.

Stevens, S.S. (1957), On the psychophysical law, *Psychol. Rev.*, 64(2), 153–181.

Streeter, H.W. (1926), the rate of atmospheric reaeration of sewage polluted streams, *Trans. ASCE (Am. Soc. Civil Eng.)*, 89, 1351–1364.

Stuetz, R. and F.-B. Frechen (eds.) (2001), *Odours in Wastewater Treatment— Measurement, Modelling and Control*, IWA Publishing, London, p. 437.

Stuetz, R.M., R.A. Fenner, S.J. Hall, I. Stratful, and D. Loke (2000), Monitoring of wastewater odours using an electronic nose, *Water Sci. Technol.*, 41(6), 41–47.

Stumm, W. and J.J. Morgan (1996), *Aquatic Chemistry: Chemical Equilibria and Rates in Natural Waters*, 3rd edn, John Wiley & Sons, New York, p. 1022.

Taghizadeh-Nasser, M. (1986), Gas–liquid mass transfer in sewers (in Swedish) [Materieöverföring gas-vätska i avloppsledningar], *Chalmers Tekniska Högskola*, Göteborg, Publikation, 3:86 (Licentiatuppsats).

Tchobanoglous, G., F.L. Burton, and H.D. Stensel (2003), *Wastewater Engineering—Treatment and Reuse*, 4th edn, McGraw-Hill, New York, USA, p. 1819.

Thibodeaux, L.J. (1996), *Environmental Chemodynamics*, John Wiley & Sons, New York, p. 593.

Thistlethwayte, D.K.B. (ed.) (1972), *The Control of Sulfides in Sewerage Systems*, Butterworth, Sydney, Australia.

Thistlethwayte, D.K.B. and E.E. Goleb (1972), The composition of sewer air, in: *Proceedings from the 6th International Conference on Water Pollution Research*, Israel, June 1972, pp. 281–289.

Tsivoglou, E.C. and L.A. Neal (1976), Tracer measurement of reaeration: III. Predicting the reaeration capacity of inland streams, *J. Water Pollut. Control Fed.*, 48(12), 2669.

U.S. National Research Council, Division of Medical Sciences (1979), Hydrogen sulfide, report by Committee on Medical and Biological Effects of Environmental Pollutants, Subcommittee on $H_2S$.

Vincent, A. and J. Hobson (1998), *Odour Control, CIWEM (Chartered Institution of Water and Environmental Management) Monograph of Best Practice No. 2*, Terence Dalton Publishing, London, p. 32.

WEF (2004), *Control of Odors and Emissions from Wastewater Treatment Plants*, WEF Manual of Practice No. 25, Water Environment Federation (WEF), Alexandria, VA, USA, 537 pp.

WERF (1998), *Modeling the Stripping and Volatilization of VOCs in Wastewater Collection and Treatment Systems*, Final Report, Project 91-TFT-1, Water Environment Research Foundation, Alexandria, VA, USA. ISBN 0-9662553-6-4.

Wilke, C.R. and P. Chang (1955), Correlation of diffusion coefficients in dilute solutions, *J. AIChE*, 1(2), 264–270.

Yongsiri, C., J. Vollertsen, M. Rasmussen, and T. Hvitved-Jacobsen (2004a), Air–water transfer of hydrogen sulfide: an approach for application in sewer networks, *Water Environ. Res.*, 76(1), 81–88.

Yongsiri, C., J. Vollertsen, and T. Hvitved-Jacobsen (2004b), Effect of temperature on air–water transfer of hydrogen sulfide, *J. Environ. Eng.*, 130, 104–109.

Yongsiri, C., J. Vollertsen, and T. Hvitved-Jacobsen (2005), Influence of wastewater constituents on hydrogen sulfide emission in sewer networks, *J. Environ. Eng.*, 131, 1676–1683.

# 5 Aerobic and Anoxic Sewer Processes
## *Transformations of Organic Carbon, Sulfur, and Nitrogen*

This chapter deals with the aerobic and anoxic transformations as they occur in a sewer network. These transformations are relevant in both the water phase and in the biofilm. The aerobic heterotrophic microbial processes dealing with the transformations of wastewater organic matter are in focus. The aerobic and anoxic oxidation of sulfide and their relation to the formation of sulfide under anaerobic conditions in the biofilm are also discussed. The anoxic transformations involving nitrate and nitrite as electron acceptors are likewise covered in this chapter. In this respect, it is important to realize that anoxic conditions in a sewer mainly occur artificially when nitrate is added for odor and sulfide control purposes.

In several ways, the contents of this chapter are a continuation of the subjects covered in Chapters 2 through 4. The basic chemical and biological aspects of sewer processes are focused on in Chapters 2 and 3, respectively, and the reaeration process is dealt with in Chapter 4. These subjects are crucial for a deeper understanding of the contents of this chapter. The core objective of Chapter 5 is on the basic understanding to conceptualize the description of the processes in view of a real sewer system with its different process-related subsystems in terms of the water phase, the biofilm, and the sediments. Chapter 5 thereby forms a central basis for what is needed when modeling sewer processes (cf. Chapters 8 and 9).

## 5.1 AEROBIC, HETEROTROPHIC MICROBIAL TRANSFORMATIONS IN SEWERS

The aerobic, heterotrophic microbial processes in sewers are often occurring under semioptimal conditions, and the corresponding rates of transformation may be relatively high. In aerobic sewer networks, having a relatively long transport and residence time, significant changes in the wastewater quality will occur. These transformations are, for example, seen as a reduced amount of biodegradable substrate and production of biomass. Such quality changes affect the subsequent treatment processes. In the case of nutrient removal, the aerobic transformations of the wastewater in a sewer may therefore cause a reduced capacity for denitrification and biological phosphorous removal. On the other hand, when primary treatment is required, a positive interaction may occur because of a reduced amount of soluble and biodegradable

organic matter and an increased amount of biomass, i.e., production of less biodegradable particulate organic matter. The sewer is therefore a reactor for processes
that interact in either a positive or a negative way with the downstream processes in
what, in traditional terms, is named the "treatment plant."

The extent of the aerobic transformations of organic matter in a sewer depends
on the presence of an active heterotrophic biomass, electron donors (biodegradable
organic matter), and an electron acceptor (dissolved oxygen [DO]). The continuous supply of the electron acceptor is, in this respect, crucial. The reaeration process often limits the transformations and is therefore a key process (cf. Chapter 4).
Traditionally, aerobic conditions are characterized by the presence of DO. However,
more fundamentally, it is the thermodynamic characteristics expressed by the redox
potential that define the process conditions (cf. Section 2.1.3).

Aerobic, heterotrophic microbial processes play a dominating role for transformation of organic matter in wastewater treatment plants. When dealing with sewer
processes, it is important to realize the differences between the microbial community in a sewer and in an activated sludge treatment plant. Typically, the following
are, in this respect, essential:

- The ratio of heterotrophic biomass to easily biodegradable substrate is
  lower in a sewer than in an activated sludge treatment plant.
- The biomass in a sewer is often in an exponential growth phase (cf. Section
  2.2.1.4).
- The residence time of the wastewater is often lower in a sewer than in an
  activated sludge treatment plant where it is typically in the order of 1 day.
- The biomass in a sewer exists mainly in suspension as single cells, whereas
  in activated sludge systems it forms flocs.
- The biomass in biofilms and deposits typically play a central role in sewers.
- The composition and the concentration levels of the process-relevant COD
  (chemical oxygen demand) fractions of wastewater in a sewer and in activated sludge are different.

It is particularly important to note that the process-related role of organic matter compounds in the wastewater of a sewer is rather different compared with that
in an activated sludge. Although important similarities exist, this fact calls for a
characterization of parameters that are differently expressed for the two systems (cf.
Section 3.2.6). In general, it is not simple to "translate" knowledge originating from
wastewater treatment to knowledge on sewer processes.

The balance between aerobic and anaerobic conditions in the wastewater of a
sewer system is like walking on a tightrope. Depending on the temperature, the
solubility of DO in the wastewater of a sewer is typically 8–11 g $O_2$ m$^{-3}$ (cf. Section
4.4.1). This value should be assessed against a respiration rate that—depending on
the quality (availability of easily degradable organic matter) and the temperature—
may reach a value as high as 20–40 g $O_2$ m$^{-3}$ h$^{-1}$ under DO nonlimiting conditions. The reaeration rate in a gravity sewer pipe is typically rather small compared
with this range. It is therefore not surprising that anaerobic conditions are generally

occurring in pressure mains and are common in gravity sewers, particularly in warm climates.

## 5.2   ILLUSTRATION OF AEROBIC TRANSFORMATIONS IN SEWERS

The basic aspects of aerobic and anaerobic processes relevant for wastewater in sewer networks are discussed in Chapters 2 and 3. In Figure 5.1, the difference between aerobic and anaerobic processes is briefly illustrated. It is thereby exemplified how protein in a wastewater sample originating from a sewer system undergoes transformation. Under aerobic conditions, suspended protein components are significantly increased, and the soluble part is correspondingly reduced. These changes are interpreted as the result of a growth process of the bacterial biomass and a corresponding consumption of substrate for this growth. Under anaerobic conditions, no significant transformations of soluble and particulate protein take place.

The results shown in Figure 5.1 are in agreement with a rather steep increase in the oxygen uptake rate (OUR) during the initial exponential growth phase as illustrated in Figure 3.11a. These examples demonstrate that organic matter components and turnover rates of biomass in wastewater are affected by the presence of DO as the electron acceptor. They are also in agreement with the basic concept of growth and substrate utilization shown in Figure 2.2. Example 5.1 will further illustrate these transformations.

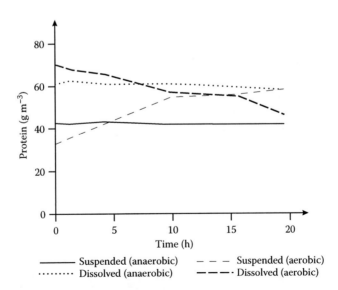

**FIGURE 5.1**   Transformation of protein in wastewater under aerobic and anaerobic conditions. (From Nielsen, P.H. et al., *Water Sci. Technol.,* 25(6), 17–31, 1992. With permission.)

## Example 5.1: Transformation of Wastewater
## Components in a Gravity Sewer

Almeida (1999) performed transformation studies of wastewater components in a gravity sewer. The sewer has a length of 7.2 km with a typical retention time of 1.5 h. An average slope equal to 0.007 m m$^{-1}$ and several drops result in a sewer where aerobic processes dominate. In addition to the organic components (COD$_{tot}$, COD$_{sol}$ and BOD$_5$), other relevant parameters [ammonia, nitrate, total suspended solids (TSS), and volatile suspended solids (VSS)] were included. The estimated average removal percentages of selected wastewater components are shown in Table 5.1.

As shown in Table 5.1, the average removal percentages for COD$_{tot}$ and COD$_{sol}$ are 6% and 19%, respectively. The interpretation of these results is that the catabolic microbial processes result in 6% removal of organic matter used for energy purposes, whereas 19−6 = 13% of the organic matter by anabolic processes is transformed to biomass (cf. Figure 2.2). The average heterotrophic biomass yield constant in this example is therefore $Y_H$ = 13/19 = 0.68 g COD biomass per g COD substrate, a value that corresponds well to what can be observed in a wastewater treatment plant.

Although experimental studies under sewer conditions are subject to high variability, it can be concluded that the removal of the components shown in the Table 5.1 is mainly attributable to the activity of the heterotrophic biomass. Theoretical considerations and several studies such as the one conducted by Almeida (1999) clearly demonstrate that the heterotrophic biomass is essential for understanding aerobic transformations. Studies by Stoyer (1970), Stoyer and Scherfig (1972), Koch and Zandi (1973), Pomeroy and Parkhurst (1973), and Green et al. (1985) have also focused on the removal of organic matter in sewers, primarily in terms of BOD (biochemical oxygen demand) and COD.

---

### TABLE 5.1
### Estimated Average Removal Percentages of Selected Wastewater Components in a 7.2-km Gravity Sewer with an Average Retention Time of about 1.5 h

| Component | Average Removal (%) | Number of Observations | Correlation Coefficient |
|---|---|---|---|
| COD$_{tot}$ | 6 | 80 | 0.94 |
| COD$_{sol}$ | 19 | 80 | 0.95 |
| BOD$_5$ | 7 | 20 | 0.95 |
| NH$_3$/NH$_4^+$ | 6 | 79 | 0.98 |

*Source:* Data from Almeida, M., Pollutant transformation processes in sewers under aerobic dry weather flow conditions, Ph.D. thesis, Imperial College of Science, Technology and Medicine, UK, p. 422, 1999. With permission.

*Note:* The sewer processes are predominantly aerobic.

---

The relative importance for transformation in a sewer in terms of limitations of biomass, substrate (electron donor), or electron acceptor ($O_2$) may vary. It is often seen that the heterotrophic biomass in wastewater compared with substrate is the limiting factor for transformations in upstream sections of a sanitary sewer, e.g., a collecting sewer or a lateral sewer. In this case, the availability of readily biodegradable substrate may result in a biomass being in a growth phase. However, in downstream sections of a sewer network (e.g., a trunk sewer or an intercepting sewer), the readily biodegradable substrate is typically—relative to the biomass—a limiting factor for the aerobic transformations. In both cases, however, it is crucial to note that an overall limiting factor is often caused by the supply of the electron acceptor (DO), i.e., the reaeration process. In contrast to what has been proposed by several authors, it would therefore serve no useful purpose to inject biomass (sludge) in a sewer line to enhance a "treatment process" for organic matter as known from treatment plants (cf. Example 5.2).

## Example 5.2: Reaeration and Aerobic Transformations in a Sewer

It is by this rather simplistic formulated example illustrated how the consumption of oxygen is related to the degradation of organic matter in a sewer, a central occurring transformation under aerobic conditions. Basically, the example quantifies the catabolism as shown in Figure 2.2. What is related to anabolism, i.e. production of new biomass, is of course also important but basically not expressed in terms of a change in a COD value.

The example refers to transformation of organic matter in a gravity sewer concrete pipe with a diameter $D = 0.5$ m and a slope $s = 0.003$ m m$^{-1}$. The wastewater in the pipe is flowing half-full under steady-state conditions, and the DO concentration is assumed constant and equal to 0.3 g $O_2$ m$^{-3}$. The sewer is an interceptor and serves a separate sewer catchment. The wastewater originates from domestic sources and has a temperature of $T = 15°C$. Only aerobic processes proceed in the water phase, and biofilm and sediments are considered absent.

The characteristics of the wastewater are approximately as depicted in Figure 3.10, and the process rate for the aerobic transformation of organic matter is limited by neither the biomass nor the substrate. Under these conditions, the limiting process for the removal of organic matter in terms of COD is the supply rate of oxygen, i.e., the reaeration rate. Equation 6 in Table 4.3 is used to calculate the reaeration rate. The COD removal can be expressed in units of g COD m$^{-3}$ h$^{-1}$ or g COD m$^{-3}$ km$^{-1}$.

The first step in the calculation of the reaeration rate is to determine the hydraulic conditions in the sewer pipe. The Manning formula (Equation 1.7), estimating a Manning coefficient of roughness, $n = 0.014$ s m$^{-1/3}$, is used to calculate the average flow velocity, $u$:

$$u = \frac{Q}{A} = \frac{1}{n}R^{2/3}s^{1/2}$$

where the hydraulic radius, $R$, is:

$$R = \frac{A}{P} = \frac{\frac{\pi}{8}D^2}{0.5\pi D} = \frac{D}{4} = 0.125 \text{ m}$$

Based on the Manning formula, the flow velocity and volumetric flow rate of the wastewater for the half-filled pipe are as follows:

$$u = 0.987 \text{ m s}^{-1}$$

$$Q = uA = 96.9 \text{ L s}^{-1}$$

The hydraulic mean depth of the water phase is:

$$d_m = \frac{\dfrac{\pi}{8}D^2}{D} = \frac{\pi}{8}D = \frac{\pi}{8}0.5 = 0.196 \text{ m}$$

Referring to Section 4.4.1, the DO saturation concentration, $S_{OS}$, at 15°C is 10 g $O_2$ m$^{-3}$. At this temperature, the rate of oxygen supply at the air–water interface is (cf. Equation 4.37 and Equation 6 in Table 4.3):

$$F = 0.86\,(1 + Fr^2)(su)^{3/8}d_m^{-1}1.024^{T-20}(S_{OS} - S_O)$$

$$= 0.86\left(1 + 0.2\frac{0.987^2}{9.81 \times 0.196}\right)(0.003 \times 0.987)^{3/8}(0.196)^{-1}1.024^{-5}(10 - 0.3)$$

$$= 4.7 \text{ gO}_2 \text{ m}^{-3} \text{ h}^{-1}$$

That is, 4.7 g COD m$^{-3}$ h$^{-1}$ is removed.

The transformation per unit length of the sewer is:

$$\frac{4.7}{0.987 \times 10^{-3} \times 3600} = 1.32 \text{ g COD m}^{-3}\text{km}^{-1}$$

From this example, it is readily seen that the reaeration rate in the half-filled sewer results in a rather low rate of transformation of the organic matter. However, under DO nonlimiting conditions, i.e., in practice a DO concentration above 2–4 g $O_2$ m$^{-3}$, the quality of the wastewater could probably support a rate of transformation that is 2- to 5-fold as high (cf. Figure 3.11). Under such conditions, the process rate for the aerobic transformation would be limited by either the biomass or the substrate, and a decrease in the DO concentration would soon be observed because the reaeration process could not fully support the need for DO. This simple example has therefore a clear message:

> Even a rather simple phenomenon is basically complex, and only a conceptual understanding of a problem or phenomenon and a formulation hereof expressed in terms of a model can fully observe the dynamics of a sewer process.

A pipe in a combined sewer network is typically designed having a relatively low dry-weather volumetric flow rate compared with its capacity. Under such conditions, an enhanced oxygen supply rate per unit volume of wastewater could support a corresponding increased removal rate of COD assuming a sufficiently high quality of the wastewater. Example 5.2 therefore also illustrates that, aside from the wastewater quality characteristics, the sewer system characteristics are also crucial in determining the extent of the transformations.

## 5.2  A CONCEPT FOR AEROBIC TRANSFORMATIONS OF WASTEWATER IN SEWERS

In the preceding part of this book, in particular Chapters 2 and 3, the understanding of the sewer processes from their microbiological, chemical, and physicochemical basis has been in focus. From this basis, it is possible to combine the different elements into a concept for the description of the transformations that occur in sewer networks and that may result in corresponding impacts and effects.

In brief, the concept for sewer processes constitutes a comprehensive and integrated understanding expressed in quantitative terms. Furthermore, it is important to stress that a concept allows description at different levels of detail and that it is open for future scientific advances. The concept dealt with will be described in general terms, and it can be applied irrespective of physical constraints and structural details of the sewer network. In order to support the understanding of the basic characteristics, the concept is—as a first attempt—restricted to aerobic microbial transformations of wastewater in sewer networks. However, it is crucial to note that minor extensions of the concept may lead to the inclusion of, e.g., anoxic and anaerobic transformations, biofilm processes, and processes at the sewer walls exposed to the air phase. This will be explained in the following sections and chapters of this book.

Finally, it is important to note that a concept is not a model but it is a prerequisite for a following expression of the sewer processes in model terms. As such, a concept is central and indispensable.

### 5.2.1  CONCEPTUAL BASIS FOR AEROBIC SEWER PROCESSES

Traditionally, when transformations of wastewater organic matter in sewers have been dealt with, bulk parameters such as BOD and COD are used as measures for organic matter and organic matter removal. Experimentally determined removal percentages have been used for prediction purposes (Stoyer 1970; Koch and Zandi 1973; Pomeroy and Parkhurst 1973; Green et al. 1985; Almeida 1999). Raunkjaer et al. (1995) extended this approach by investigating the transformation of organic matter fractions under aerobic conditions in a gravity sewer. These investigations included the transformation pattern of soluble and particulate fractions of proteins, carbohydrates, lipids, and volatile fatty acids (VFAs) (cf. Figure 3.9). The methods for analyses of these fractions were directly developed for their applicability to a wastewater matrix (Raunkjaer et al. 1994).

Although the investigations of both Raunkjaer et al. (1995) and Almeida (1999) showed that removal of the dissolved COD fraction took place in aerobic sewers, a total COD removal was more difficult to identify (cf. Example 5.1). From a process viewpoint, it is clear that total COD is a parameter with fundamental limitations, because it does not reflect the transformation of dissolved organic fractions of substrates into particulate biomass. The dissolved organic fractions (i.e., VFAs and parts of the carbohydrates and proteins) are—from an analytical point of view and under aerobic conditions—considered to be useful indicators of microbial activity and substrate removal. The kinetics of the removal or transformations of these specific compounds, however, cannot express the basic characteristics of the microbial transformations: substrate removal for energy purposes and buildup of biomass. As an example, removal of dissolved carbohydrates can be empirically described in terms of first-order kinetics, but a conceptual formulation of a theory of the microbial activity in a sewer in this way is not possible. The fact is that theoretical limitations and methodological problems are major obstacles for characterization of microbial processes in sewers based on bulk parameters such as COD, or parameters determined as specific chemical or physical fractions of organic matter.

The biology of microorganisms is complex. It is therefore important to identify a rather simple—albeit, generally well-accepted—concept for the sewer processes that can also be extended when further knowledge on the transformations exists. Although details may be missing in a simple description, advantages in terms of possibilities for a sound use of the concept as a modeling basis increase because less components, processes, and parameters are required to be experimentally determined. Different approaches may arise when determining the narrow gap between what is a simple and informative description of a system and what is more scientifically correct.

As a first step, it is important to determine to which extent details of substrates and biomass should be included in the conceptual description. From an application point of view, it is considered unrealistic to include all important, chemically different substances although it is evident that their kinetic characteristics are different. Even a distinction between carbohydrates, protein, and fats seems at present not realistic. Correspondingly, the biomass must also be included as rather crude groups of microorganisms, e.g., as heterotrophic biomass and autotrophic biomass. The description of organic matter fractions at a level as shown in Table 3.7 is therefore the pragmatic approach. A concept for sewer processes thereby includes both soluble and particulate components and, indirectly, different characteristics related to the occurrence of organic components and corresponding different kinetic characteristics of these two groups of substances. From the very beginning, the concept is therefore not designed to observe a short-term variability and a microbial succession, i.e., selection and sequential development of different populations of microorganisms. However, the concept should be well suited to describe an "average" behavior of the processes in a sewer network.

It is furthermore important in a concept to distinguish between the different phases where the substances occur, i.e., the water phase, the biofilm, the sewer sediments, the solid surfaces of the sewer, and the sewer air phase. Mass transfer between such phases can thereby be taken into account (cf. Figures 1.6 and 5.2). As the first step, the aerobic microbial activity and the transformations of organic matter will be focused on in the description of the concept. It is, however, important to stress that the concept is expanded to account for anoxic and anaerobic processes. Furthermore, nitrogen- and sulfur-containing substances will also be included.

The concept for description of aerobic, microbial transformations of wastewater organic matter in a sewer should include processes that are central for the description of wastewater quality. The heterotrophic biomass, the organic substrates, and the relevant electron acceptor (DO) are therefore all central in this respect. The fundamental understanding of the concept is based on the generally accepted fact that substrate utilization for growth of biomass takes place parallel with its removal for energy purposes by the electron acceptor (cf. Figures 2.3 and 5.3).

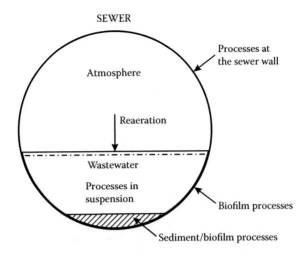

**FIGURE 5.2**  Different phases of a sewer where the microbial and chemical processes proceed and between which exchange processes occur.

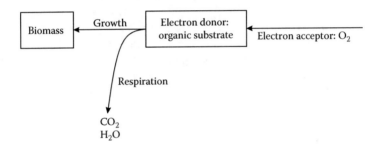

**FIGURE 5.3**  Biomass and substrate relations as applied to aerobic wastewater processes in sewer systems.

The traditional BOD and COD removal concept considers organic matter as "degradable" in a fictitious "removal process." In contrast, the concept described has moved to highlight biomass as being the real active component, depending on the nature and availability of organic substrates and electron acceptors. The heterotrophic biomass is, therefore, in terms of its activity, the central component of such a concept.

### 5.2.2  A Concept for Microbial Transformations in Sewers

Microbial processes in terms of sludge (biomass) production, BOD, and nutrient removal have been focused on when dealing with biological wastewater treatment. Therefore—and because the sewer is the input system for a wastewater treatment plant—there is a basis as well as a perspective to establish a process concept for the sewer that can be integrated with the activated sludge processes.

In the description of activated sludge processes in wastewater treatment plants, organic matter is subdivided into a number of fractions. The major components for the aerobic, heterotrophic processes of activated sludge have been identified as being heterotrophic biomass and different fractions of biodegradable and inert organic matter (cf. Section 3.2.6). Several researchers have contributed to the fundamental steps toward this understanding, e.g., Kountz and Forney (1959), McKinney and Ooten (1969), Gaudy and Gaudy (1971), Marais and Ekama (1976), Gujer (1980), and Dold et al. (1980). These early findings were the basis for a comprehensive description of activated sludge processes formulated in model terms by Grady et al. (1986) and as the "Activated Sludge Model No. 1" described by Henze et al. (1987). Further developments on the kinetics of organic carbon transformations and associated methodologies in activated sludge have taken place (e.g., Sollfrank and Gujer 1991; Henze et al. 1995, 2000).

Differences between the microbial transformations in a sewer and in a wastewater treatment plant should be considered in a process concept for two major reasons. First, it must be assessed which process-related aspects are relevant for each of the two systems. Second, it is important that different conditions for the microbial processes exist in activated sludge compared with wastewater in a sewer.

A treatment plant is, in terms of its microbial and chemical processes, in general a controlled and an optimized system. This is not the case for a sewer network. In addition, it is crucial that the sewer process concept for aerobic processes can be extended with anoxic and anaerobic processes and transformation in other phases than just a suspended water phase. The two fundamental aspects or criteria for development of a process concept will in the following be further discussed. In this discussion, it is crucial to assess a potential use of elements from the activated sludge process concept, but also to focus on those specific characteristics of sewer processes that are needed in a sewer process concept.

*Which process-related aspects are relevant in a concept?*

Wastewater treatment is, as far as microbial processes are concerned, assessed by the ability and efficiency of organic carbon, nitrogen, and phosphorous removal from the water phase. Furthermore, it is crucial that the removed excess sludge can be further treated (e.g., in terms of dewatering). In the sewer, however, the heterotrophic organic carbon transformations are particularly relevant to focus on in terms of wastewater quality changes such as production of odorous volatile substances and the availability of organic matter as a substrate for other heterotrophic processes such as sulfide production under anaerobic conditions (e.g., in the biofilm). In contrast to a removal of organic matter, it is—if advanced treatment of the wastewater is required—important to preserve readily biodegradable organic matter during transport in a sewer. The point is that this fraction of organic matter is essential for a subsequent denitrification (removal of nitrate from the water phase and its transformation to gaseous, molecular nitrogen) and biological phosphorous removal in the treatment plant. The sulfur cycle and fermentation processes in sewers are of interest when anaerobic conditions prevail, and formation of hydrogen sulfide and volatile odorous substances take place during transport of wastewater. A process concept must therefore reflect the different objectives and aspects that are potentially relevant for both sewer processes and processes in wastewater treatment plants. Although similarities exist, it seems therefore evident that these two process concepts should be differently expressed.

*Which process conditions should be included in a concept?*

As discussed in Section 3.2.6, the autotrophic biomass, i.e., the nitrifying biomass, and the phosphorous-accumulating biomass are—in contrast to wastewater treatment plants—expected to be close to zero in sewer networks. The following comparison between the process conditions in sewers and wastewater treatment plants concern the heterotrophic processes. The activated sludge processes proceed typically in a system characterized by a high concentration of microbial flocs (sludge) including both biologically active biomass and nonactive particulate organic matter. From this point of view, the activated sludge flocs consist primarily of living heterotrophic and autotrophic biomass, slowly hydrolyzable organic substrate, and biologically inert organic matter (cf. Table 3.6 and Figure 3.10). This mixture of organic matter is a result of a relatively high residence time of the sludge in the treatment plant. The amount of active heterotrophic biomass is dominating in activated sludge, and it typically exists under substrate-limited growth conditions (Section 2.2.1.4). Compared with this rather dense microbial floc system, the wastewater in a sewer is characterized by a relatively low concentration of active heterotrophic biomass in suspension that in upstream sewer sections often exists in an exponential growth phase (i.e., under nonlimited substrate conditions). On the other hand, a fast change to substrate-limited conditions occurs when readily biodegradable substrate is depleted and the availability of the electron acceptor (DO) is furthermore crucial. Under such conditions, hydrolysis of entrapped particulate substrate becomes the source for supply of a readily biodegradable substrate. Compared with a rather stable microbial system in a wastewater treatment plant, the sewer system is highly dynamic. The sewer process concept must reflect the dynamic relation between the biomass, the substrate, and the electron acceptor. The dynamics of the microbial system is typically seen when comparing upstream and downstream locations of a sewer and is influenced by both system-dependent conditions and changes in composition caused by the transformations that occur along a sewer reach. Furthermore, biofilm and sediment processes in the sewer will add to the differences between the process conditions that prevail in sewers and treatment plants.

The highlighted two questions define as a starting point the criteria for development of a sewer process concept. In particular, it is important to note the dynamic behavior of sewer processes, the fact that the processes occur both within and between different phases, and the potential need for including different system and process conditions. The first attempt to establish a conceptual understanding and corresponding model for microbial carbon transformations of wastewater under aerobic conditions in a sewer was proposed and applied by Bjerre et al. (1995, 1998a, 1998b). The starting point was the existing concept developed for the aerobic, heterotrophic transformations in activated sludge (cf. Figure 5.4). The sewer process

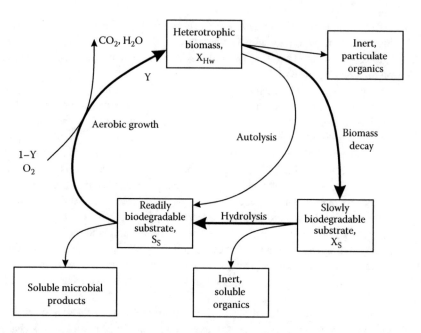

**FIGURE 5.4** Flow of substrate and biomass according to activated sludge concept for aerobic, heterotrophic transformations. Three major pathways of the flow, i.e., aerobic biomass growth, hydrolysis, and biomass decay, were included in first attempt to describe corresponding processes in the sewer. Compounds are defined in Section 3.2.6.

concept included as a first approach the biomass growth, the hydrolysis of particulates, and the decay of biomass as central processes. Experimental results and modeling confirmed that the concept, in principle, could be applied for the description of processes in the water phase when extended to include two to three fractions of hydrolyzable substrate characterized by their different rates of hydrolysis (Bjerre et al. 1998b). Biofilm and sediment processes were included as simple surface flux descriptions based on both batch experiments and field investigations (Bjerre et al. 1998a). This first attempt to establish a sewer process concept was important because it confirmed some kind of a "similarity" between processes in sewers and treatment plants; however, it also showed that a different approach was needed.

Methods for characterization of wastewater components in terms of different COD fractions and parameters for stoichiometric and kinetic description of the microbial processes were primarily developed based on procedures for activated sludge characterization. However, as a consequence of different dynamics for growth of the heterotrophic biomass in activated sludge and in wastewater of a sewer, a modified procedure and a different interpretation of the OUR versus time was used (cf. Figure 3.11 and Section 10.1.3) (Bjerre et al. 1995, 1998b).

Based on this slightly modified activated sludge concept and the corresponding procedures for estimation of parameters, it was possible to produce acceptable model simulation results for the water-phase processes of the heterotrophic carbon transformations in sewers. However, problems were identified for the description of the heterotrophic biomass decay, also named endogenous respiration. A major problem was the magnitude of the first-order decay rate constant with respect to the biomass concentration. Henze et al. (1987) and Kappeler and Gujer (1992) assumed this constant to be 7% to 10% and 5% of the maximum specific growth rate ($\mu_H$) for activated sludge, respectively. Bjerre et al. (1995) found the decay rate constant for wastewater from a sewer system to be 15% of $\mu_H$, and Vollertsen and Hvitved-Jacobsen (1998) estimated values in the order of 40% to 60% of $\mu_H$ for suspended sewer sediments. Such decay rates for bacteria are unrealistic (Kurland and Mikkola 1993). Vollertsen and Hvitved-Jacobsen (1998) therefore concluded that the concept that has been adopted until now should be reconsidered.

Taking this fact into consideration and based on new experiments, Vollertsen and Hvitved-Jacobsen (1998) concluded that a nongrowth-related substrate removal process should be included in the concept. Results originating from previous investigations also confirmed that it is a relevant approach (Tempest and Neijssel 1984; Russel and Cook 1995). This nongrowth process can be interpreted as a maintenance energy requirement of the heterotrophic biomass and not as endogenous respiration. Compared with this process, the decay of biomass (endogenous respiration) is considered less important for wastewater and is therefore omitted as an element of the heterotrophic processes in a sewer. The existence of a maintenance energy requirement was also confirmed based on observations of the heterotrophic biomass activity under varying aerobic and anaerobic conditions (Tanaka and Hvitved-Jacobsen 1998). Only if the readily biodegradable substrate produced by hydrolysis is insufficient to support the maintenance energy requirement of the biomass does it become relevant that—for mass balance reasons—biomass can undergo endogenous respiration resulting in a reduction of biomass. Results from addition of readily biodegradable substrate to sewer solids subject to aerobic conditions for several days confirmed that this process is likely to take place (Vollertsen and Hvitved-Jacobsen 1999). The rate constant for the maintenance energy requirement is constant and does not depend on, e.g., the availability of substrate.

Although the maintenance energy requirement of the biomass defined as a nongrowth substrate utilization process is considered real, the overall concept of biomass-related processes still reflects a pragmatic approach of reality. It is, however, important to state that in addition to the general requirement of using a relatively simple concept including only the major microbial processes, the COD mass balance must be observed. It is also important to note that any change in a concept will affect the value of the parameters in a sewer process model based on the concept. This means that because the biomass decay is replaced by a biomass maintenance energy requirement, in addition to the increased number of hydrolyzable substrate fractions, the parameters of a sewer process model are different compared with what is typically recommended and accepted for the activated sludge model. As an example, a biomass yield constant based on the sewer process concept must be defined differently compared to what is the case for the activated sludge model concept. These facts call for theoretically based and experimentally reliable methodologies to determine process parameters, a subject that will be focused on in Chapter 10. Finally, it is important to stress that the concept and the methodologies to determine the corresponding compounds and parameters must be in agreement. If it is not the case, the concept is nothing but theoretical.

Based on the findings and considerations discussed, the concept for the aerobic, heterotrophic biomass processes in wastewater of a sewer is depicted in Figure 5.5. The difference between this concept and the concept behind the activated sludge model as shown in Figure 5.4 is evident. It is not a point for discussion which concept is "correct," but it is important to realize that wastewater and activated sludge are different matrices and therefore by nature also behave differently in terms of the microbiological processes. The important point is to consider the validity of a concept formulated in relatively simple terms under the given constraints. The differences between sewer processes and processes in treatment plants are, however, no obstacle for the integration of these two systems. What is important is the fact that COD fractions, in terms of their role in biotransformation processes, are compatible and well defined at the interfaces between the sewer and the treatment plant.

The concept shown in Figure 5.5 is in accordance with the basic criterion and characteristics of redox reactions that have been highlighted and finally depicted in Figure 5.3.

Referring to Figure 5.5, it is readily seen that the biomass is the central compound that, by its activity, is the driving force for the changes in the organic compounds in wastewater. It seems relevant to deal with microbial transformations based on this fundamental statement. However, it also creates basic problems. It is somewhat problematic that the biomass in terms of its activity is determined based on an OUR measurement, at the same time that it is materialized as a COD fraction. This fundamental conflict, mixing biomass activity and biomass as something being materialized, is both the strength and the weakness of the concept. Explicitly expressed methodologies for determination of the biomass (e.g., applying genetic engineering

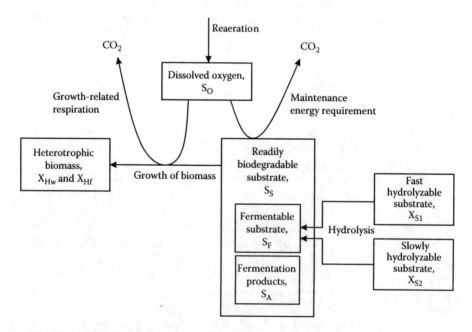

**FIGURE 5.5** Concept for aerobic, heterotrophic transformations of wastewater organic matter in a sewer. Compounds are defined in Section 3.2.6.

methods) are at present not realistic alternatives, although a future potential may exist (Vollertsen et al. 2000). Only a careful validation of the concept performed under varying external conditions for the biomass can make this dual situation meaningful. The concept including methods for determination of related characteristics is—when this basic requirement is observed—considered theoretically sound and, at the same time, operational and useful from a practical point of view.

The aerobic, heterotrophic concept for processes in a sewer as shown in Figure 5.5 can be extended according to specific objectives. An extension including aerobic processes of the sulfur cycle according to Figure 4.7 is shown in Figure 5.6.

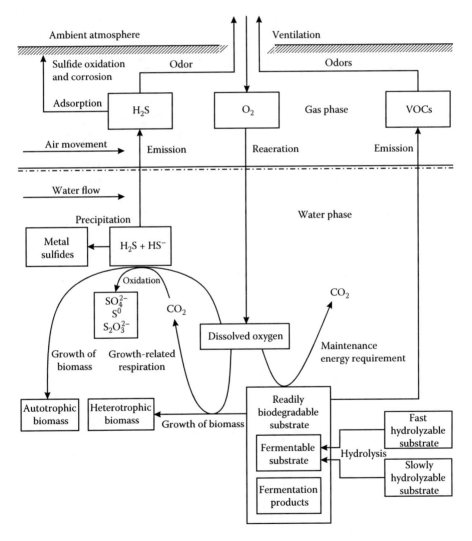

**FIGURE 5.6** Extended version of concept shown in Figure 5.5 by including aerobic processes of sulfur cycle and sewer gas phase.

Furthermore, this extended concept includes the sewer atmosphere with water–gas exchange processes and ventilation (cf. Figure 4.5). It is important to note that Figure 5.6 includes sulfur-related processes that are directly relevant for aerobic transformation, whereas the formation of sulfide occurring under anaerobic conditions is not shown. The sewer process concept as shown in Figure 5.6 is, for example, relevant in an aerobic gravity sewer receiving inflow of anaerobic wastewater from a force main with sulfide production. The occurrence of sulfide and VOCs in the gravity sewer is therefore basically a result of a transport from an upstream location. This example thereby demonstrates that effects of sewer processes may occur at locations different from where the sewer processes proceed (Nielsen et al. 2008). Further details of the sulfur cycle under aerobic conditions (sulfide oxidation) are discussed in Section 5.5.

Referring to Figure 5.6, it should be noted that the oxidation of sulfide to sulfate, elemental sulfur, and thiosulfate occurs both biologically by autotrophic bacteria and as a chemical process with the biological process as the most important (Nielsen et al. 2006). Both types of processes are included in the concept (although not visualized by double arrows).

## 5.3 FORMULATION IN MATHEMATICAL TERMS OF AEROBIC, HETEROTROPHIC PROCESSES IN SEWERS

Chapter 5 deals with both aerobic and anoxic processes in sewers. However, in this section, only the aerobic, heterotrophic processes in terms of their formulation in mathematical terms will be dealt with. The concept as shown in Figure 5.5 defines which processes are central for this description.

### 5.3.1 EXPRESSIONS FOR SEWER PROCESSES: OPTIONS AND CONSTRAINTS

The aerobic, heterotrophic microbial transformations of wastewater described in the concept shown in Figure 5.5 include COD compounds as defined in Section 3.2.6. The figure shows the central transformations of organic matter (the electron donors) in two subsystems of the sewer: the suspended wastewater phase and the sewer biofilm. The air–water oxygen transfer (reaeration) provides the aerobic microbial processes with the electron acceptor (cf. Section 4.4). Sediment processes are omitted in the concept but are indirectly taken into account in terms of a biofilm that not only exists at the sewer walls but also at the sediment surface. Water phase/biofilm exchange of electron donors and DO is part of the concept.

All subsystems influence the integrated picture of the in-sewer processes. Reaeration is a process that, at low DO concentrations in the water phase, may limit the rate of the aerobic processes. The relative importance of the processes in the suspended water phase and in the biofilm may vary, e.g., determined by the magnitude of the area/volume ratio, i.e., the ratio of biofilm area to bulk water volume in the sewer pipe. In this way, it is clear that system-related aspects also influence the extent of chemical and biological processes.

When considering the details of the sewer process concept, the processes of the subsystems are typically described at different levels. The most detailed description is normally done for the water phase by including the biomass–substrate relationship. Fewer details in terms of a description at an empirical level are included in the expression of the reaeration and the processes in the biofilm, although theoretical information is available for a detailed description of the biofilm processes (Characklis 1990; Gujer and Wanner 1990). However, a fundamental requirement to establish applicable experimental procedures for determination of substances and process parameters typically delimits the use of details in the formulation of the processes. A rather simple description of the biofilm processes in terms of 1/2-order or 0-order kinetics as shown in Section 2.2.2.1 is often more appropriate to use.

The dilemma related to which details should be selected for a process description is fundamental. In general, it is recommended to select a level where a process description and the experimental potential for its quantification are in agreement and optimal. The criteria for this optimum level rely on the need for quantification of those processes relevant for the sewer performance being discussed. Furthermore, it is important that the processes are expressed with sufficient accuracy, applicable under real conditions, and at the same time in agreement with sound theoretical knowledge and methods for experimental determination of model parameters.

The following section will highlight the mathematical expressions for the processes included in the concept shown in Figure 5.5.

### 5.3.2 Mathematical Expressions for Aerobic, Heterotrophic Processes in Sewers

Mathematical expressions for central aerobic, heterotrophic processes according to the concept shown in Figure 5.5 are described in the following four subsections. The main objective is to reach a formulation and a level appropriate for model use (cf. Chapter 8).

We focus on expressions for the microbial processes. The reaeration process is described in Section 4.4 and is therefore not included here. The nomenclature applied in these mathematical expressions is for the different substances shown in Section 3.2.6. For a more comprehensive list of terms and symbols, see the nomenclature given in Appendix A.

#### 5.3.2.1 Heterotrophic Growth of Suspended Biomass and Growth-Related Oxygen Consumption

Growth of suspended biomass under the limiting conditions of organic substrate and DO follows the classical Monod formulation as described in Section 2.2.1 (Monod 1949). The correctness of this formulation was confirmed based on laboratory and field experiments with wastewater under gravity sewer conditions (Bjerre et al. 1995, 1998a). The formulation of the heterotrophic growth rate of suspended biomass in the water phase, $r_{grw}$, therefore follows the concept of heterotrophic growth in the "Activated Sludge Model No. 1" (Henze et al. 1987):

$$r_{grw} = \mu_{H,O_2} \frac{S_F + S_A}{K_{Sw} + (S_F + S_A)} \frac{S_O}{K_O + S_O} X_{Hw} \alpha_w^{(T-20)} \qquad (5.1)$$

where:

    $r_{grw}$ = growth rate of heterotrophic biomass in suspension (g COD m$^{-3}$ day$^{-1}$)

    $\mu_{H,O_2}$ = maximum specific growth rate (day$^{-1}$)

    $S_F$ = readily (fermentable) biodegradable substrate (g O$_2$ m$^{-3}$)

    $S_A$ = volatile acids, i.e., fermentation products (g O$_2$ m$^{-3}$)

    $K_{Sw}$ = saturation constant for readily biodegradable substrate (g O$_2$ m$^{-3}$)

    $S_O$ = DO concentration in bulk water phase (g O$_2$ m$^{-3}$)

    $K_O$ = saturation constant for DO (g O$_2$ m$^{-3}$)

    $X_{Hw}$ = heterotrophic biomass concentration in the water phase (g COD m$^{-3}$)

    $\alpha_w$ = temperature coefficient for the water phase process (–)

    $T$ = temperature (°C)

It should be noted that organic matter in terms of COD is expressed in units of oxygen consumption by its degradation, i.e., the two units g COD and g O$_2$ are identical.

Suspended biomass growth is basically a transformation of readily biodegradable substrate into new biomass. A yield constant, $Y_{Hw}$, typically about 0.55 g COD biomass produced per g COD substrate consumed, has been observed (Bjerre et al. 1998a). The corresponding energy-producing degradation of organic matter therefore results in a DO consumption ratio of $1 - Y_{Hw}$.

### 5.3.2.2   Maintenance Energy Requirement of Suspended Biomass

Considerations behind the introduction of nongrowth-related consumption of substrate and corresponding uptake of DO are discussed in Section 5.2.2. The concept outlined in Figure 5.5 accounts for a maintenance energy requirement of the biomass, which—in addition to the growth or yield-related energy requirement—removes readily biodegradable substrate and, if it is not available, the biomass itself by cell lysis. The latter process is needed to observe the fundamental COD mass balance. The maintenance energy requirement concept is considered reasonable in a sewer network where the biomass often exists under organic carbon conditions of unlimited growth and where biomass decay in the suspended water phase is considered less relevant.

The maintenance energy requirement rate, $r_{maint}$ of the suspended biomass following Monod kinetics is:

$$r_{maint} = q_m \frac{S_O}{K_O + S_O} X_{Hw} \alpha_w^{(T-20)} \qquad (5.2)$$

where:

    $r_{maint}$ = rate of maintenance energy requirement for biomass in suspension (g COD m$^{-3}$ day$^{-1}$)

    $q_m$ = maintenance energy requirement rate constant (day$^{-1}$)

### 5.3.2.3   Heterotrophic Growth and Respiration of Sewer Biofilms

Biofilms develop at the solid sewer surfaces exposed to both the water phase and the humid air phase. In the context of the heterotrophic sewer processes, it is the exchange of electron donors, electron acceptor, and active biomass between the biofilm and the wastewater phase that is relevant. Referring to Figure 5.5, the heterotrophic biomass in the biofilm, $X_{Hf}$, indicates that the biofilm is included in the concept for aerobic, heterotrophic transformations of wastewater organic matter.

Before deciding the level of description for aerobic heterotrophic growth in the biofilm, it is important to consider the role of the biomass. First, a biofilm is not homogeneous, and it is typically developed with a rather fluffy surface. The "surface" of a biofilm is therefore not well defined, and the exchange of both particulate mass and soluble substances between the biofilm and the surrounding water phase is complex. As shown by Characklis and Marshall (1990) and Huisman and Gujer (2002), it is possible to describe a biofilm mathematically with aerobic, anoxic, and anaerobic zones where the corresponding processes proceed. However, difficulties arise when it comes to selection of realistic parameter values for this description. It is therefore evident that the complex biofilm structure and mix of processes are severe obstacles for process description in mathematical terms.

From a practical point of view, it must be realized that details concerning sewer biofilm growth and activity are not available to the same extent as is the case for the suspended biomass. A simple expression, compared to other well-known deterministic biofilm models (e.g., as described by Gujer and Wanner 1990; Huisman and Gujer 2002), is therefore selected for a combined biofilm growth and respiration.

An important finding based on laboratory and mixed field–laboratory studies has confirmed that the half-order kinetics for DO biofilm surface removal rates is a reasonable approximation (Raunkjaer et al. 1997; Bjerre et al. 1998b). This observation fits well with the fact that a rather thick aerobic biofilm with a relatively high process rate is typically partly penetrated by oxygen (cf. Section 2.2.2.1). The results from these studies also showed the influence of readily biodegradable substrate. Furthermore, temperature dependency limited by diffusion is included in the process description according to Nielsen et al. (1998). The following equation for the aerobic growth rate was therefore selected:

$$r_{grf} = k_{1/2} S_O^{0.5} \frac{Y_{Hf}}{1 - Y_{Hf}} \frac{S_F + S_A}{K_{sf} + (S_F + S_A)} \frac{A}{V} \alpha_f^{(T-20)} \tag{5.3}$$

where:

$r_{grf}$ = growth rate of heterotrophic biomass in a biofilm (g COD m$^{-3}$ day$^{-1}$)

$k_{1/2}$ = 1/2-order rate constant $\left( g\ O_2^{0.5}\ m^{-0.5}\ day^{-1} \right)$

$Y_{Hf}$ = biofilm yield constant [g COD, biomass (g COD, substrate)$^{-1}$]

$K_{Sf}$ = saturation constant for readily biodegradable substrate in biofilm (g COD m$^{-3}$)

$A/V$ = wetted sewer pipe surface area divided by the water volume, i.e., $R^{-1}$, where $R$ is the hydraulic radius (m$^{-1}$)

$\alpha_f$ = temperature coefficient for the biofilm process (–)

It should be noted that the biofilm growth rate for biomass (Equation 5.3) is differently expressed compared with Equation 5.1 for the suspended growth rate. Equation 5.1 directly expresses the growth by referring to a maximum specific growth rate, $\mu_{H,O_2}$, and the corresponding substrate limitation. Without the term $Y_{Hf}/(1 - Y_{Hf})$, Equation 5.3 is basically a 1/2-order expression for DO uptake in the biofilm. With a yield constant, $Y_{Hf}$, the fraction of DO uptake in terms of respiration is $1 - Y_{Hf}$ (cf. Figure 2.3). The term $Y_{Hf}/(1 - Y_{Hf})$ is therefore included in the expression to convert the expression from units of DO consumption to units of corresponding heterotrophic biomass production. It is also noteworthy that the 1/2-order description of the heterotrophic biofilm processes is rather simple compared with the description of the processes in the bulk water phase.

The important advantage of a rather simple process description is that the corresponding parameters can be determined experimentally. The growth expression for the biofilm requires a minimum of kinetics and stoichiometric coefficients to be determined, and no hydraulic or mass transport details are included. The dynamics of sewer biofilm detachment are not quantitatively known, and a steady-state biofilm with a biomass release to the bulk water phase, equal to the biomass growth within the biofilm, is therefore a pragmatic estimate. Example 5.3 will illustrate the process-related behavior of an aerobic biofilm.

### Example 5.3: Aerobic Processes in Sewer Biofilms

The description of biomass growth in a biofilm as expressed by Equation 5.3 includes the influence of both the electron donor (organic substrate) and the electron acceptor (DO). These two dependencies are expressed by the Monod type of limitation and by 1/2-order kinetics, respectively (cf. Sections 2.2.1.4 and 2.2.2.1).

Investigations were performed to exemplify the DO surface removal rates from biofilms grown on different types of wastewater (Bjerre et al. 1998b). Such investigations may show if Equation 5.3 can be considered an appropriate description of the aerobic activity. The wastewater for these studies originates from an open sewer system, the Emscher river, Germany. The results of the experiments are outlined in Table 5.2. Further details are shown in Figures 5.7 and 5.8.

---

### TABLE 5.2
### Surface Removal Rates for DO in Sewer Biofilms. Wastewater from the Sewer was Continuously Supplied to the Biofilm

| Distance from Wastewater Treatment Plant (km) | Typical COD Concentration (g COD m$^{-3}$) | Average DO Removal Rate, $r_O$ (g O$_2$ m$^{-2}$ h$^{-1}$) | Coefficient of Determination, $R^2$ | DO Concentration (g O$_2$ m$^{-3}$) | Biofilm Thickness (µm) |
|---|---|---|---|---|---|
| 52 | 70 | $r_O = 0.07\ S_O^{0.46}$ | 0.98 | 0–1.5 | 80 |
| 52 | 70 | $r_O = 0.085$ | – | 1.5–6 | 80 |
| 36 | 130 | $r_O = 0.11\ S_O^{0.53}$ | 0.83 | 0.2–5.5 | 90–180 |
| 23 | 280 | $r_O = 0.08\ S_O^{0.45}$ | 0.93 | 0.2–8 | 130–230 |
| 10 | 280 | $r_O = 0.10\ S_O^{0.35}$ | 0.86 | 0.2–6.5 | 130–250 |

*Source:* Data from Bjerre, H. L. et al., *Water Environ. Res.*, 70(6), 1151–1160, 1998b. With permission.
*Note:* The experiments were carried out under organic substrate nonlimiting conditions.

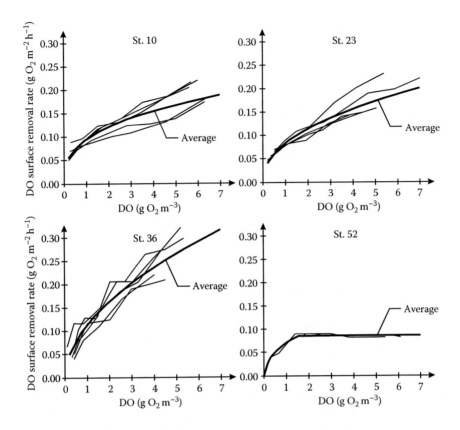

**FIGURE 5.7** DO surface removal rates for sewer biofilms (cf. Table 5.2).

The results of the experiments, as outlined in Figure 5.7, are performed under organic substrate nonlimiting conditions. Figure 5.8 shows similar experiments; however, these are performed under stepwise, aerated conditions without the addition of organic substrate. The experiments show the influence of organic matter on the aerobic biofilm activity in terms of the biodegradability of the wastewater.

The results of this study outlined in Table 5.2 and Figures 5.7 and 5.8 follow Equation 5.3. This equation observes theoretically sound concepts and is, therefore, found to be an appropriately simple expression for the estimation of heterotrophic biofilm activity.

### 5.3.2.4 Hydrolysis

Hydrolysis of particulate substrates produces readily biodegradable substrate important for the growth of the biomass (cf. Figure 5.6 and Section 3.2.3). The kinetics of the hydrolysis follows in principle the concept of Activated Sludge Model No. 1 as described in Section 5.2.2.

Hydrolysis in a sewer proceeds in both the suspended water phase and in the biofilm. The following pragmatic interpretation for the hydrolysis in each of these phases is considered: the hydrolyzable particulate substrate available in the bulk water phase originates mainly from households and the food industry, whereas that

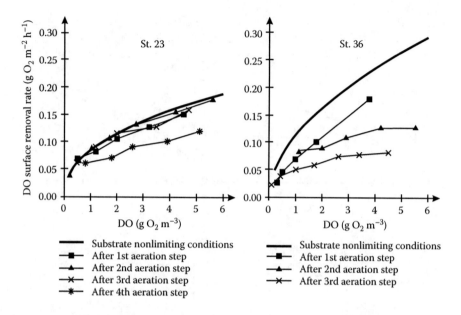

**FIGURE 5.8** DO surface removal rates for sewer biofilms showing organic substrate and a DO limitation.

found in the biofilm is more or less inactive biomass. It is therefore reasonable to assume a reduced rate of hydrolysis in the biofilm compared with that in the water phase. Under these conditions, the rate of hydrolysis, $r_{hydr}$, for each of the hydrolyzable fractions, $n$, is as follows:

$$r_{hydr} = k_{hn} \frac{X_{Sn}/X_{Hw}}{K_{Xn} + X_{Sn}/X_{Hw}} \frac{S_O}{K_O + S_O} \left( X_{Hw} + \varepsilon X_{Hf} \frac{A}{V} \right) \alpha_w^{(T-20)} \qquad (5.4)$$

where:

$r_{hydr}$ = rate of hydrolysis (g COD m$^{-3}$ day$^{-1}$)

$k_{hn}$ = hydrolysis rate constant, fraction $n$ (day$^{-1}$)

$X_{Sn}$ = hydrolyzable substrate, fraction # $n$; $\Sigma n$ typically equals 2 or 3: $n = 1$ (fast degradable), $n = 2$ (slowly degradable); $n = 1$ (fast degradable), $n = 2$ (medium degradable), $n = 3$ (slowly degradable) (g COD m$^{-3}$)

$K_{Xn}$ = saturation constant for hydrolysis of fraction $n$ (g COD g COD$^{-1}$)

$\varepsilon$ = relative efficiency constant for hydrolysis of the biofilm biomass (–)

$X_{Hf}$ = heterotrophic biomass in the biofilm (g COD m$^{-2}$)

The contribution to the hydrolysis rate as expressed by Equation 5.4 is typically mainly caused by the water phase hydrolysis. It is therefore also reasonable to assume that $\alpha_f$ can be replaced by $\alpha_w$.

The existence of different fractions of particulate substrates in terms of their specific hydrolysis rates is an important finding that originates from investigations of wastewater and resuspended sediments (Bjerre et al. 1995, 1998a; Vollertsen and Hvitved-Jacobsen 1998; Tanaka and Hvitved-Jacobsen 1998). Typically, two to three fractions must be considered to interpret the rate of hydrolysis when the process occurs in the wastewater of sewer systems (cf. Section 3.2.6). Three fractions are generally needed for relatively big sewer networks with a corresponding varying quality of the wastewater.

### 5.3.2.5 Final Comments

The mathematical expressions of the aerobic, heterotrophic processes in sewers as presented in Sections 5.3.2.1 through 5.3.2.4 are all designed to be included in a quantification of the concept as depicted in Figures 5.5 and 5.6. This description can furthermore be extended with other relevant sewer processes (cf. Chapter 6 and Sections 5.5 and 5.6). The final integrated formulation of the concept in model terms and in principle based on a set of coupled differential equations will be discussed in detail in Chapter 8.

The conceptual description as formulated in this book is included in the WATS (Wastewater Aerobic/anaerobic Transformations in Sewers) sewer process model. The description of WATS and examples of its use are given in Chapters 9 and 10. The sulfide odor and corrosion management model, SEWEX, is another example of a sewer process model (Yuan et al. 2010).

## 5.4  DO MASS BALANCES AND VARIATIONS IN GRAVITY SEWERS

DO is, without comparison, the central component and parameter for the course of aerobic processes in wastewater of sewers: it is the electron acceptor for the aerobic microbial processes and, as such, it controls the redox conditions for their activities. From a practical point of view, it is important that reliable equipment is available for the monitoring of DO and that solid scientific knowledge exists for calculation of its transfer (reaeration) across the air–water interface (cf. Chapter 4). In this section, a number of examples will be given to illustrate the quantification of a DO mass balance, its level, and its variations under the conditions that exist in sewer networks.

The overall DO mass balance in a sewer network is shown in Figure 5.9, formulated by Matos and de Sousa (1996). The figure outlines the central role of the DO concentration in the bulk water phase and the different elements that determine its variation. The mass balance is formulated in general terms: the flow of oxygen occurs into the suspended water phase with the reaeration as the driving force, and the DO consuming processes have their source of oxygen in the water phase. The DO concentration in the water phase is therefore central, and measurements of the DO concentration also typically take place in this phase.

The DO mass balance in a sewer can be established at different levels. The following expression is generally expressed and stated in relatively simple terms (Parkhurst and Pomeroy 1972; Jensen and Hvitved-Jacobsen 1991; Matos and de Sousa 1991, 1996):

$$\frac{dS_O}{dt} = \text{input} - \text{output} + K_L a (S_{OS} - S_O) - (r_w + r_f) \qquad (5.5)$$

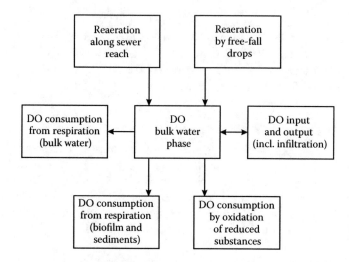

**FIGURE 5.9** Main processes, sources, and sinks involved in DO mass balance for wastewater in a sewer network (formulation after Matos and de Sousa 1996).

where

$S_O$ =  dissolved oxygen concentration in bulk water phase (g m$^{-3}$)

$t$ =  time (h)

$K_La$ =  overall oxygen transfer coefficient, also named reaeration constant (h$^{-1}$)

$S_{OS}$ =  dissolved oxygen saturation concentration in bulk water phase (g m$^{-3}$)

$r_w$ =  rate of DO consuming processes in the flowing wastewater (g O$_2$ m$^{-3}$ h$^{-1}$)

$r_f$ =  rate of DO consuming processes in the biofilm, in units transferred to the water volume (g O$_2$ m$^{-3}$ h$^{-1}$)

For a sewer network with self-cleansing conditions (without sediments), without drops, and without considerable amounts of reduced substances (e.g., produced by anaerobic processes in the deeper parts of the biofilm), the mass balance in Equation 5.5 follows what is outlined in Figure 5.9. Equation 5.5 can be solved by numerical methods or methods analytically corresponding to different conditions of DO consumption (Matos and de Sousa 1996).

Different levels of description of the rate expressions for $r_w$ and $r_f$ can be selected when using a simple mass balance as expressed in Equation 5.5. Matos and de Sousa (1996) propose a temperature-dependent and DO-nondependent value for $r_w$:

$$r_w = r_w (20) \, \alpha^{(T - 20)} \tag{5.6}$$

It is possible to include a DO dependency in Equation 5.6 by adding a Monod-type of expression as shown in Equations 5.1 and 5.2.

The value of $r_w(20)$ depends on the quality of the wastewater and is therefore subject to considerable variability. It can be determined from simple laboratory experiments. The values shown in Table 5.3 originate from different investigations

## TABLE 5.3

### Experimentally Determined Values of the DO Consumption Rate, $r_w$, for Different Types of Wastewaters

| Source | Characteristics | $r_w$ (g O$_2$ m$^{-3}$ h$^{-1}$) |
|---|---|---|
| Boon and Lister (1975) | Wastewater subject to anaerobic conditions | 11–16 |
| USEPA (1985) | Young wastewater | 2–3 |
| Matos and de Sousa (1991) | Small sewers | 0.3–6 |
| Huisman et al. (1999) | Main sewers, 15°C | 0.5 (night)–3 (day) |

of wastewater from sewers and clearly show that local knowledge is required for selection of a value for $r_w(20)$.

A simple empirical description of the DO consumption rate in the biofilm, $r_f$ (in units of g O$_2$ m$^{-3}$ h$^{-1}$), is proposed by Parkhurst and Pomeroy (1972):

$$r_f = 5.3 \, S_O \, (su)^{0.5} \, R^{-1} \qquad (5.7)$$

where:

$s =$ slope of sewer line (m m$^{-1}$)
$u =$ average flow velocity (m s$^{-1}$)
$R =$ hydraulic radius, i.e., the cross-sectional area of the water volume divided by the wetted perimeter (m)

Equation 5.7 shows a linear dependency in the DO concentration that is not in agreement with the results shown in Figure 5.7. Matos (1992) also found a discrepancy between Equation 5.7 and experimental results, and substituted the term 5.3$S_O$ in Equation 5.7 with a constant equal to 10.9. This constant depends on biofilm and wastewater characteristics and should be determined from local measurements.

In addition to the information given in Example 5.3 based on Bjerre et al. (1998b), measured values of surface respiration rates for sewer biofilms are shown in Table 5.4.

It is important to note that the empirical expressions 5.6 and 5.7 are simple, and in contrast to the conceptually expressed Equations 5.1 through 5.4, do not include the dynamics of wastewater quality changes taking place in sewers. However, the simple

## TABLE 5.4

### Experimentally Determined Values of the DO Surface Consumption Rate, $r_f$, for Different Sewer Biofilms

| Source | Characteristics | $r_f$ (g O$_2$ m$^{-2}$ h$^{-1}$) |
|---|---|---|
| Boon and Lister (1975) | 15°C | 0.7 |
| Matos and de Sousa (1991) | Small sewers, 20°C | 0.9–2.6 |
| Norsker et al. (1995) | 20°C | 1.2–1.8 |
| Huisman et al. (1999) | Main sewers, 15°C | 0.17–0.25 |

DO mass balance expressed in Equation 5.5 may give useful information, as shown in Example 5.4. In general, it is crucial that both simple and more complex models exist. A model most appropriate to use will always depend on the actual objective and the information available.

### Example 5.4: DO Concentration Profiles in a Gravity Sewer Pipe under Varying Conditions of Water Depth/Diameter Ratio

The objective is to exemplify the use of Equation 5.5 as also illustrated by Matos and de Sousa (1996). DO concentration profiles, i.e., the varying DO concentration along the length of the sewer, were calculated based on this equation. Results are shown in Figure 5.10 based on the following parameters:

$S_O = 4.5 \ g \ O_2 \ m^{-3}$, DO concentration at the inflow to the sewer pipe section
$K_La = $ oxygen transfer coefficient ($h^{-1}$) was calculated based on Equation 3 in
 Table 4.3
$r_w = 5 \ g \ O_2 \ m^{-3} \ h^{-1}$
$r_f = 10.9 \ (s \ u)^{0.5}$
$u = $ average flow velocity ($m \ s^{-1}$) was calculated according to the Manning
 formula, Equation 1.7
$s = 0.005 \ m \ m^{-1}$
$T = 20°C$
$D = 0.3 \ m$, sewer diameter

The curves in Figure 5.10 illustrate the importance of the water depth/diameter ratio ($y/D$) on the DO concentration. Although Equation 5.5 and thereby Figure 5.10 include DO consumption processes, it is interesting to compare Figure 5.10 with Figure 4.8. This figure depicts a corresponding influence of the $y/D$ ratio on the magnitude of the reaeration, i.e., the supply with oxygen. As also illustrated in

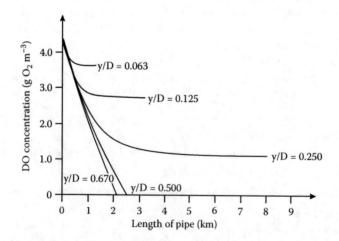

**FIGURE 5.10**   Calculated DO concentration profiles along a gravity sewer line for varying values of water depth/diameter ratio ($y/D$).

Example 5.2, these different aspects of the DO mass balance show that reaeration often plays a central role in establishing aerobic conditions.

Figure 5.10 also shows that DO equilibrium conditions with DO $> 0$ g $O_2$ m$^{-3}$ are only established at relatively low values of $y/D$.

The processes in the DO mass balance, and thereby the DO concentration, are subject to considerable variability. In sewers with low DO concentrations, in particular, this variability may ultimately result in varying aerobic/anaerobic conditions, a case that will be further dealt with in Chapter 6. At a specific location, the flow, the wastewater quality, and the temperature vary over time, daily and annually, and thereby result in substantial changes in the DO concentration. Example 5.4 illustrates such relations with emphasis on the flow conditions. Example 5.5 will further exemplify these aspects and especially focus on the impact of the temperature and the wastewater quality on the DO concentration.

### Example 5.5: DO Concentration Profiles in a Gravity Sewer Pipe Subject to Daily Variations in Flow, Wastewater Quality, and Temperature

This example supplements and extends what is illustrated in Example 5.4 related to the variability of DO in sewer networks. Example 5.4 was based on a simple DO mass balance as expressed in Equation 5.5. This example will make use of the conceptual descriptions of aerobic sewer process as expressed by Equations 5.1 through 5.4, and the expression for the reaeration as shown in Equation 6 in Table 4.3. The detailed way of performing a computation with a sewer process model based on these equations will be described in Chapter 8.

As a starting point, Figure 5.11 shows measured values of the DO concentration in a gravity sewer. The example originates from an investigation in an intercepting gravity sewer (diameter $D = 0.5$ m and slope $s = 0.0023$ m m$^{-1}$). The DO

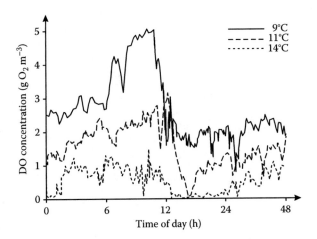

**FIGURE 5.11** Measured variability over day and night of DO concentration in a gravity sewer. Variability during 3 days and nights corresponds to different temperatures of wastewater.

variability over day and night is affected by the flow and the quality of wastewater (Gudjonsson et al. 2001). Around noon, the wastewater produced and discharged to the upstream part of the network in the morning hours results in an increased depleting effect on the DO concentration. Furthermore, a comparison between the curves measured over three different days from April through June shows the impact of the temperature on the rate of DO consumption. Figure 5.12 includes all measurements during this period and indicates the effect of a temperature increase from 9°C to 14°C on the daily duration of anaerobic conditions defined as a DO concentration <0.1 g $O_2$ m$^{-3}$.

The observations shown in Figure 5.11 are interpreted by applying a sewer process model based on Equations 5.1 through 5.4 and the reaeration (cf. Chapter 8).

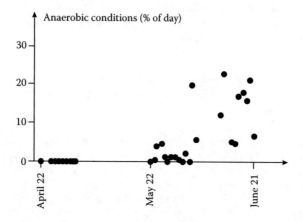

**FIGURE 5.12** Duration of anaerobic conditions in an intercepting gravity sewer over a certain period with a wastewater temperature that increases from 9°C to 14°C.

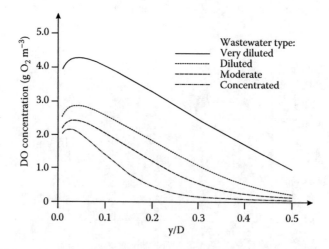

**FIGURE 5.13** Steady-state equilibrium concentrations of DO at a wastewater temperature of 11°C. The sewer is identical with the one used as an example in Figure 5.11 and 5.12. Simulation results are based on a sewer process model (cf. Chapter 8).

Simulation results are shown in Figure 5.13. The variability in DO concentration is, in this case, with given sewer systems parameters, determined by the flow (represented by the water depth/diameter ratio, $y/D$) and the quality (strength in terms of COD) of the wastewater varying between diluted in the night and concentrated during daytime.

Figure 5.13 illustrates that the flow conditions—here expressed as $y/D$—and the quality of the wastewater affect the DO concentration. According to the daily variations in wastewater quality and the $y/D$ ratio (0.14–0.18), the simulated DO concentration for concentrated wastewater and diluted wastewater varies between 0.5 and 2.5 g $O_2$ m$^{-3}$, respectively. This result fits well with the observations shown in Figure 5.11.

## 5.5 AEROBIC SULFIDE OXIDATION

The importance of hydrogen sulfide in terms of its impact on odor, toxicity, and corrosion in sewer networks means that sulfur plays a specific role and therefore must be focused on in all its relevant aspects. The main subject in this section is the aerobic sulfide oxidation in the water phase and the associated wetted biofilm. Aerobic sulfide oxidation is thereby a potential sink for hydrogen sulfide. Sulfide oxidation in sewers also occurs at the sewer walls exposed to the atmosphere where it results in concrete corrosion. This specific corrosion-related aspect of sulfide oxidation is dealt with in Section 6.5.2.2.

The occurrence of sulfide in an aerobic water phase of a sewer is typically caused by one of the following two situations:

1. Sulfide is at the location produced in the anaerobic part of the sewer biofilm or in the sediments, and diffuses from there into the aerobic part of the biofilm and possibly into the water phase.
2. Sulfide is produced in an upstream part of the sewer network and is transported to an aerobic section. Typically, this situation occurs when sulfide is produced in a pressure sewer located upstream of a gravity sewer section.

Sulfide oxidation in the water phase—or, alternatively, in the aerobic part of a biofilm—is therefore of importance as a sink process for sulfide. The kinetics and stoichiometric aspects of sulfide oxidation are central for the quantification of the phenomenon.

### 5.5.1 SULFIDE OXIDATION IN WASTEWATER OF SEWERS

Sulfide oxidation in wastewater occurs both biologically and chemically (Wilmot et al. 1988; Nielsen et al. 2003). It is likely that biological sulfide oxidation processes are chemoautotrophic, i.e., the oxidation of sulfide provides energy for microorganisms that utilize inorganic carbon for growth (cf. Section 3.1.3).

The major products and intermediates that have been identified by sulfide oxidation are elemental sulfur ($S^0$), thiosulfate $\left(S_2O_3^{2-}\right)$, sulfite $\left(SO_3^{2-}\right)$, and sulfate $\left(SO_4^{2-}\right)$. The corresponding reactions that determine the stoichiometry of the oxidation are:

$$2HS^- + O_2 \rightarrow 2S^0 + 2OH^- \tag{5.8}$$

$$2HS^- + 2O_2 \rightarrow S_2O_3^{2-} + H_2O \tag{5.9}$$

$$2HS^- + 3O_2 \rightarrow 2SO_3^{2-} + 2H^+ \tag{5.10}$$

$$2HS^- + 4O_2 \rightarrow 2SO_4^{2-} + 2H^+ \tag{5.11}$$

### 5.5.1.1 Stoichiometry of Sulfide Oxidation

Stoichiometric investigations of aerobic sulfide oxidation in wastewater systems were performed under conditions relevant for sewer networks (Nielsen et al. 2003). Such conditions include oxidation at relatively low DO concentrations.

The stoichiometry of sulfide oxidation is quantitatively expressed in terms of a reaction coefficient, $R_C$, defined as follows:

$$R_C = \Delta S(-II)/\Delta O_2 \tag{5.12}$$

where:
$R_C$ = reaction coefficient (mol S (mol $O_2$)$^{-1}$)
$\Delta S(-II)$ = transformed amount of dissolved sulfide (mol S)
$\Delta O_2$ = consumed amount of DO (mol $O_2$)

A reaction coefficient can be determined for chemical sulfide oxidation and biological sulfide oxidation in wastewater expressed as $R_{Cwc}$ and $R_{Cwb}$, respectively (Nielsen et al. 2003). Experimental results leading to values for these parameters can be summarized, as shown in the following subsections.

#### 5.5.1.1.1 Chemical Sulfide Oxidation

The reaction coefficient, $R_{Cwc}$, was typically found within the interval of 0.8–0.9 mol S (mol $O_2$)$^{-1}$. A single reaction scheme corresponding to Equations 5.8 through 5.11 cannot explain this result. However, it seems likely that both thiosulfate ($R_C = 2/2 = 1$ mol S (mol $O_2$)$^{-1}$; Equation 5.9) and sulfite ($R_C = 2/3 = 0.67$ mol S (mol $O_2$)$^{-1}$; Equation 5.10) are major intermediates. Although these intermediates undergo transformations, further oxidation of these compounds apparently proceeds at a low rate. These compounds can therefore be considered rather stable during the relatively short period that a certain volume of wastewater is detained in a sewer network.

#### 5.5.1.1.2 Biological Sulfide Oxidation

Biological oxidation of sulfide in wastewater generally results in its oxidation to elemental sulfur. The oxidation process is therefore as shown in Equation 5.8, and consequently is $R_{Cwb}$ equal to 2 mol S (mol $O_2$)$^{-1}$. During the residence time of wastewater in a sewer, further oxidation of $S^0$ can normally be neglected.

### 5.5.1.2 Kinetics of Sulfide Oxidation in Wastewater

The reaction kinetics with respect to both sulfide and DO is at constant temperature and pH for both biological and chemical sulfide oxidation described by power functions (cf. Section 2.2.1.3). When appropriate, the indices $c$ and $b$ are used to describe chemical oxidation and biological oxidation, respectively.

$$r_{S(-II)} = k_{S(-II)} (C_S)^{n1}(C_O)^{n2} \tag{5.13}$$

where:
  $r_{S(-II)}$ = rate of sulfide oxidation (g S m$^{-3}$ day$^{-1}$)
  $k_{S(-II)}$ = rate constant (unit depends on the values of n1 and n2)
  $C_S$ = concentration of dissolved sulfide (g m$^{-3}$)
  $C_O$ = concentration of DO (g m$^{-3}$)
  $n = n1 + n2$ = reaction order (unit dependent on n1 and n2)

The kinetics of sulfide oxidation is complex. Equation 5.13 is, in this respect, simple since—for biological oxidation—it does not include the biomass. It is already indicated by the fact that different processes, Equations 5.8 through 5.11, may govern the oxidation of sulfide and that both chemical and biological oxidation proceed. Furthermore, pH, temperature, and catalysis are important for the rate of sulfide oxidation.

Numerous authors have published valuable information on the kinetics of sulfide oxidation (Chen and Morris 1972; Wilmot et al. 1988; Buisman et al. 1990). However, when it comes to the specific conditions that prevail in sewer networks, it becomes difficult to apply the knowledge that has been gained. It is therefore crucial that further investigations under sewer conditions have been performed (Nielsen et al. 2003, 2006, 2008). Experimental results show that both chemical and biological sulfide oxidation are important processes. The main findings important for adding details to the general expressed rate equation, Equation 5.13, are discussed in the following.

### 5.5.1.2.1 Chemical and Biological Sulfide Oxidation, Reaction Order

The reaction rate was found to be relatively independent of the DO concentration. Sulfide oxidation is therefore important even at low DO concentrations (Nielsen et al. 2003). As an example, the chemical oxidation rate in the wastewater phase of sewers was at a DO concentration of about 0.1 g $O_2$ m$^{-3}$, about 65% of the rate at 1 g $O_2$ m$^{-3}$. For both chemical and biological sulfide oxidation, it was consequently found that n1 >> n2 (cf. Equation 5.13). Nielsen et al. (2003, 2006) report typical values of n1 (both n1$_c$ and n1$_b$) ranging from 0.8 to 1.0 and those for n2 (both n2$_c$ and n2$_b$) ranging from 0.1 to 0.2.

### 5.5.1.2.2 Chemical Sulfide Oxidation, Effect of pH on Reaction Rate

The molecular form of sulfide ($H_2S$) is oxidized with a lower rate than the ionic form ($HS^-$). This fact must be included in the formulation of the overall rate for chemical sulfide oxidation. In mathematical terms, this rate is therefore proportional with the sum of the oxidation rates for each of these two compounds. The pH effect on the sulfide oxidation rate in wastewater is therefore closely related to the distribution between $H_2S$ and $HS^-$ as shown in Figure 2.7. Referring to Equation 5.13, the rate constant for chemical oxidation, $k_{S(-II)c,pH}$, at varying pH values is formulated as follows (Nielsen et al. 2004, 2006):

$$k_{S(-II)c,pH} = \frac{k_{H_2Sc} + k_{HS^-c} \dfrac{K_{a1}}{10^{-pH}}}{1 + \dfrac{K_{a1}}{10^{-pH}}} \tag{5.14}$$

where:

$k_{S(-II)c,\ pH}$ = pH-dependent rate constant for chemical sulfide oxidation $((g\ S\ m^{-3})^{1-nlc}$ $(g\ O_2\ m^{-3})^{-n2c}\ h^{-1})$

$k_{H_2Sc}$ = rate constant for chemical sulfide oxidation of molecular sulfide, $H_2S$ $((g\ S\ m^{-3})^{1-nlc}\ (g\ O_2\ m^{-3})^{-n2c}\ h^{-1})$

$k_{HS^-c}$ = rate constant for chemical sulfide oxidation of ionic sulfide, $HS^-$ $((g\ S\ m^{-3})^{1-nlc}$ $(g\ O_2\ m^{-3})^{-n2c}\ h^{-1})$

$K_{a1}$ = $10^{-7.1}$ $(25°C)$ = the first dissociation constant for sulfide (cf. Table 2.1)

The average values of the two rate constants, $k_{H_2Sc}$ and $k_{HS^-c}$, are found in the order of 0.04 and 0.5, respectively, and both values are expressed in units of $(g\ S\ m^{-3})^{0.1}$ $(g\ O_2\ m^{-3})^{-0.2}\ h^{-1}$ (Nielsen et al. 2006).

### 5.5.1.2.3  Biological Sulfide Oxidation, Effect of pH on Reaction Rate

In general, biological processes are pH dependent and typically have an optimum pH value for activity. Similar characteristics are found for biological sulfide oxidation. Mathematically, it is formulated in terms of an equation that simulates a pH-dependent activity range around a pH value $(pH_{opt})$ at which biological sulfide oxidation proceeds at a maximum rate (Nielsen et al. 2006). Referring to Equation 5.13, the pH dependent rate constant is:

$$k_{S(-II)b,pH} = k_{S(-II)b,pH_{opt}} f_{S(-II),pH}$$

$$= k_{S(-II)b,pH_{opt}} \frac{\omega_{S(-II)b}}{\omega_{S(-II)b} + 10^{|pH_{opt}-pH|} - 1} \tag{5.15}$$

where:

$k_{S(-II)b,\ pH}$ = pH dependent rate constant for biological sulfide oxidation $((g\ S\ m^{-3})^{1-nlb}$ $(g\ O_2\ m^{-3})^{-n2b}\ h^{-1})$

$k_{S(-II)b,pH_{opt}}$ = maximum rate constant for biological sulfide oxidation at the $pH_{opt}$ value $((g\ S\ m^{-3})^{1-nlb}\ (g\ O_2\ m^{-3})^{-n2b}\ h^{-1})$

$pH_{opt}$ = optimum pH value for activity of sulfide oxidation (–)

$f_{S(-II),\ pH}$ = factor for relative pH dependency (–)

$\omega_{S(-II)b}$ = constant that determines the shape of the activity curve for sulfide oxidation versus pH (–)

It is expected that a biomass, in terms of its activity level, is adapted to the ambient environmental conditions. It is therefore often found that $pH_{opt}$ in wastewater is within the pH range of 7.5–8.5. Experiments have shown that a value of $\omega_{S(-II)b} = 25$ can describe an average activity level of a pH-dependent biological sulfide oxidation (Nielsen et al. 2006).

### 5.5.1.2.4 Chemical and Biological Sulfide Oxidation, Effects of Temperature

The temperature dependency of sulfide oxidation follows the general Arrhenius equation as expressed by Equation 2.38. For chemical sulfide oxidation, a temperature coefficient $\alpha = 1.07$ is a realistic value (Millero et al. 1987; Nielsen et al. 2004). For short-term changes in temperature, $\alpha = 1.10$ was found for biological sulfide oxidation (Nielsen et al. 2006).

### 5.5.1.3 Sulfide Oxidation under Field Conditions

Field investigations performed in an aerobic gravity sewer with a relatively high pH (pH = 8.3) revealed that the major decrease (approximately 90%) in sulfide inflow from a pressure main located upstream was caused by sulfide oxidation. Only a small fraction entered the sewer atmosphere. This observation shows that sulfide oxidation is a sewer process that potentially must be considered when assessing and simulating the fate of sulfide.

The kinetics of both chemical and biological sulfide oxidation is complex and subject to significant variability among different wastewaters. It is therefore important to note that the kinetic parameters for sulfide oxidation in wastewater are site specific (Wilmot et al. 1988; Nielsen et al. 2006). Kinetic parameters should accordingly be selected with caution, and results from locally performed investigations are therefore crucial. Table 5.5 exemplifies the level of rate constants for both chemical and biological sulfide oxidation found for three different municipal wastewaters.

The variability of the reaction rates shown in Table 5.5 is significant although the corresponding wastewaters are all considered "normal" municipal types. In contrast to sewer #1 and #3, the sewer network upstream the monitored section of sewer #2 has no pressurized pipes with potential sulfide production. It is therefore possible

### TABLE 5.5
### Average Values of Rate Constants for Chemical and Biological Sulfide Oxidation of Different Wastewaters at pH 8 and 25°C

| Sewer Network # | $k_{S(-II)c}$ $(g\ S\ m^{-3})^{1-n1c}$ $(g\ O_2\ m^{-3})^{-n2c}\ h^{-1}$ | $k_{S(-II)b}$ $(g\ S\ m^{-3})^{1-n1b}$ $(g\ O_2\ m^{-3})^{-n2b}\ h^{-1}$ |
|---|---|---|
| 1 | 0.35 | 0.67 |
| 2 | 0.09 | 0.04 |
| 3 | 0.42 | 0.99 |

*Source:* Nielsen, A.H. et al., *Water Environ. Res.*, 78(3), 275–283, 2006. With permission.

that an active sulfide oxidizing biomass in sewer #2 is only being developed to a minor degree, resulting in a relatively low value of $k_{S(-II)b}$. It is not possible to explain why the value of $k_{S(-II)c}$ in this sewer is also low.

Finally, it is important to note that the heterotrophic DO removal rate under DO nonlimiting conditions typically varies between 2 and 20 g $O_2$ m$^{-3}$ h$^{-1}$, corresponding to low and high biological activity, respectively. The magnitude of the DO removal rate from sulfide oxidation is typically found in the lower end of this range; however, this is still significant in terms of DO consumption.

## 5.5.2 Sulfide Oxidation in Sewer Biofilms

Biofilms (slimes) that develop in gravity sewers are typically rather thick (more than 1–2 mm), aerobic at the surface, and anaerobic in the deeper layers of the biofilm (cf. Section 2.2.2.1). The thickness of the aerobic layer of the biofilm is typically in the order of 0.5 mm but depends on the DO concentration in the water phase and the degradability of organic matter. Sulfide is typically produced in these anaerobic regions from where it diffuses into the aerobic part of the biofilm. The occurrence of both sulfide and DO creates, in this zone, an environment favorable for growth of sulfide-oxidizing bacteria, resulting in an internal sulfur cycle (cf. Figure 5.14). Furthermore, sulfide produced in pressure pipes located upstream a gravity sewer section may diffuse into the biofilm and undergo oxidation.

Investigations of the stoichiometry of sulfide oxidation in biofilms have shown that elemental sulfur is the main product and that sulfate and thiosulfate are either not produced or only are found in minor quantities (Nielsen et al. 2005). According to the results obtained for sulfide oxidation in the wastewater phase, it is therefore

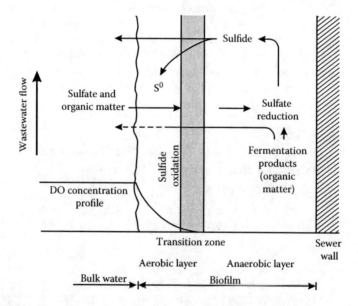

**FIGURE 5.14** Aerobic and anaerobic process interactions in a gravity sewer biofilm showing internal sulfur cycle.

likely that biological sulfide oxidation is dominating and that chemical sulfide oxidation in general can be excluded (cf. Section 5.5.1).

Biological sulfide oxidation follows the half-order kinetics in terms of both sulfide and DO in the bulk water phase (cf. Section 2.2.2.1) (Nielsen et al. 2005). At constant temperature and pH, the rate equation is therefore formulated as follows:

$$r_{S(-II),ox,f} = k_{S(-II),ox,f} S_{S(-II)}^{0.5} S_O^{0.5} \tag{5.16}$$

where:

$r_{S(-II),ox,f}$ = biofilm surface specific sulfide oxidation rate (g S $m^{-2}$ $h^{-1}$)
$k_{S(-II),ox,f}$ = rate constant for sulfide oxidation ((g S $m^{-3}$)$^{0.5}$ (g $O_2$ $m^{-3}$)$^{-0.5}$ m $h^{-1}$)
$S_{S(-II)}$ = dissolved sulfide concentration (g S $m^{-3}$)
$S_O$ = DO concentration (g $O_2$ $m^{-3}$)

The temperature dependency of the sulfide oxidation rate in biofilms follows the Arrhenius equation (Equation 2.38) with a temperature coefficient $\alpha = 1.03$. As also observed for the heterotrophic oxidation of organic matter, the temperature coefficient in biofilms is relatively low (limited by diffusion) compared with the corresponding value in the suspended water phase (cf. Sections 2.2.3 and 5.5.1.2.4).

The pH dependency of sulfide oxidation in biofilms is expressed in terms of a pH dependency factor, $f_{S(-II),pH}$ (cf. Equation 5.15). The values of both pH$_{opt}$ and $\omega_{S(-II)b}$ is assumed to follow those for sulfide oxidation in the water phase (cf. Section 5.5.1.2.3).

## 5.6   ANOXIC TRANSFORMATIONS IN SEWERS

Anoxic transformations in wastewater of sewer networks require availability of nitrate or other oxidized nitrogen compounds as electron acceptors for the microbial processes, and the DO concentration must be extremely low (cf. Sections 4.1 and 2.1). The concentration level of nitrate, $NO_3^-$, in wastewater is low, often 0–0.5 g N $m^{-3}$, and anoxic conditions are therefore typically not established. Under conditions where DO in wastewater is depleted and the redox potential is being reduced, the transition from the aerobic state to anaerobic conditions is in practice directly occurring without any intermediate anoxic state.

It is, however, important that anoxic conditions in sewers are artificially established when nitrate is used to control sulfide problems (cf. Section 7.3.1.3). In this case, kinetic as well as stoichiometric characteristics for anoxic transformations are therefore important for efficient sulfide control. As an example, a low nitrate uptake rate (NUR) is preferred to keep the consumption of nitrate low for economic reasons, and an overdose of nitrate in the sewer network may result in further treatment requirements at the wastewater treatment plant.

### 5.6.1   RELATIONS BETWEEN ANOXIC AND AEROBIC SEWER PROCESSES

The anoxic microbial processes are in several ways identical with the aerobic processes. The anoxic growing microorganisms are—like the aerobic heterotrophic microorganisms—chemoheterotrophs, and they use organic matter as their source of energy and organic substances as their carbon source. It is furthermore important that the electron transport pathway—except for the final step—for anoxic and aerobic transformations is the same and that several types of microorganisms can therefore switch between DO and nitrate as electron acceptors. For this and other reasons, it is important to compare aerobic and anoxic transformations. Example 5.6 illustrates this in terms of the stoichiometry.

## Example 5.6: Comparison between the Stoichiometry of Aerobic and Anoxic Transformations

Aerobic and anoxic degradation of organic substances are, by nature, occurring via transfer of electrons from the organic matter to $O_2$ and $NO_3^-$ as electron acceptors, respectively (cf. Section 2.1.4). The stoichiometric comparison of aerobic and anoxic transformations requires that the electron acceptors $O_2$ and $NO_3^-$ be considered by using electrons as a common and suitable unit for the redox processes. The fundamental reactions for this comparison are shown in Examples 2.4 and 2.5:

$$\frac{1}{4}O_2 + H^+ + e^- \rightarrow \frac{1}{2}H_2O$$

$$\frac{1}{5}NO_3^- + \frac{6}{5}H^+ + e^- \rightarrow \frac{1}{10}N_2 + \frac{6}{10}H_2O$$

From these two reduction processes (consumption of electrons), it follows that the transformation of 1/5 mol $NO_3$-N equals the transformation of 1/4 mol $O_2$. The oxygen/nitrate ratio for comparison of the two substances as electron acceptors is therefore as follows:

$$R_{O_2,N} = \frac{5M_{O_2}}{4A_N} = \frac{5 \times 32}{4 \times 14} = 2.86 \text{ g } O_2 \text{ (g } NO_3\text{-N)}^{-1}$$

where:
$M_{O_2}$ = molar weight of $O_2$ (g $O_2$ mol$^{-1}$)
$A_N$ = atomic weight of N (g N mol$^{-1}$)

It is important to note that the oxygen/nitrate ratio, $R_{O_2,N}$, is valid for the relation between DO and nitrate consumption as electron acceptors in aerobic and anoxic processes, respectively. These processes thereby refer to a corresponding transformation of organic matter as electron donor, and the ratio thereby refers to redox processes. Organic matter is also used for the production of the heterotrophic biomass. However, this growth process is, in this context, a quite different type of process.

## 5.6.2 Anoxic Transformations in the Water Phase

Anoxic transformations are well known in wastewater treatment plants where they are typically known under the name denitrification because of the impact on the removal of nitrogen from the water phase. Denitrification in activated sludge systems, i.e., the transformation of nitrate into atmospheric (molecular) nitrogen, is intensively studied and well described (Henze et al. 2002; Tchobanoglous et al. 2003). However, when it comes to anoxic transformations relevant for sewer systems, only a few investigations have been performed.

### 5.6.2.1 Heterotrophic Anoxic Processes

Although all oxidized nitrogen compounds that occur in wastewater can undergo anoxic transformations (denitrification), nitrate is the most important and relevant compound. In the case of hydrogen sulfide control, it is also nitrates that are artificially added to the wastewater. The following sequence of N transformations may proceed corresponding to a gradual change in the oxidation level, $OX_N$, from +5 to 0:

$$NO_3^- \rightarrow NO_2^- \rightarrow N_2O \rightarrow N_2 \qquad (5.17)$$

Abdul-Talib et al. (2002) and Yang et al. (2004) have, based on laboratory studies of anoxic transformations in wastewater, found that nitrite is accumulated as an intermediate until the concentration of nitrate is reduced to a very low concentration level (cf. Figure 5.15). Based on these findings, the transformation of nitrate can be described in two steps:

$$\text{Step I:} \quad NO_3^- \rightarrow NO_2^- \rightarrow N_2 \qquad (5.18)$$

$$\text{Step II:} \quad NO_2^- \rightarrow N_2 \qquad (5.19)$$

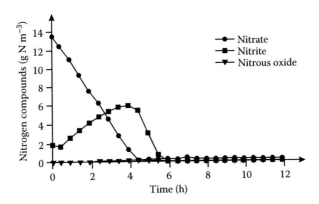

**FIGURE 5.15** Example of nitrate, nitrite, and nitrous oxide ($N_2O$) transformations under anoxic conditions in bulk water phase of wastewater.

During step I, nitrate is used with a significant accumulation of nitrite, i.e., the rate of nitrite production is faster than its transformation to molecular nitrogen. Step II occurs when nitrate is depleted and the accumulated nitrite is used as electron acceptor.

Although laboratory studies show that nitrite accumulates in wastewater under anoxic conditions, it should be noted that a similar observation so far has not been reported from real sewer systems. For this reason, until further knowledge on the transformation is available, it is recommended to consider the utilization of nitrate as electron acceptor in one step:

$$NO_3^- \rightarrow N_2 \qquad (5.20)$$

Corresponding to an OUR as described in Section 3.2.6 and shown in Figure 3.11, an NUR, also named nitrate utilization rate, can be defined for an anoxic process:

$$NUR = -dS_{NO3}/dt \qquad (5.21)$$

where:
  NUR = nitrate uptake rate (g N m$^{-3}$ h$^{-1}$)
  $S_{NO3}$ = concentration of nitrate (g N m$^{-3}$)

Figure 5.16 shows an example of the variability over time for NUR in the bulk water face of wastewater compared with an OUR variation for the same wastewater type.

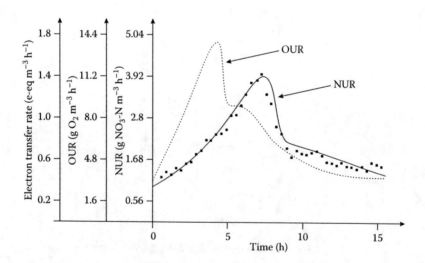

**FIGURE 5.16**  Measured NUR (nitrate uptake rate) curve compared with the measured OUR curve for the same wastewater sample. Transformation from units of e-eq m$^{-3}$ h$^{-1}$ (for both curves) to units of g O$_2$ m$^{-3}$ h$^{-1}$ and g NO$_3$-N m$^{-3}$ h$^{-1}$ for aerobic and anoxic conditions, respectively, are shown (cf. Section 2.1.4.2).

The two ratios between the uptake rates of DO and nitrate relative to the corresponding electron equivalents, respectively, are shown in Figure 5.16. According to Example 5.6, these are calculated as follows:

Aerobic process:

$$R_{ae} = M_{O_2}/4 = 32/4 = 8 \text{ g O}_2 \text{ m}^{-3} \text{ h}^{-1} (\text{e-eq m}^{-3} \text{ h}^{-1})^{-1}$$

Anoxic process:

$$R_{an} = A_N/5 = 14/5 = 2.8 \text{ g NO}_3\text{-N m}^{-3} \text{ h}^{-1} (\text{e-eq m}^{-3} \text{ h}^{-1})^{-1}$$

Once again, it is seen that $R_{O_2,N} = R_{ae} / R_{an} = 2.86$ g $O_2$ (g $NO_3$-N)$^{-1}$ (cf. Example 5.6).

The electron equivalent unit e-eq m$^{-3}$ h$^{-1}$ used in Figure 5.16 for both OUR and NUR expresses that these two rates are quantitatively comparable in their ability as electron acceptors for aerobic and anoxic degradation of organic matter, respectively (cf. Section 2.1.4.2). It is also therefore seen in Figure 5.16 that the unlimited anoxic heterotrophic growth that takes place during the first 7 h of this experiment proceeds with a significant lower rate than is the case for the unlimited aerobic degradation of organic matter. This example reflects the general observation from kinetic studies of wastewater that the anoxic heterotrophic biomass growth occurs with a maximum rate that is about 40% of the corresponding aerobic growth rate (Yang et al. 2004). Furthermore, in this study it is observed that the saturation constant, $K_{S,NO_3}$, is in the order of 1–2.5 g $NO_3$-N m$^{-3}$, i.e., nitrate is crudely a limiting factor for anoxic heterotrophic processes below a nitrate concentration of 2–5 g N m$^{-3}$. These findings are important because the consumption rate for nitrate for sulfide control has economic consequences (cf. Section 7.3.1.3).

### 5.6.2.2 Autotrophic Anoxic Sulfide Oxidation

Investigations have shown that biological—and not chemical—sulfide oxidation is possible under anoxic conditions in wastewater (Einarsen et al. 2000; Yang et al. 2005). Sulfide is oxidized to elemental sulfur ($S^0$), and nitrate is reduced to nitrite $\left(NO_2^-\right)$ or molecular nitrogen ($N_2$). Although elemental sulfur potentially can be further oxidized to sulfate, $S^0$ is apparently rather stable. The formation of $S^0$ is therefore considered the most important, and this first step of oxidation seems to govern the rate of sulfide removal. Sulfide is thereby the energy source, i.e., the electron donor, and inorganic carbon the carbon source for biomass growth. Anoxic sulfide oxidation is therefore a chemoautotrophic process (cf. Section 3.1.3).

The stoichiometry of anoxic sulfide oxidation is not well understood. It is difficult to quantify in wastewater because a simultaneously occurring heterotrophic nitrate reduction proceeds with a higher rate. The following reaction schemes are therefore just two of the several potential transformations (but likely occurring processes):

$$H_2S + NO_3^- \rightarrow S^0 + NO_2^- + H_2O \tag{5.22}$$

$$3H_2S + 2NO_2^- + 2H^+ \rightarrow 3S^0 + N_2 + 4H_2O \tag{5.23}$$

The processes are based on experiments observing that $H^+$ is consumed (Yang et al. 2005).

Yang et al. (2005) measured the anoxic sulfide oxidation rates in wastewater and found maximum values of these rates in the order of 0.48 and 0.62 g S m$^{-3}$ h$^{-1}$ at pH values of 7.0 and 8.5, respectively. These sulfide oxidation rates are, in general, lower than aerobic sulfide oxidation rates (cf. Section 5.5.1.3).

## 5.6.3  ANOXIC HETEROTROPHIC TRANSFORMATIONS IN BIOFILMS

The structure of an anoxic biofilm is like an aerobic biofilm (cf. Section 3.2.7). It is typically 1–3 mm thick, fluffy, and may also be affected by $N_2$ gas formation.

Unpublished results show that an anoxic biofilm is typically partly penetrated by nitrate and that the surface removal rate consequently follows 1/2-order kinetics (cf. Section 2.2.2.1). If the shear stress on the biofilm surface is not too small, i.e., larger than about 1 N m$^{-2}$, the following 1/2-order reaction rate was found typical:

$$r_{NO_3} = k_{1/2}S_{NO_3}^{0.5} = 0.08S_{NO_3}^{0.5} \tag{5.24}$$

where:

$r_{NO_3}$ = biofilm surface flux of nitrate (g $NO_3$-N m$^{-2}$ h$^{-1}$)

$k_{1/2}$ = 1/2-order reaction rate constant per unit area of biofilm surface (g $NO_3$-$N^{0.5}$ m$^{-0.5}$ h$^{-1}$)

$S_{NO_3}$ = nitrate concentration in the bulk water phase (g $NO_3$-N m$^{-3}$)

Poulsen (1997) investigated the anoxic transformations of wastewater in biofilms originating from a biofilter and found maximum NUR values of 0.025–0.055 g $NO_3$-N m$^{-2}$ h$^{-1}$ at 20°C. For relatively low concentration levels (i.e., concentrations less than about 1 g N m$^{-3}$), this interval corresponds well with the findings according to Equation 5.24. Poulsen (1997) observed 1/2-order kinetics when the concentration of nitrate was less than about 3 gN m$^{-3}$. The biomass yield constant, was according to Poulsen (1997), about 0.38 g COD g COD$^{-1}$.

Aesoey et al. (1997) found an NUR from a 1- to 2-mm-thick sewer biofilm to be 0.15–0.18 g $NO_3$-N m$^{-2}$ h$^{-1}$ at 15°C. They showed that 4.8 g COD (g $NO_3$-N)$^{-1}$ is removed when readily biodegradable substrate is available.

## 5.6.4  PREDICTION OF NITRATE REMOVAL UNDER ANOXIC CONDITIONS

Based on the findings from Sections 5.6.2.1 and 5.6.3, a simple empirical approach for the prediction of the nitrate removal rate under heterotrophic anoxic conditions is formulated. The following equation includes both nitrate removal in the water phase and in the biofilm of a sewer pipe:

$$r_{NO_3} = -\left( r_{w,max} \frac{S_{NO_3}}{K_{S,NO_3} + S_{NO_3}} \alpha_{w,an}^{T-20} + k_{1/2}S_{NO_3}^{0.5} \frac{A}{V} \alpha_{f,an}^{T-20} \right) \tag{5.25}$$

where:

$r_{NO_3}$ = overall nitrate removal rate in the water phase and in the biofilm (g $NO_3$-N $m^{-3}$ $h^{-1}$)

$r_{w,max}$ = nitrate removal rate at 20°C in the water phase under substrate unlimited conditions (g $NO_3$-N $m^{-3}$ $h^{-1}$)

$S_{NO_3}$ = nitrate concentration in the water phase (g $NO_3$-N $m^{-3}$)

$K_{S,NO_3}$ = saturation constant for nitrate (g $NO_3$-N $m^{-3}$)

$k_{1/2}$ = 1/2-order reaction rate constant per unit area of biofilm surface (g $NO_3$-$N^{0.5}$ $m^{-0.5}$ $h^{-1}$)

$A$ = area of the wetted pipe surface ($m^2$)

$V$ = volume of water ($m^3$)

$\alpha_{w,an}$ = temperature coefficient for the water phase (–)

$\alpha_{f,an}$ = temperature coefficient for the biofilm (–)

$T$ = temperature (°C)

Equation 5.25 describes a situation where the nitrate concentration is a limiting factor for nitrate removal in the water phase, and the equation includes a Monod expression for this limitation (cf. Section 2.2.1.4). The information from Section 5.6.2.1 indicates that the saturation constant for nitrate, $K_{S,NO_3}$, is in the order of 1–2.5 g $NO_3$-N $m^{-3}$.

Values for $r_{w,max}$ and $k_{1/2}$ can be estimated based on the information given in Sections 5.6.2.1 and 5.6.3, respectively. The temperature coefficients for the anoxic transformations are not well known; however, $\alpha_{w,an} = 1.07$ and $\alpha_{f,an} = 1.03$ are considered relevant estimates for the water phase and the biofilm, respectively (cf. Section 2.2.3).

The nitrate removal in a sewer pipe can be calculated based on Equation 5.25:

$$\Delta C_{NO_3} = r_{NO_3} t_r \tag{5.26}$$

where:

$\Delta C_{NO_3}$ = nitrate removal (g $NO_3$-N $m^{-3}$)

$t_r$ = anoxic residence time in the pipe (h)

### 5.6.5   CONCEPT FOR HETEROTROPHIC ANOXIC TRANSFORMATIONS

As discussed in Section 5.6.1, there are numerous shared characteristics of the electron transport pathway for anoxic and aerobic transformations. The difference is basically just the terminal electron acceptor being DO and nitrate (or oxidized, inorganic nitrogen compounds) for aerobic and anoxic processes, respectively. It is therefore not surprising that a concept for the heterotrophic, anoxic transformations is very similar to the aerobic, heterotrophic concept. Figure 5.17 is therefore—except for the electron acceptor part of the concept—rather identical with Figure 5.5.

The anoxic concept as shown in Figure 5.17 is formulated in the same terms as the corresponding aerobic concept. The formulation is based on the findings outlined in the preceding portion of Section 5.6. The concept is—as is also the case

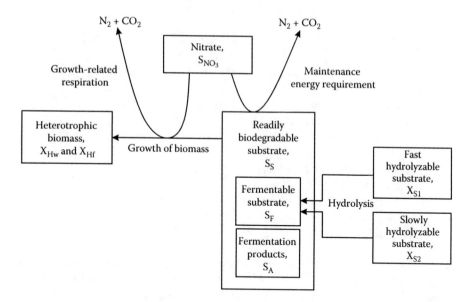

**FIGURE 5.17** A concept for heterotrophic transformations of wastewater organic matter under anoxic conditions in a sewer. Compounds are defined in Section 3.2.6.

for the aerobic concept—in several ways expressed in simple terms. As an example, the heterotrophic biomass is—in contrast to the real situation in a sewer—expressed via two parameters only, with each having uniform growth characteristics, namely, $X_{Hw}$ in the water phase and $X_{Hf}$ in the biofilm. Furthermore, nitrite is not expected to accumulate as an intermediate and is therefore not included as a "stable" electron acceptor (cf. Section 5.6.2.1).

It is important to note that although the anoxic, heterotrophic concept—as is also the case for the aerobic concept—seems simple in several ways, it does include details that in terms of the biological wastewater system have a scientifically sound foundation: biomass growth, biomass maintenance requirement of substrate, hydrolysis of substrate, and consumption of an electron acceptor. This "balance" between a simple and yet scientifically sound description is the strength of the concept. When applying the concept as the basis for sewer process modeling, relatively few process parameters are required, and such parameters can be experimentally determined (cf. Chapter 10).

A rather detailed explanation of the formulation in mathematical terms of the aerobic, heterotrophic processes in wastewater of sewers is given in Section 5.3 in terms of Equations 5.1 through 5.4. Similar mathematical expressions for the corresponding anoxic processes will be presented shortly. However, as a first approach, it is important to introduce a mathematical term that controls the switch from an aerobic process to the corresponding anoxic process when the DO concentration approaches a value close to 0. This switch term, $x_{switch}$, is formulated as follows:

$$x_{switch} = \frac{K_O}{K_O + S_O} \tag{5.27}$$

where:

$K_O$ = saturation constant for DO (g $O_2$ m$^{-3}$)
$S_O$ = DO concentration in bulk water phase (g $O_2$ m$^{-3}$)

With a typical value of $K_O$ in the order of 0.1 g $O_2$ m$^{-3}$, it is seen that $x_{switch}$ varies between 0 at high DO concentrations and 1 when the DO concentration is close to 0. The switch term selected is therefore appropriate to include in anoxic process expressions for their activation or deactivation.

The similarities between aerobic and anoxic processes means that the formulation of the aerobic processes, Equations 5.1 through 5.4, with minor corrections can be used for the corresponding anoxic processes as shown by Equations 5.28 through 5.31. All rates, $r_{index}$, are expressed in units of g COD m$^{-3}$ day$^{-1}$.

The anoxic growth rate of heterotrophic biomass in suspension:

$$r_{grw,an} = \mu_{H,NO_3} \frac{S_F + S_A}{K_{Sw} + (S_F + S_A)} \frac{S_{NO_3}}{K_{NO_3} + S_{NO_3}} \frac{K_O}{K_O + S_O} X_{Hw} \alpha_w^{(T-20)} \quad (5.28)$$

The rate of anoxic maintenance energy requirement for biomass in suspension:

$$r_{maint,an} = q_{m,NO_3} \frac{S_{NO_3}}{K_{NO_3} + S_{NO_3}} \frac{K_O}{K_O + S_O} X_{Hw} \alpha_w^{(T-20)} \quad (5.29)$$

The anoxic growth rate of heterotrophic biomass in a biofilm:

$$r_{grf,an} = k_{1/2,NO_3} S_{NO_3}^{0.5} 2.86 \frac{Y_{Hf,NO_3}}{1 - Y_{Hf,NO_3}} \frac{S_F + S_A}{K_{sf} + (S_F + S_A)} \frac{K_O}{K_O + S_O} \frac{A}{V} \alpha_f^{(T-20)} \quad (5.30)$$

The constant, 2.86, in Equation 5.30 is in units of g $O_2$ (g $NO_3$-N)$^{-1}$. It is the oxygen/nitrate ratio that accounts for the transformation from nitrate utilization to DO consumption as the corresponding electron acceptor (cf. Example 5.6). Equation 5.30 is thereby, in terms of a mass balance, directly comparable with the corresponding growth rate of an aerobic biofilm as expressed by Equation 5.3. The anoxic biofilm growth rate, $r_{grf,an}$, is therefore expressed in units of g COD m$^{-3}$ day$^{-1}$.

The rate of anoxic hydrolysis in the water phase and in the biofilm is:

$$r_{hydr,anox} = \eta_{h,anox} k_{hn} \frac{X_{Sn}/X_{Hw}}{K_{Xn} + X_{Sn}/X_{Hw}} \frac{S_{NO_3}}{K_{NO_3} + S_{NO_3}} \left( X_{Hw} + \varepsilon X_{Hf} \frac{A}{V} \right) \frac{K_O}{K_O + S_O} \alpha_w^{(T-20)}$$

$$(5.31)$$

The parameter $\eta_{h,anox}$ in Equation 5.31 is an efficiency constant for anoxic hydrolysis in comparison with aerobic hydrolysis (cf. Equation 5.4).

The parameters in Equations 5.28 through 5.31 are in terms of nomenclature either identical with the parameters already explained for Equations 5.1 through 5.4 or changed, as denoted by the index of the parameter: "$O_2$" for an aerobic parameter and "$NO_3$", "an" or "anox" for an anoxic parameter. An overview of the nomenclature for the parameters is given in Appendix A.

## REFERENCES

Abdul-Talib, S., T. Hvitved-Jacobsen, J. Vollertsen, and Z. Ujang (2002), Anoxic transformations of wastewater organic matter in sewers—process kinetics, model concept and wastewater treatment potential, *Water Sci. Technol.*, 45(3), 53–60.

Aesoey, A., M. Storfjell, L. Mellgren, H. Helness, G. Thorvaldsen, H. Oedegaard, and G. Bentzen (1997), A comparison of biofilm growth and water quality changes in sewers with anoxic and anaerobic (septic) conditions, *Water Sci. Technol.*, 36(1), 303–310.

Almeida, M. (1999), Pollutant transformation processes in sewers under aerobic dry weather flow conditions, Ph.D. thesis, Department of Civil and Environmental Engineering, Imperial College of Science, Technology and Medicine, UK, p. 422.

Bjerre, H.L., T. Hvitved-Jacobsen, B. Teichgräber, and D. te Heesen (1995), Experimental procedures characterizing transformations of wastewater organic matter in the Emscher river, Germany, *Water Sci. Technol.*, 31(7), 201–212.

Bjerre, H.L., T. Hvitved-Jacobsen, S. Schlegel, and B. Teichgräber (1998a), Biological activity of biofilm and sediment in the Emscher river, Germany, *Water Sci. Technol.*, 37(1), 9–16.

Bjerre, H.L., T. Hvitved-Jacobsen, B. Teichgräber, and S. Schlegel (1998b), Modelling of aerobic wastewater transformations under sewer conditions in the Emscher river, Germany, *Water Environ. Res.*, 70(6), 1151–1160.

Boon, A.G. and A.R. Lister (1975), Formation of sulphide in rising mains and its prevention by injection of oxygen, *Prog. Water Technol.*, 7(2), 289–300.

Buisman, C., P. Ijspeert, A. Janssen, and G. Lettinga (1990), Kinetics of chemical and biological sulphide oxidation in aqueous solutions, *Water Res.*, 24(5), 667–671.

Characklis, W.G. (1990), Kinetics of microbial transformations, in: W.G. Characklis and K.C. Marshall (eds.), *Biofilms*, John Wiley & Sons, New York, pp. 233–264.

Characklis, W.G. and K.C. Marshall (eds.) (1990), *Biofilms*, John Wiley & Sons, New York, pp. 796.

Chen, K.Y. and J.C. Morris (1972), Kinetics of oxidation of aqueous sulfide by $O_2$, *Environ. Sci. Technol.*, 6(6), 529–537.

Dold, P.L., G.A. Ekama, and G. v. R. Marais (1980), A general model for the activated sludge process, *Prog. Water Technol.*, 12, 47–77.

Einarsen, A.M., A. Aesoey, A.-I. Rasmussen, S. Bungum, and M. Sveberg (2000), Biological prevention and removal of hydrogen sulphide in sludge at Lillehammer Wastewater Treatment Plant, *Water Sci. Technol.*, 41(6), 175–182.

Gaudy, A.F. Jr. and E.T. Gaudy (1971), Biological concepts for design and operation of the activated sludge process, *Water Pollution Research Series, Rep. no. 17090*, FQJ, 09/71, US Environmental Protection Agency, Washington, DC.

Grady, C.P.L. Jr., W. Gujer, M. Henze, G. v. R. Marais, and T. Matsuo (1986), A model for single-sludge wastewater treatment systems, *Water Sci. Technol.*, 18, 47–61.

Green, M., G. Shelef, and A. Messing (1985), Using the sewerage system main conduits for biological treatment, *Water Res.*, 19(8), 1023–1028.

Gudjonsson, G., J. Vollertsen, and T. Hvitved-Jacobsen (2001), Dissolved oxygen in gravity sewers—measurement and simulation, in: *Proceedings from the 2nd International Conference on Interactions between Sewers, Treatment Plants and Receiving Waters in Urban Areas (INTERURBA II)*, Lisbon, Portugal, February 19–22, 2001, pp. 35–43.

Gujer, W. (1980), The effect of particulate organic material on activated sludge yield and oxygen requirement, *Prog. Water Technol.*, 12, 79–95.

Gujer, W. and O. Wanner (1990), Modelling mixed population biofilms, in: W.G. Characklis and K.C. Marshall (eds.), *Biofilms,* John Wiley & Sons, New York, pp. 397–443.

Henze, M., C.P.L. Grady Jr., W. Gujer, G. v. R. Marais, and T. Matsuo (1987), *Activated Sludge Model No. 1, Scientific and Technical Report No. 1,* IAWPRC (International Association on Water Pollution Research and Control), London.

Henze, M., W. Gujer, T. Mino, T. Matsuo, M.C. Wentzel, and G. v. R. Marais (1995), *Activated Sludge Model No. 2, Scientific and Technical Report No. 3,* IAWQ (International Association on Water Quality), London, p. 32.

Henze, M., W. Gujer, T. Mino, and M.v. Loosdrecht (2000), *Activated Sludge Models ASM1, ASM2, ASM2d and ASM3, Scientific and Technical Report No. 9,* IWA (International Water Association), London, p. 121.

Henze, M., P. Harremoës, J. la Cour Jansen, and E. Arvin (2002), *Wastewater Treatment— Biological and Chemical Processes*, 3rd edn, Springer-Verlag, Berlin, p. 430.

Huisman, J.L., C. Gienal, M. Kühni, P. Krebs, and W. Gujer (1999), Oxygen mass transfer and biofilm respiration rate measurement in a long sewer, evaluated with a redundant oxygen balance, in: I.B. Joliffe and J.E. Ball (eds.), *Proceedings of the 8th International Urban Storm Drainage Conference,* Sydney, Australia, August 30–September 3, 1999, vol. 1, pp. 306–314.

Huisman, J.L. and W. Gujer (2002), Modelling wastewater transformation in sewers based on ASM3, *Water Sci. Technol.,* 45(6), 51–60.

Jensen, N.A. and T. Hvitved-Jacobsen (1991), Method for measurement of reaeration in gravity sewers using radiotracers, *Res. J. WPCF,* 63(5), 758–767.

Kappeler, J. and W. Gujer (1992), Estimation of kinetic parameters of heterotrophic biomass under aerobic conditions and characterization of wastewater for activated sludge modelling, *Water Sci. Technol.,* 25(6), 125–139.

Koch, C.M. and I. Zandi (1973), Use of pipelines as aerobic biological reactors, *J. Water Pollut. Control Fed.,* 45, 2537–2548.

Kountz, R.R. and C. Forney, Jr. (1959), Metabolic energy balances in a total oxidation activated sludge system, *J. Water Pollut. Control Fed.,* 31, 819–826.

Kurland, C.G. and R. Mikkola (1993), The impact of nutritional state on the microevolution of ribosomes, in: S. Kjelleberg (ed.), *Starvation in Bacteria*, Plenum Press, New York, pp. 225–237.

Marais, G. v. R. and G.A. Ekama (1976), The activated sludge process part 1—steady state behaviour, *Water SA,* 2(4), 163–200.

Matos, J.S. (1992), Aerobiose e septicidade om sistemas de drenagem de águas residúais, Ph.D. dissertation, IST, Lisbon, Portugal.

Matos, J.S. and E.R. de Sousa (1991), Dissolved oxygen in small wastewater collection systems, *Water Sci. Technol.,* 23(10–12), 1845–1851.

Matos, J.S. and E.R. de Sousa (1996), Prediction of dissolved oxygen concentration along sanitary sewers, *Water Sci. Technol.,* 34(5–6), 525–532.

McKinney, R.E. and R.J. Ooten (1969), Concepts of complete mixing activated sludge, *Transactions of the 19th Sanitary Engineering Conference,* University of Kansas, pp. 32–59.

Millero, F.J., S. Hubinger, M. Fernandez, and S. Garnett (1987), Oxidation of $H_2S$ in seawater as a function of temperature, pH and ionic strength, *Environ. Sci. Technol.,* 21(5), 439–443.

Monod, J. (1949), The growth of bacterial cultures, *Annu. Rev. Microbiol.,* 3, 371–394.

Nielsen, A.H., J. Vollertsen, and T. Hvitved-Jacobsen (2003), Determination of kinetics and stoichiometry of chemical sulfide oxidation in wastewater of sewer networks, *Environ. Sci. Technol.,* 37(17), 3853–3858.

Nielsen, A.H., J. Vollertsen, and T. Hvitved-Jacobsen (2004), Chemical sulfide oxidation of wastewater—effects of pH and temperature, *Water Sci. Technol.*, 50(4), 185–192.

Nielsen, A.H., T. Hvitved-Jacobsen, and J. Vollertsen (2005), Kinetics and stoichiometry of sulfide oxidation by sewer biofilms, *Water Res.*, 39, 4119–4125.

Nielsen, A.H., J. Vollertsen, and T. Hvitved-Jacobsen (2006), Kinetics and stoichiometry of aerobic sulfide oxidation in wastewater from sewers—effects of pH and temperature, *Water Environ. Res.*, 78(3), 275–283.

Nielsen, A.H., J. Vollertsen, H.S. Jensen, H.I. Madsen, and T. Hvitved-Jacobsen (2008), Aerobic and anaerobic transformations of sulfide in a sewer system—field study and model simulations, *Water Environ. Res.*, 80(1), 16–25.

Nielsen, P.H., K. Raunkjaer, N.H. Norsker, N.A. Jensen, and T. Hvitved-Jacobsen (1992), Transformation of wastewater in sewer systems—a review, *Water Sci. Technol.*, 25(6), 17–31.

Nielsen, P.H., K. Raunkjaer, and T. Hvitved-Jacobsen (1998), Sulfide production and wastewater quality in pressure mains, *Water Sci. Technol.*, 37(1), 97–104.

Norsker, N.-H., P.H. Nielsen, and T. Hvitved-Jacobsen (1995), Influence of oxygen on biofilm growth and potential sulfate reduction in gravity sewer biofilm, *Water Sci. Technol.*, 31(7), 159–167.

Parkhurst, J.D. and R.D. Pomeroy (1972), Oxygen absorption in streams, *J. Sanit. Eng. Div., ASCE*, 98(SA1), 121–124.

Pomeroy, R.D. and J.D. Parkhurst (1973), Self-purification in sewers. Advances in ater pollution research, in: *Proceedings of the 6th International Conference, Jerusalem, June 18–23, 1972*, Pergamon Press, Elmsford, NY.

Poulsen, B.K. (1997), Anoxisk omsætning af organisk stof i biofiltre [Anoxic transformations of organic matter in biofilters], MSc thesis, Department of Environmental Technology, Technical University of Denmark (in Danish), p. 56.

Raunkjaer, K., T. Hvitved-Jacobsen, and P.H. Nielsen (1994), Measurement of pools of protein, carbohydrate and lipid in domestic wastewater, *Water Res.*, 28(2), 251–262.

Raunkjaer, K., T. Hvitved-Jacobsen, and P.H. Nielsen (1995), Transformation of organic matter in a gravity sewer, *Water Environ. Res.*, 67(2), 181–188.

Raunkjaer, K., P.H. Nielsen, and T. Hvitved-Jacobsen (1997), Acetate removal in sewer biofilms under aerobic conditions, *Water Res.*, 31, 2727–2736.

Russel, J.B. and G.M. Cook (1995), Energetics of bacterial growth: balance of anabolic and catabolic reactions, *Microbiol. Rev.*, 59(1), 48–62.

Sollfrank, U. and W. Gujer (1991), Characterisation of domestic wastewater for mathematical modelling of the activated sludge process, *Water Sci. Technol.*, 23(4–6), 1057–1066.

Stoyer, R.L. (1970), The pressure pipe wastewater treatment system. Presented at the 2nd Annual Sanitary Engineering Research Laboratory Workshop on Wastewater Reclamation and Reuse, Tahoe City, CA.

Stoyer, R. and J. Scherfig (1972), Wastewater treatment in a pressure pipeline, *Am. City*, 87(10), 84–93.

Tanaka, N. and T. Hvitved-Jacobsen (1998), Transformations of wastewater organic matter in sewers under changing aerobic/anaerobic conditions, *Water Sci. Technol.*, 37(1), 105–113.

Tchobanoglous, G., F.L. Burton, and H.D. Stensel (2003), *Wastewater Engineering—Treatment and Reuse*, 4th edn, McGraw-Hill, New York, p. 1819.

Tempest, D.W. and O.M. Neijssel (1984), The status of YATP and maintenance energy as biological interpretable phenomena, *Annu. Rev. Microbiol.*, 38, 459–486.

USEPA (1985), *Odor and Corrosion Control in Sanitary Sewerage Systems and Treatment Plants*, US Environmental Protection Agency, EPA 625/1-85/018, Washington, DC.

Vollertsen, J. and T. Hvitved-Jacobsen (1998), Aerobic microbial transformations of resuspended sediments in combined sewers—a conceptual model, *Water Sci. Technol.*, 37(1), 69–76.

Vollertsen, J. and T. Hvitved-Jacobsen (1999), Stoichiometric and kinetic model parameters for microbial transformations of suspended solids in combined sewer systems, *Water Res.*, 33(14), 3127–3141.

Vollertsen, J., A. Jahn, J.L. Nielsen, T. Hvitved-Jacobsen, and P.H. Nielsen (2001), Determination of microbial biomass in wastewater, *Water Res.*, 35(7), 1649–1658.

Wilmot, P.D., K. Cadee, J.J. Katinic, and B.V. Kavanagh (1988), Kinetics of sulfide oxidation by dissolved oxygen, *J. Water Pollut. Control Fed.*, 60(7), 1264–1270.

Yang, W., J. Vollertsen, and T. Hvitved-Jacobsen (2004), Anoxic control of odour and corrosion from sewer networks, *Water Sci. Technol.*, 50(4), 341–349.

Yang, W., J. Vollertsen, and T. Hvitved-Jacobsen (2005), Anoxic sulfide oxidation in wastewater of sewer networks, *Water Sci. Technol.*, 52(3), 191–199.

Yuan, Z., K.R. Sharma, O. Gutierrez, R. Rootsey, and J. Keller (2010), Corrosion and odour management in sewers: recent advances and key knowledge gaps, 6th International Conference on Sewer Processes and Networks, November 7–10, 2010, Gold Coast, Australia.

# 6 Anaerobic Sewer Processes

## Hydrogen Sulfide and Organic Matter Transformations

Wastewater is a matrix with potential impacts and risks in terms of waterborne diseases and deterioration of the environment (cf. Sections 1.2.5 and 1.2.6). Careful management of wastewater is needed when collecting and conveying it to a location for "detoxification"—named as "treatment"—before its discharge to the adjacent environment can be considered feasible. Details of impacts on humans such as diseases caused by pathogenic organisms and environmental effects caused by wastewater substances are not discussed in this book. The impacts and control of the different compounds in terms of chemical, physicochemical, and biological transformations onto the "sewer environment" are, however, central objectives.

Without comparison, it is the anaerobic conditions (wastewater septicity) in collection systems that cause the majority of process-related problems for the sewer network stakeholders, the sewer personnel, and the public. Anaerobic microbial-induced transformations of wastewater occur in general when dissolved oxygen (DO) is absent. The problems are primarily associated with the formation of hydrogen sulfide and volatile organic compounds (VOCs). The corresponding problems appear as concrete and metal corrosion degrading the sewer network, health-related impacts on the sewer personnel, and malodors. Such in-sewer process-related problems have been known for centuries (cf. Section 1.2). However, it was not until the beginning of the twentieth century that the problems were identified in terms of a cause–effect relationship (Olmsted and Hamlin 1900). A deliberate management of the problems based on a sound scientific and technical knowledge started about 80 years ago (Bowlus and Banta 1932; Parker 1945a, 1945b; Pomeroy and Bowlus 1946). The concept of today's sewer process management has its starting point in these studies.

Although the "sulfide problem" generally is considered superior to other problems associated with anaerobic conditions in sewers, the production and occurrence of malodors in terms of VOCs are, in principle, equally important. Furthermore, it is important that the treatment of wastewater is also affected by the anaerobic production of low molecular organics, i.e., readily biodegradable organics (Tanaka and Hvitved-Jacobsen 1998, 1999). Microbiologically active organic matter (readily biodegradable and fast hydrolyzable substrate) is important as substrate in wastewater

215

treatment plants designed for N and P removal by enhancing denitrification and biological phosphorous removal. On the other hand, such substrates may turn out to be
a problem in mechanical wastewater treatment plants.

Anaerobic sewer processes—related to both the sulfur and the carbon cycles—
are, therefore, central to this discussion. The interaction between these processes and
the aerobic transformations of wastewater are furthermore important in the performance of urban wastewater systems.

## 6.1  HYDROGEN SULFIDE IN SEWERS: A WORLDWIDE OCCURRING PROBLEM

In Section 1.2, we briefly described how several historical facts have indicated that
human excreta—although largely unrealized—have caused problems related to both
human health and deterioration of the environment. Hydrogen sulfide problems have
also been known as phenomena. However, they have been rather lately deliberately
identified in terms of their cause–impact relations. Example 6.1 describes two small sto-
ries on how things may develop when such process aspects are not taken into account.

### Example 6.1: Sulfide Problems in Sewer Networks: Examples from Denmark and Portugal

It is interesting to note that sulfide problems related to conveyance of wastewater
in sewers were not generally considered important in Denmark—a country with
a temperate climate—until the mid-1980s. Traditionally, Danish gravity sewers—and, in particular, the combined sewer networks—were designed to have a
relatively low dry-weather flow compared with their capacity. Under such conditions, the reaeration is normally sufficient to prevent the negative effects of sulfide,
although its formation may take place within the sewer slimes.

The establishment of the centralized wastewater treatment concept in Denmark
during the 1980s with few relatively big treatment plants changed this situation
drastically. In particular, several pressure mains were required to transport the
wastewater over long distances. Pumping stations and concrete gravity sewer
networks located downstream from pressure mains with anaerobic conditions of
the wastewater were in several municipalities often substantially degraded after
2 to 4 years of operation owing to sulfide formation. As a result, the first investigation in Denmark on sulfide formation and its prediction was initiated by the
National Environmental Protection Agency as late as in 1986 (Miljøstyrelsen 1988).
Recent investigations have furthermore shown that considerable sulfide formation
also takes place in pressure mains during winter periods with wastewater temperatures of about 6°C to 8°C (Nielsen et al. 1998). Furthermore, the negative impact
of sulfide on floc stability in activated sludge treatment plants was documented
(Nielsen and Keiding 1998).

In Portugal, a country in Southern Europe, temperatures of about 25°C are
common in wastewater of sewer networks during summer. Since the beginning
of the 1960s, the collapse of concrete and fiber–cement gravity sewers located
downstream pressurized sewer systems have been recognized. These sulfide-
initiated problems particularly arose in relatively long sanitary sewer networks in
the coastal zone, where trunk sewers were constructed to avoid contamination

of the bathing waters. After just 6 to 8 years of operation, the general observation was that these gravity sewers were no longer functioning properly. The gravity sewers were gradually replaced by rising mains with air injection for sulfide control.

These two examples illustrate the climatic widespread occurrence and the complexity of sulfide formation in sewers. It is evident that detailed process knowledge and models for prediction of sulfide formation that can be used in the design phase of the sewers are sorely needed. Efficient sulfide control measures are also required.

## 6.2 OVERVIEW OF BASIC KNOWLEDGE ON SULFUR-RELATED PROCESSES

Because of the importance of hydrogen sulfide in sewer networks, basic knowledge on sulfur-related processes is described in the preceding chapters of this book. In several cases, sulfur is selected as an appropriate and relevant compound to illustrate the general phenomena and processes occurring in sewers. It is considered relevant in this chapter, where hydrogen sulfide plays a dominating role, to provide the reader with a brief overview on where these sulfur-related subjects can be found in this book (see Table 6.1). It should be noted that only sulfur-related information is referred to.

**TABLE 6.1**
**Overview of Sulfur-Related Information Given in Preceding Chapters**

| Location in the Text | Subject |
|---|---|
| Section 1.2 | A brief historical overview |
| Section 2.1.3.3 | $E$–pH diagram for sulfur and $C$–pH diagram for hydrogen sulfide |
| Example 2.6 | Sulfate reduction (anaerobic half-reaction) |
| Section 2.4.1 | Speciation of iron and sulfide |
| Section 2.4.2 | Sulfide control by addition of iron salts |
| Section 3.2.2 | Anaerobic microbial processes |
| Section 4.1.3 | Dissociation of hydrogen sulfide |
| Example 4.4 | Emission constant for hydrogen sulfide |
| Figure 4.7 | The sulfur cycle in sewers |
| Section 4.3.5 | Odor and health-related problems of sulfide |
| Section 4.4.3 | Air–water mass transfer of hydrogen sulfide |
| Figure 5.6 | The integrated aerobic sulfur and carbon cycle in a sewer |
| Section 5.5 | Aerobic sulfide oxidation |
| Section 5.6.2.2 | Autotrophic anoxic sulfide oxidation |

## 6.3 INTRODUCTION TO HYDROGEN SULFIDE IN SEWER NETWORKS

The sulfur cycle in sewers is relevant under both anaerobic and aerobic conditions (cf. Figures 4.7 and 5.6). In a sewer system with changing aerobic and anaerobic conditions, the corresponding sulfur cycle is therefore a particularly interesting—albeit rather complex—topic. The sulfur cycle becomes complex because the processes proceed in a multiphase system: the biofilm, the sewer deposits, the water phase, and at the air–solid surfaces.

The release of odorous compounds in terms of water–air mass transfer is dealt with in Chapter 4, and the behavior of sulfur (hydrogen sulfide) is, in this respect, exemplified. Figures 4.7 and 5.6 not only illustrate the release phenomena but also provide an overall view of the pathways and sinks of sulfur compounds in a sewer.

### 6.3.1 BASIC PRINCIPLES OF SULFUR CYCLE IN SEWERS

Details of the sulfur cycle shown in Figure 4.7 are depicted in Figure 6.1, showing the biological transformations related to the sulfur cycle. It is important to note that Figure 6.1 includes both aerobic (oxidation of sulfur) and anaerobic (reduction of sulfur) processes. Biomass growth and degradation are mainly relevant in an aerobic environment. The anaerobic, sulfate-reducing bacteria (principally *Desulfovibrio* and *Desulfotomaculum*) are slow growing and therefore subject to being washed out of the sewer system if they occur in the water phase. However, in the biofilm (sewer slimes) and sewer sediments (deposits), they may be retained. As a consequence, sulfate reduction primarily takes place in the biofilm and the sediments. However, detached (anaerobic) biofilm may result in a minor sulfide production in the wastewater phase, typically less than 10% of the total amount. The importance of

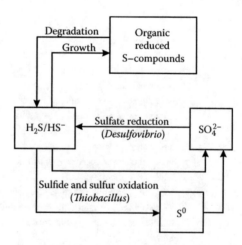

**FIGURE 6.1** Overview of biological sulfur cycle related to both aerobic and anaerobic processes in a sewer.

the biofilm for sulfide production in anaerobic pressure mains is evident. In gravity sewers, it is supported by the fact that the deep part of a rather thick biofilm is often permanently anaerobic, whereas the conditions in the water phase may vary between aerobic and anaerobic depending on the extent of reaeration and the sulfate reduction rate (cf. Figure 5.14).

The aerobic and anoxic oxidation of sulfide to mainly elemental sulfur ($S^0$) and sulfate ($SO_4^{2-}$) is detailed in Sections 5.5 and 5.6.2.2, respectively. If sulfide is produced in the deep part of a biofilm in a gravity sewer, it can be oxidized in an aerobic upper layer of the biofilm or in the water phase (cf. Figure 5.14). Although the final step of this oxidation process is sulfate, elemental sulfur is typically relatively stable and often governing the stoichiometry and kinetics of the transformation. Hydrogen sulfide emitted to the sewer atmosphere and following oxidized at the sewer walls with a potential effect in terms of concrete corrosion is dealt with in Section 6.5.

In Figure 6.1, it is readily seen that there are basically two types of processes for production of sulfide: sulfate reduction and degradation of sulfur-containing organic matter. Quantitatively, only sulfate reduction is important. However, it should be noticed that degradation of sulfur-containing organic matter (certain types of proteins) under anaerobic conditions results in formation of odorous volatile sulfur compounds, for example, mercaptans (cf. Table 4.1). Further details on the biological sulfur cycle are found in Brüser et al. (2000).

Referring to the interaction between aerobic and anaerobic processes in sewers, it is interesting to note that DO consumption proceeds by oxidation of reduced compounds. Several of these reduced compounds, for example, sulfide and low molecular organics, are produced under anaerobic conditions.

In general, the yield of biomass production for anaerobic microbial processes including fermentation is relatively low, typically with a yield constant, $Y$, in the order of 0.05–0.1 g COD g COD$^{-1}$ (cf. Equations 6.19 and 6.20). The result of Example 3.1 that deals with the aerobic biomass production based on the total DO consumption rate of a biofilm may therefore overestimate the biomass production.

Models and procedures for the estimation of sulfide formation in sewer networks are a major subject of this chapter (cf. Section 6.4.3 and Chapters 8 and 9). A very simple illustration of how to predict sulfide formation is shown in Example 6.2.

### Example 6.2: Hydrogen Sulfide Formation in a Pressure Main

Hydrogen sulfide is produced in the biofilm of a pressure main with a constant flux rate, $r_a = 0.10$ g S m$^{-2}$ h$^{-1}$. No significant production takes place in the water phase. The pipe length is 4000 m, and the volumetric flow rate of the wastewater is $Q = 100$ m$^3$ h$^{-1}$. The DO concentration and the sulfide concentration at the inlet to the pipe are both 0.

Two different scenarios for transport of the wastewater will be compared with respect to the sulfide concentration in the wastewater phase at the outlet of the pipe. Transport will take place in a pipe (pipe #1) with a diameter $D_1 = 0.3$ m and in a pipe (pipe #2) with a diameter $D_2 = 0.4$ m. Pipe #1 results in the shortest residence time. However, it has a higher biofilm area/volume ratio than pipe #2 and, therefore, a higher volumetric production rate for sulfide.

The residence time in each of the two pipes is as follows:

$$\text{Pipe \#1:} \quad t_1 = \frac{\frac{\pi}{4} \cdot D_1^2 \cdot L}{Q} = \frac{\frac{\pi}{4} \cdot 0.3^2 \cdot 4000}{100} = 2.83 \text{ h}$$

$$\text{Pipe \#2:} \quad t_2 = \frac{\frac{\pi}{4} \cdot D_2^2 \cdot L}{Q} = \frac{\frac{\pi}{4} \cdot 0.4^2 \cdot 4000}{100} = 5.03 \text{ h}$$

The hydraulic radius is:
$$R = \frac{V}{A} = \frac{\frac{\pi}{4}D^2}{\pi \cdot D} = \frac{D}{4}$$

The volumetric rate of hydrogen sulfide formation in each of the two pipes multiplied with the residence time results in the hydrogen sulfide concentration in the wastewater at the outlet:

$$\text{Pipe \#1:} \quad c_1 = \frac{r_{a,1}}{R_1} t_1 = \frac{0.10}{0.3/4} 2.83 = 3.8 \text{ g S m}^{-3}$$

$$\text{Pipe \#2:} \quad c_2 = \frac{r_{a,2}}{R_2} t_2 = \frac{0.10}{0.4/4} 5.03 = 5.0 \text{ g S m}^{-3}$$

The example shows that residence time (flow rate conditions) and pipe diameter (system related conditions) affect the resulting sulfide concentration in the wastewater of a force main. Although this impact has opposite directions, a fast transport rate results in the lowest sulfide concentration.

### 6.3.2 Basic Aspects and Stoichiometry of Hydrogen Sulfide Formation

Organic sulfur compounds such as certain proteins and amino acids will produce hydrogen sulfide when degraded (cf. Section 6.3.1 and Figure 6.1). The following hydrolysis of cysteine, an amino acid, exemplifies this process:

$$CH_2(SH)CH(NH_2)COOH + H_2O \rightarrow CH_3COCOOH + NH_3 + H_2S \quad (6.1)$$
$$\text{Cysteine}$$

As briefly mentioned in Section 6.3.1, the degradation of organic sulfur compounds in wastewater as exemplified by Equation 6.1 can normally be disregarded as a source of sulfides. The main process for formation of sulfide is the sulfate respiration by sulfate-reducing heterotrophic bacteria such as *Desulfovibrio* and *Desulfotomaculum* (cf. Figure 6.1). Schematically, and without taking into account the formation of low molecular organics as shown in Table 3.3, the sulfide-producing heterotrophic respiration process is briefly as follows:

$$SO_4^{2-} + \text{organic carbon} \rightarrow HCO_3^- \text{ (carbon dioxide)} + H_2S \quad (6.2)$$

Sulfate is the electron acceptor in Equation 6.2, and the reduction process is as follows (cf. Example 2.6):

$$\frac{1}{8}SO_4^{2-} + \frac{5}{4}H^+ + e^- \rightarrow \frac{1}{8}H_2S + \frac{1}{2}H_2O \qquad (6.3)$$

To establish the stoichiometry of sulfide formation, Equation 6.3 must be combined with the oxidation process for the organic matter that is the electron donor for the heterotrophic sulfate-reducing bacteria. The procedure for the combination of the oxidation and the reduction process steps is the same as shown in Section 2.1.4. If organic matter is considered simply as $CH_2O$, the combination of the oxidation process as depicted in Example 2.3 and the reduction reaction for sulfate shown in Equation 6.3 result in the following redox process:

$$SO_4^{2-} + 2CH_2O + 2H^+ \rightarrow 2H_2O + 2CO_2 + H_2S \qquad (6.4)$$

This equation is the stoichiometrically detailed version of Equation 6.2 with $CH_2O$ as an example of the composition of organic matter.

Equation 6.4 shows that, for 1 mol of $H_2S$ produced, i.e., 32 g $H_2S$-S, 2 mol of organic carbon—considering organic matter as $CH_2O$—are oxidized. From an electron transfer point of view, this corresponds with the fact that $\Delta OX_C = 4$ for oxidation of carbon in $CH_2O$, and $\Delta OX_S = 8$ for reduction of sulfur in $SO_4^{-2}$ (cf. Section 2.1.4).

If organic matter cannot be considered as simple as $CH_2O$, another formula should be used to establish the stoichiometry. For an average quality of organic matter in wastewater, the oxidation level for carbon, $OX_C$, is normally low. As an example, COD (chemical oxygen demand) and TOC (total organic carbon) for medium strength wastewater are 500 g $O_2$ m$^{-3}$ and 150 g C m$^{-3}$, respectively (cf. Table 3.2). According to Equation 2.8, the average oxidation level of carbon in this type of wastewater is $OX_C = 4 - 1.5 \ 500/150 = -1$.

### 6.3.3 CONDITIONS AFFECTING FORMATION AND BUILDUP OF SULFIDE

Anaerobic conditions, i.e., the absence of DO and nitrate or other oxidized inorganic nitrogen compounds, are crucial for the formation of sulfide. Although sulfate reduction requires strict anaerobic conditions, it is important that sulfide can be transported to and temporarily can also exist at those parts of the sewer where the DO concentration is not zero. As an example, sulfide is produced in the inner part of a sewer biofilm and from there transported to an upper aerobic environment or into the bulk water phase (cf. Figure 5.14).

In principle, anaerobic conditions occur in full-flowing gravity sewers and rising mains. In cases where aerobic wastewater, typically with moderate DO concentration levels, flows into such sewers, the DO concentration is typically fast depleted, often after 5–30 min, depending on the level of the DO concentration and the aerobic respiration rate of the wastewater. Although sulfide problems in sewer networks are particularly widespread, occurring in climates with high temperatures, it may also exist in pressure mains during winter under temperate climate conditions, i.e., at wastewater temperatures around 5°C to 12°C (Hvitved-Jacobsen et al. 1995; Nielsen et al. 1998). At such low temperatures, the sulfide production rate is generally low, and the anaerobic residence time should typically exceed 0.5 to 2 h before sulfide production becomes significant.

Sulfide production in gravity sewers typically takes place in slow-flowing water ($<0.3$ m s$^{-1}$), large diameter pipes with insufficient reaeration, and at relatively high temperatures ($>15–20°C$). The DO mass balance as dealt with in Section 5.4 is crucial for the net formation of sulfide (in the biofilm) and for the occurrence of sulfide in the wastewater phase. The permanent presence of sulfide in bulk water of gravity sewers in Northern Europe is not common, owing to a moderate temperature even during summer periods. However, as seen in Figures 5.10 through 5.13, changing aerobic and anaerobic conditions may occur. In warm climates, for example, Central and Southern Europe, the southern United States, Africa, and Australia, sulfide problems under gravity sewer conditions are common (Meyer and Hall 1979; ASCE 1989).

It is not possible to elaborate simple and general valid criteria that indicate the extent of sulfide-related problems although certain parameters may give some sign. According to USEPA (1974), sulfide is usually not present in the bulk water phase of a gravity sewer at DO concentrations $>1$ g O$_2$ m$^{-3}$. Furthermore, experience and literature seem to indicate that if the DO concentration exceeds $0.2–0.5$ g O$_2$ m$^{-3}$, sulfide problems are not typically occurring. When sulfide problems are identified, sulfide concentration in the bulk water is a simple indicator for the corresponding problems (in networks with a free water surface). Sulfide concentrations of 0.5, 3, and 10 g S m$^{-3}$ are considered as low, moderate and high, respectively, in terms of the problems typically reported. In countries with long pressure mains or high temperature of the wastewater in gravity sewers, significantly higher concentrations than 10 g S m$^{-3}$ are reported (Thistlethwayte 1972; Pomeroy and Parkhurst 1977).

The occurrence of sulfide and corresponding problems in a sewer network relates to a complex balance between several wastewater quality parameters. Furthermore, the extent of the problem depends on the design of the sewer system and the local climate. The level of the DO concentration in the bulk water phase, the magnitude of the relevant microbial process rates, and interfacial exchange rates for oxygen and sulfide outline, in general, the nature of the central factors. It is crucial to "organize" such factors in a framework that can be applied when predicting the sulfide problem. In the following, seven main quality and system factors that influence the occurrence of sulfide will be briefly discussed.

### 6.3.3.1 Sulfate

Sulfate is typically found in all types of municipal wastewaters in concentrations larger than 5–15 g S m$^{-3}$. Such concentration levels are not limiting sulfide formation in relatively thin biofilms due to full penetration (cf. Figure 2.11) (Nielsen and Hvitved-Jacobsen 1988). In sewer sediments, however, where sulfate may penetrate into deeper sediment layers, the potential for sulfate reduction may increase with increasing sulfate concentration in the bulk water phase. Under specific conditions, for example, industrial wastewater, it is important that oxidized sulfur compounds, for example, thiosulfate and sulfite, in addition to sulfate may act as sulfur sources for sulfate-reducing bacteria (Nielsen 1991). In case of intensive sulfide formation, it is of course possible that sulfate has been depleted below the actual concentration for nonlimited growth.

### 6.3.3.2 Quality and Quantity of Biodegradable Organic Matter

Biodegradable organic matter is needed as substrate for biomass growth and as electron donor for the sulfate-reducing bacteria. Wastewater from, for example, food industries, with typically relatively high concentrations of readily biodegradable organics are good substrates for the sulfate-reducing bacteria, and the sulfate reduction rate can be higher than in wastewater from households. However, also in high strength domestic wastewater, readily biodegradable and fast hydrolyzable COD may lead to a high potential for sulfide formation. Several specific organics, for example, formate, lactate, and ethanol, have been identified as particularly suitable substrates for sulfate-reducing bacteria (cf. Table 3.3) (Nielsen and Hvitved-Jacobsen 1988).

### 6.3.3.3 Temperature

The temperature dependency of the sulfate reduction rate for single sulfate-reducing bacteria is high, corresponding to a temperature coefficient, $\alpha$, of about 1.13, which corresponds to a change in the rate with a factor $Q_{10} = 3.4$ per 10°C of temperature increase (cf. Section 2.2.3). Because diffusion of substrate into biofilms or sediments is typically limiting the sulfide formation rate, the temperature coefficient is reduced to about 1.03 (Nielsen et al. 1998). The development of sulfate-reducing species that are adapted to low and to high temperatures, respectively, may result in an apparent reduced temperature dependency between winter and summer. This long-term temperature dependency may have resulted in the rather minor impact of temperature on the sulfide production rates observed by Nielsen et al. (1998) in pressure mains at wastewater temperatures between 5°C and 12°C.

It is furthermore interesting to note that the sulfate reduction rate observed in studies of sewer systems in Japan and in the United Arab Emirates have been found relatively high at "normal" wastewater temperatures (Tanaka et al. 2000a; Vollertsen et al. 2011a). The reason for that is not known but may be attributable to sulfate-reducing bacteria having specific characteristics, also in terms of their temperature-related activity.

### 6.3.3.4 pH

Sulfate-reducing bacteria mainly exist between pH 6 and 9. A significant inhibition of these bacteria will, however, not take place below a pH value of about 10. The relative distribution between the two sulfide components, $H_2S$ and $HS^-$, with an oxidation level of $-2$, depends on the pH. This aspect and the importance hereof in terms of the air–water mass transfer characteristics are detailed described in Chapter 4 (cf. the overview in Table 6.1). It is furthermore important that the oxidation kinetics of sulfide is also pH-dependent (cf. Section 5.5).

### 6.3.3.5 Area/Volume Ratio of Sewer Pipes

Sulfide is primarily produced in the biofilm—and in the sewer sediment if it permanently occurs—and the corresponding water phase concentration of sulfide is caused by diffusion of sulfide from the biofilm. The observed water phase concentration is therefore related to the biofilm area/volume ($A/V$) ratio. In a pressure main and a full-flowing gravity sewer, the governing parameters correspond to the surface area and the volume of the pipe. In a partly flowing gravity sewer pipe, $A$ and $V$ are the wetted pipe surface area and the bulk water volume, respectively. As an example and

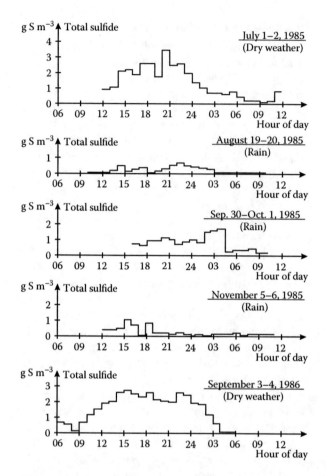

**FIGURE 6.2** Diurnal variation of sulfide concentration of a 4-km pressure main located in North Jutland, Denmark. Typical variation of residence time during a dry weather period is between 6 and 14 h, corresponding to daily wastewater flow. During a rain event, residence time can be as low as about 1 h.

considering unchanged residence time, small-diameter pressure pipes having an $A/V$ ratio larger than bigger diameter pipes, typically result in higher sulfide concentrations in the water phase (cf. Example 6.2).

### 6.3.3.6 Flow Velocity

The potential production of sulfide depends on the anaerobic thickness of the biofilm. If the flow velocity in a pressure main frequently exceeds 0.8–1 m s$^{-1}$, the corresponding biofilm is relatively smooth and thin, typically 100–300 μm. However, high flow velocities also reduce the diffusion layer at the biofilm–water interface and thereby the resistance against transport of substrates and products. In total, a high flow velocity will normally—considering constant residence time—reduce the potential for sulfide formation in pressure mains. Furthermore, the flow conditions in gravity sewers affect the air–water exchange processes, for example, the emission of hydrogen sulfide and the reaeration (cf. Chapter 4).

### 6.3.3.7  Anaerobic Residence Time

The anaerobic residence time of the wastewater during its transport is a factor that affects the time effective for sulfide formation and thereby the level of sulfide concentration. The residence time in a specific pressure pipe is determined by the magnitude of wastewater inflow compared with the volume (length and diameter) of the pipe. The level of sulfide formation in a given pipe is therefore typically subject to the diurnal variation of the inflowing wastewater and to the precipitation pattern in combined sewer catchments (cf. Figure 6.2).

## 6.4  PREDICTING MODELS FOR SULFIDE FORMATION

Traditionally, the prediction of sulfide formation in both force mains and gravity sewers is based on relatively simple empirical models including various fundamental relations as summarized in Section 6.3.3. These models have gradually been developed and correspondingly improved. In spite of the simplicity—or maybe therefore—these models are still used, in particular as a first attempt of predicting a sulfide formation level for single and simple constructed pipe sections. For the more complex constructed and large sewer networks, and when modeling impacts of and control options for sulfide, these simple models cannot be used as predicting tools and must be replaced by conceptually formulated models. This complex and extended approach of modeling sewer processes is in detail dealt with in Chapters 8 and 9.

In Sections 6.4.1–6.4.3, the different prediction approaches and simple models for sulfide formation will be discussed in a historical perspective as a convenient framework for their origin and applicability.

### 6.4.1  SULFIDE AS A SEWER PHENOMENON: 1900–1940

It is likely that human wastes collected in sewers have always caused a nuisance. We know that the systems built 200 to 300 years ago accumulated extensive amounts of deposits that must have established conditions for sulfate reduction (Bechmann 1900; Bertrand-Krajewski 2003). However, the evidence of sulfide-related problems is a rather late discovery. The first details of sulfide-based concrete corrosion were probably reported for an outfall sewer that was built in Los Angeles, CA, in 1895 (Olmsted and Hamlin 1900). The authors observed that the mortar used in the joints of the brickwork swelled during its disintegration to 2 to 3 times its original bulk. The decomposed cement was chemically analyzed and comparison with the original composition of Portland cement showed that 80% of the original amount of lime was lost and "that the balance has been changed to sulphates." The authors also noticed "this fact indicates that the corrosive or destructive agent is sulphuric acid." However, they refer the problem to "the usual gases given off by sewage during putrefaction, viz., nitrogen, carbureted hydrogen, light carbureted hydrogen, carbonic acid and carbonic oxide." It is interesting that although Olmsted and Hamlin

(1900) give several precise and detailed descriptions of the corrosion process, they never mentioned the word "hydrogen sulfide" in their report.

A similar observation was done in Madras, India (James 1917). In 1902, the Madras Municipality decided to use cement pipes up to 9 in. in diameter. About 5 years later, these pipes were discovered to be in a "very curious condition: that portion of the pipes which were generally covered by sewage seemed to have stood satisfactorily, but the part exposed to sewer gas had corroded to the extent of collapse." James (1917) refers to Colonel J. van Guyzel, Chemical Engineer to Government, giving the following explanation based on an analysis (showing mainly silica and calcium sulfate) of the soft and easily removable substance in the corroded parts of the pipes: "The calcium silicate is one of the important ingredients of cement. When the sulphuretted hydrogen, which is present in the sewage, comes in contact with the portion of the cement pipe, which is not covered by the sewage, it decomposes the calcium silicate and forms calcium sulphide and free silica. The calcium sulphide again in the presence of moisture gets into calcium sulphate and in the process deposits a small quantity of free sulphur. This reaction accounts for the presence of the soft substance, the chemical composition of which I have already given." We are hereby approaching an understanding of the sewer corrosion phenomenon, but it is evident that there is still a long way to go until it is correctly described in quantitative terms.

Further steps in the understanding of the role of sulfide in sewers in the following period took place in California. The Los Angeles County Sanitation Districts started in 1928 a 2-year construction program for wastewater treatment that included more than 350 km of trunk sewers. High temperatures of the wastewater (20°C–25°C) and high strength industrial wastes promptly initiated generation of hydrogen sulfide. In 1931, the Districts' staff started a systematic survey of sulfides throughout the trunk sewer system. Focus in these investigations was on methods for sulfide control and chlorination was found to be effective (Bowlus and Banta 1932). Sulfide formation was considered associated with sewer deposits and the design criterion for sewer pipes to avoid sulfide buildup was basically defined in terms of a required minimum flow velocity about 0.6 m s$^{-1}$. Although the argument for the criterion was not (fully) correct, the "cure" was—from today's standpoint—not bad.

A number of other investigations, for example, on aeration and nitrate addition to control sulfide in sewer systems, were performed during the 1930s and thereby established an understanding that led to a breakthrough in the following decade. It is crucial that developing predicting tools for sulfide buildup in sewers requires a basic microbiological and chemical knowledge on sulfur transformations. This knowledge is combined with information on the construction and performance of the sewer network. Furthermore, the development of the methylene blue method for reliable and fast sulfide analysis in wastewater also contributed to the subsequent progress of understanding (Pomeroy 1936).

### 6.4.2 Toward a New Understanding of Sulfide in Sewers: 1940–1945

Today's understanding of sulfide formation in sewers is based on a number of investigations that were conducted during World War II and shortly thereafter. The combined scientific and technological understanding gained through these studies was crucial for the development of the corresponding predictive tools. A major part of these studies took place in California, but also in Australia and South Africa. In particular, the innovative works by C.D. Parker, R.D. Pomeroy, and F.D. Bowlus contributed to the development toward a new and solid understanding of the nature of sulfide formation and corrosion in sewers (Parker 1945a; Parker 1945b; Pomeroy and Bowlus 1946; Stutterheim and van Aardt 1953). Pomeroy and Bowlus (1946) clearly stated the basic concept and the conditions for sulfide formation, for example, in terms of its association with the sewer biofilm and the importance of temperature, pH, organic matter, and sulfate.

Pomeroy and Bowlus (1946) also formulated the first attempt to include the dynamics of sulfide formation in terms of biofilm scouring and reaeration by introducing the flow velocity as a central parameter: "Hence, velocity is a major factor in determining whether sulfide subtractions from the stream by oxidation and evolution will be able to keep up with sulfide additions by the slimes." Based on their work, Pomeroy and Bowlus (1946) formulated the importance of the velocity for sulfide formation as shown in Table 6.2. If the actual velocity relative to the concentration of biodegradable organic matter is below a certain value as indicated in Table 6.2, sulfide buildup can be expected. It is interesting that although they basically have all the pieces for the puzzle, they never formulated their knowledge directly in terms of a mathematical model for sulfide formation. However, they realized that "the limiting velocity is determined by both temperature and sewage strength." Therefore, they combined these conditions in a single factor, EBOD, defined as "effective BOD":

$$EBOD = BOD_5 \, 1.07^{T-20} \qquad (6.5)$$

where:

$BOD_5$ = 5-day biological oxygen demand ($g \, O_2 \, m^{-3}$)
$T$ = temperature (°C)

---

**TABLE 6.2**
**Velocity Required to Prevent Sulfide Buildup in a Gravity Sewer Pipe Flowing Less than Half-Full**

| EBOD ($g \, O_2 \, m^{-3}$) | Velocity, $u$ (ft $s^{-1}$)[a] |
|---|---|
| 55 | 1.0 |
| 125 | 1.5 |
| 225 | 2.0 |
| 350 | 2.5 |
| 500 | 3.0 |
| 690 | 3.5 |
| 900 | 4.0 |

Source:  Pomeroy, R.D., Bowlus, F.D., *J. Sewage Works*, 18(4), 1946. With permission.

[a]  1 ft $s^{-1}$ = 0.305 m $s^{-1}$.

The values given in Table 6.2 basically imply that the minimum velocity ($u$) to prevent sulfide formation in a gravity sewer pipe is related to the EBOD value according to the equation:

$$EBOD = 55u^2 \tag{6.6}$$

With the knowledge we have today, it is easy to give several comments on the limitations and conditions associated with such a simple correlation. However, the fundamental dynamics related to sulfide formation is understood and taken into account.

In addition to the work by Pomeroy and Bowlus, it is important to acknowledge that C.D. Parker's work contributed significantly to the understanding of the microbiology of sewers (Parker 1945a, 1945b). In particular, his work on the concrete corrosion process is unique.

### 6.4.3 EMPIRICAL SULFIDE PREDICTION AND EFFECT MODELS: 1945–1995

During the 50-year span from 1945 to 1995, a great number of different tools for prediction of sulfide formation and associated effects in sewer networks were developed. Extensive studies were in several cases conducted as the experimental basis. These tools can all be grouped under the term "empirical models" or, more simply, "empirical equations" (cf. Section 8.1.2). Compared with a conceptual model characterized by rather detailed mathematical formulations of the governing physical, chemical, and biological phenomena and processes, an empirical model is briefly the outcome of a statistical analysis of well-designed experiments. Empirical models are therefore in general not valid outside their "area of definition," determined by the conditions under which the experiments were conducted.

As a first estimate, the empirical models that were developed during the period 1945 through 1995 can be grouped in three main types based on their "degree of complexity" corresponding to their intended use:

1. Type I sulfide prediction models
   Rather simple formulated models valid for gravity sewers. Type I models can be characterized as "risk models."
2. Type II sulfide prediction models
   Models that predict sulfide in the water phase of both pressure mains and gravity sewers. These models typically include three to six central parameters characterizing the nature of the sewer system and the most important processes relevant for the occurrence of sulfide.
3. Type III sulfide prediction models
   These models predict the sulfur cycle in gravity sewers. This type of model tends to include a number of equations for both formation and sink of sulfide in the water and biofilm phases and principally also in the sewer gas phase including sulfide removal at the concrete surfaces. These empirical models are designed and applied in terms of a series of consecutive steps for calculation of sulfide formation and removal.

The overall conclusion is that Type I and Type II models were the most successful, particularly those of Type II that have been developed and used in different versions. This statement is based on several reports from projects carried out worldwide under quite different external conditions.

Relatively few empirical models belong to the Type III model group. An example is a series of models developed by Thistlethwaite (1972) for stepwise calculation of sulfide occurrence in the water phase taking into account different sinks (sulfide oxidation, emission, precipitation, and gas-phase absorption), and finally followed by an estimated pipe corrosion rate (cf. Figure 4.7). The different steps included in the calculations are basically relevant, but are associated with problems. First, the problem of calculations is the complexity of the system compared with the possibility to select relevant values of the empirical model parameters. Second, a fundamental problem exists because it is basically not possible to produce reliable results from a series of calculation steps that, per definition, do not reflect the dynamics of the real system. It is the authors' conclusion that the Type III empirical models cannot be recommended for prediction of sulfide buildup in sewer networks. These attempts are therefore not further dealt with.

### 6.4.3.1   Type I Sulfide Prediction Models

The Type I models can be characterized as screening or risk models that tend to express if a sulfide problem may exist or not.

Figure 6.3 is an example of how corrosion control was predicted for gravity sewer design (USEPA 1974; ASCE 1982; ASCE 1989). The design criterion is a very simple formulation of a complex problem. However, it is—in terms of both simplicity and fundamental insight in the nature of sulfide occurrence and effect—an outstanding example of a guideline for risk assessment, i.e., it is a Type I model. The two curves A and B in Figure 6.3 correspond to the production of maximum 0.1 to 0.2 g S m$^{-3}$ of sulfide and "several tenths" of g S m$^{-3}$ of sulfide, respectively. The risk indicated by Figure 6.3 is based on what is likely to take place for wastewater with EBOD = 500 g m$^{-3}$ during a 3-month summer period (southeastern US states), at an average flow rate at the highest 6-h daily period and at water depths not exceeding 2/3 of the pipe diameter.

Another example of a Type I model originates from the work of Davy (1950). He stated that the occurrence of sulfide problems in gravity sewers was related to Reynolds number, pipe flow characteristics, and EBOD of the wastewater. This early work was the basis for what Pomeroy (1970)—after several years of consideration—presented as the "Z formula":

$$Z = \text{EBOD} \, (s^{0.5} Q^{0.33})^{-1} \frac{P}{b} \qquad (6.7)$$

where:

$s$ = slope of pipe (m per 1000 m)
$Q$ = pipe flow (ft$^3$ s$^{-1}$); 1 m$^3$ = 35.314 ft$^3$
$P$ = wetted pipe-wall perimeter (m)
$b$ = pipe width at the water surface (m)

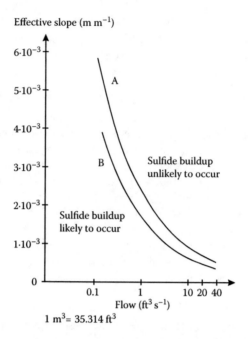

**FIGURE 6.3** Flow-to-slope relationships as guideline to sulfide forecasting and recommendation for design of gravity sewers. Effective slope of pipe is defined for average conditions at 1-h flow distance of the pipe upstream point of reckoning. (Data from USEPA, *Process Design Manual for Sulfide Control in Sanitary Sewerage Systems*, USEPA 625/1-74-005, 1974.)

Although the Z formula by Pomeroy is simple, it reflects by applying few integrated parameters, a fundamental understanding of which factors and processes are limiting sulfide buildup in gravity sewers: the reaeration of the water phase followed by a competition for DO by sulfide being produced in the biofilm and the availability of biodegradable organic matter in the water phase.

The Z formula presented in Equation 6.7 is equivalent to the one originally formulated by Pomeroy (1970). Different versions of the formula, for example, with other units of flow and slope, have since been used by other authors, in some cases resulting in different numerical values of Z for assessment (Boon 1995). Pomeroy (1970) assessed the Z formula based on an array of 50 tests and found in general the results as shown in Table 6.3. This interpretation was recommended for sewer pipe design.

The assessment of the Z formula as shown in Table 6.3 is not applicable to very small flows where it may show sulfide buildup that, in practice, will not occur. The tests performed by Pomeroy (1970) also showed that $Z > 10,000$ did not necessarily result in sulfide buildup. Such cases could usually be explained as being due to toxic substances or extreme values of pH in the wastewater. From the standpoint of sewer design it is, however, often sufficient to know that a high Z value indicates a potential for sulfide buildup. Although conceptually formulated models for sulfide occurrence and buildup today are available, the low requirement for data still makes the Z formula useful as a tool for screening. The Z formula has been widely used (Pomeroy and Parkhurst 1977; ASCE and WPCF 1982; ASCE 1989).

## TABLE 6.3
## Sulfide Problems Occurring in Gravity Sewers Related to the *Z* Formula

| *Z* Formula, Numerical Value | Estimated Magnitude of the Sulfide Problem |
|---|---|
| *Z* < 5000 | Sulfide buildup is not likely to occur |
| 5000 < *Z* < 10,000 | Some risk of sulfide buildup |
| *Z* > 10,000 | Sulfide buildup is likely to occur |

*Source:* Data from Pomeroy, R.D., *Public Works*, 101(10), 93–96 and 130, 1970. With permission.

### Example 6.3: Assessment of Sulfide Buildup in Gravity Sewers Based on the Z Formula

This example concerns an assessment based on the *Z* formula of sulfide buildup risk in a half-filled, long gravity sanitary sewer pipe designed with different slopes and a corresponding need for flow capacities.

The average flow rate per person equivalent (PE) is 200 L $PE^{-1}$ $day^{-1}$. The temperature of the wastewater is 20°C and the $BOD_5$ value is 250 g $O_2$ $m^{-3}$. The diameter of the pipe is 0.6 m, and three different values of a pipe slope are assessed corresponding to different pipe capacities. The resulting *Z* values are shown in Table 6.4.

When comparing the *Z* values in Table 6.4 with the levels shown in Table 6.3, it is seen that significant sulfide problems may easily occur in trunk sewers and intercepting gravity sewers with low or moderate pipe slopes.

Other types of screening models have been developed. As an example, Polder and Mechelen (1989) published a rather simple point system procedure for sulfide corrosion risk assessment (degree of aggressiveness) of sewer networks, taking into account the levels of sulfide concentration, temperature, turbulence, and pH value of the moist sewer surface.

### 6.4.3.2 Type II Sulfide Prediction Models

Within the area of the Type II models, the development during the period 1945 to 1995 resulted in several types of empirical mathematical models for sulfide prediction in both gravity sewers and pressure mains. Some of these models became important tools for design and control of sewer networks to prevent sulfide buildup. These models have also been included in manuals produced by governmental agencies, municipalities, and institutions (Thistlethwayte 1972; USEPA 1974; USEPA 1985; ASCE

## TABLE 6.4
## Screening of Sulfide Occurrence in Selected Gravity Sewer Pipes by the *Z* Formula

| Slope, *s* (%) | Velocity, *u* (m $s^{-1}$) | Flow, *Q* (L $s^{-1}$) | Load, Number of PEs | *Z* Value |
|---|---|---|---|---|
| 0.03 | 0.32 | 45 | 19,000 | 19,370 |
| 0.1 | 0.58 | 82 | 35,000 | 8740 |
| 0.3 | 1.00 | 141 | 61,000 | 4220 |

1989). The United States and Australia were the leading countries for the devel-opment during this period, with a number of important contributions from other countries, for example, the United Kingdom, Germany, Portugal, and Denmark. The "Process Design Manual for Sulfide Control in Sanitary Sewerage Systems," pub-lished by the U.S. Environmental Protection Agency, is a relatively early outstanding example of these efforts (USEPA 1974).

Three main types of information in terms of central parameters for sulfide predic-tion are typically included in Type II models:

1. Wastewater characteristics, often relatively simple expressed. Examples are BOD or COD (in some models, also further details of the biodegradability), sulfate concentration, and temperature.
2. Combined hydraulic and network characteristics, for example, flow rate, water surface area/volume ratio, and submersed surface area/volume ratio.
3. Specific network characteristics, for example, pipe diameter, pipe length, and slope of the pipe.

Although these models from a process point of view are rather simple, it is impor-tant to stress that they indirectly reflect a level of information that exceeds what is directly expressed by the parameters. As an example, the water surface area/volume ratio, the flow rate, and the slope includes information on reaeration, and the sub-mersed surface area/volume ratio reflects the fact that sulfide is produced in a biofilm but occurs—and is emitted from—the water phase. Provided that these models have been rationally transferred (calibrated) to their specific intended use, the strong point is the rather limited requirement for data. In this respect, they are still interesting to consider for use in practice, but typically for rather simple networks and basically only valid for single pipe sections. The interactions and dynamics of the flows in a network require a quite different approach (cf. Chapters 8 and 9).

Compared with the rather complex sulfur cycle in a sewer network as shown in Figure 4.7, the main part of the Type II models only consider it in part. However, the gravity sewer prediction models by Pomeroy and Parkhurst (1977) and Matos (1992) include a simple formulated sink term.

The occurrence of sulfide in the wastewater phase as predicted by the models is in general *the* indicator of which sulfide related problems may appear. Based on an overall assessment of what is typically reported in the literature, the information in Table 6.5 is an attempt to "translate" a sulfide concentration to the "real" occurring

## TABLE 6.5
## Empirically Based "List of Translation" from a Predicted Sulfide Concentration in Wastewater of a Sewer to a Potentially Occurring Sulfide Problem

| Total Sulfide Concentration (g S m$^{-3}$) | Level of Potential Corresponding Problem |
| --- | --- |
| <0.5 | No or rather few problems observed |
| 0.5–2 | Moderate or some problems may occur |
| >2 | Sulfide problems common and widely distributed |

problem, i.e., to make a first approach of a cause–effect relation. It should be noted that although a sulfide concentration is predicted in a pressure main, the problem refers to a case with a free water surface, for example, in a gravity sewer or a manhole located downstream. The values given in Table 6.5 should be considered as both "summer values" and values that may exist in systems with rather high strength wastewater, for example, generally occurring during daytime. Finally, it is clear that a simple relation as shown in Table 6.5 should not be interpreted as generally true.

In continuation of the information given in Table 6.5, it is important to remember the following statement formulated by Pomeroy (1970): "As far as concrete pipe is concerned, a few tenths of a milligram per liter of sulfide for a few hours each day in the hottest part of the year is not serious."

Because of the importance of the air–water exchange of hydrogen sulfide, it is not surprising that Type II models for sulfide prediction have been developed differently for use in force mains (and full flowing gravity sewers) and in partly filled gravity sewers. In the following, this distinction will therefore be maintained.

### 6.4.3.2.1  Type II Sulfide Prediction Models for Force Mains

Depending on the DO concentration of the incoming wastewater and the DO depletion rate, anaerobic conditions in pressure mains typically occur after a short initial aerobic period. The main factors that affect sulfide production in a pressure sewer are dealt with in Section 6.3.3 and include both wastewater and system characteristics.

The equation for determination of the sulfide produced in terms of a resulting concentration in the bulk water phase is based on the biofilm flux rate of sulfide production:

$$\Delta C_{S(-II)} = C_{S(-II),d} - C_{S(-II),u} = r_a \frac{A}{V} t_r \qquad (6.8)$$

where:
$\Delta C_{s(-II)}$ = sulfide production rate transformed to the water phase (g S m$^{-3}$)
$C_{s(-II),d}$ = sulfide concentration at the end of the sewer section (g S m$^{-3}$)
$C_{s(-II),u}$ = sulfide concentration at the start of the sewer section (g S m$^{-3}$)
$r_a$ = area sulfide production rate of the biofilm (g S m$^{-2}$ h$^{-1}$)
$A/V$ = area/volume ratio, i.e., wetted pipe surface area/wet pipe volume (m$^{-1}$)
$t_r$ = anaerobic residence time in the pipe (h)

A number of empirical equations have been developed for the prediction of the sulfide production rate, $r_a$, in pressure pipes. All equations take into consideration that sulfide generation is affected by the concentration of the organic matter (BOD, total COD, or soluble COD) and the temperature. Table 6.6 outlines selected models for sulfide prediction in force mains and full flowing gravity sewers. These models will be discussed in the following.

The equations shown in Table 6.6 are all based on the fact that the biofilm—and not the water phase—is the productive part of the system. Sulfide in the anaerobic water phase may occur when nitrate and DO are absent. Nitrate is typically not present in municipal wastewater, and DO is in full flowing pipes, quickly rather depleted—from a few minutes to half an hour depending on the temperature and the biodegradability

---

## TABLE 6.6

### Empirical Equations (Type II Models) for Prediction of the Biofilm Surface Formation Rate of Sulfide, $r_a$ (g S m$^{-2}$ h$^{-1}$), in Force Mains and Full Flowing Gravity Sewers

| Biofilm Surface Rate, $r_a$ | References | Nomenclature |
|---|---|---|
| (Equation 1): M BOD$_5$ 1.07$^{T-20}$ | Pomeroy (1959); USEPA (1974); Pomeroy and Parkhurst (1977) | Empirical rate constant: $M = 10^{-3}$ (m h$^{-1}$) BOD$_5$ (g O$_2$ m$^{-3}$) $T$ = temperature (°C) |
| (Equation 2): $0.5 \times 10^{-3}$ u BOD$_5^{0.8}$ S$_{SO4}^{0.4}$ 1.14$^{T-20}$ | Thistlethwayte (1972) | $u$ = flow velocity (m s$^{-1}$) S$_{SO4}$ = sulfate concentration (g SO$_4$ m$^{-3}$) |
| (Equation 3): $0.23 \times 10^{-3}$ COD 1.07$^{T-20}$ | Boon and Lister (1975) | COD (g O$_2$ m$^{-3}$) |
| (Equation 4): $a$ (COD$_S$ – 50)$^{0.5}$ 1.03$^{T-20}$ | Hvitved-Jacobsen et al. (1988); Nielsen et al. (1998) | $a^a$ = empirical rate constant (g S g O$_2^{-0.5}$ m$^{-0.5}$ h$^{-1}$) COD$_s$ = soluble COD (g O$_2$ m$^{-3}$) |

$a^a$ = 0.001–0.002 domestic wastewater without sewage from food industries
$a^a$ = 0.003–0.006 wastewater from mixed domestic and industrial sources
$a^a$ = 0.007–0.010 wastewater with biodegradable organic matter from mainly foodstuff industries

---

of the wastewater. Sinks of sulfide in terms of emission and oxidation can, under the conditions of a full flowing pipe as a first estimate, be neglected for most practical purposes. Compared with gravity sewers, prediction of sulfide buildup in full flowing pipes therefore becomes significantly simpler.

Specific comments to Equations 1 through 4 shown in Table 6.6 are given in the following:

*Equation 1.*

According to Pomeroy (1959), USEPA (1974), and Pomeroy and Parkhurst (1977), the most frequent values of $M$ in Equation 1, Table 6.6, are found in the interval from $0.4 \times 10^{-3}$ to $1.3 \times 10^{-3}$ m h$^{-1}$. With a value of $M = 10^{-3}$ m h$^{-1}$, the equation is approaching a "risk model" rather than a prediction model. For design purposes it is, however, an appropriate (conservative) estimate.

*Equation 2.*

Compared with Equation 1, Equation 2 includes the sewer flow velocity, $u$, during pumping and the sulfate concentration, SO$_4^{2-}$ (Thistlethwayte 1972). The predictability of Equation 2 was not found to be reliable at flow velocities less than about 0.6 m s$^{-1}$, i.e., values that are not typically observed in operating pressure mains (Melbourne and Metropolitan Board of Works 1989). Concerning the importance of sulfate, we know today that sulfate is not typically a limiting factor for sulfate reduction at concentrations exceeding

5–15 g S m$^{-3}$, which is often the case in municipal wastewater (Nielsen and Hvitved-Jacobsen 1988). Furthermore, Thistlethwayte's equation has a temperature coefficient of 1.14. Although this value may be correct for the sulfate-reducing bacteria themselves, it turns out to be wrong in biofilms where diffusion is normally limiting the apparent transformations. If Equation 2 results in reasonable estimates of the sulfide formation rate close to 20°C, it will underestimate the rate in the low temperature region. The relevance of including the flow velocity as a model parameter in a pressure main may be correct but has been questioned (Pomeroy and Parkhurst 1977).

*Equation 3.*

Equation 3 in Table 6.6, developed by Boon and Lister (1975), follows the concept of Equation 1 except for the fact that BOD$_5$ is replaced with COD. The equation is developed from results obtained in intermittently pumped systems where the transfer of substances across the biofilm–water interface is expected to be reduced in the nonpumping periods.

*Equation 4.*

Equation 4 is developed based on results from intermittently operated pressure mains (Hvitved-Jacobsen et al. 1988; Nielsen et al. 1998). The equation follows Equations 1 and 3 as far as the basic formulation is concerned. However, it includes further details important for the sulfide production rate in terms of biofilm kinetics, wastewater quality, and temperature dependency. The substrate-limited growth of the sulfate-reducing bacteria in biofilms is taken into account in two ways: via different rate constants depending on the type of wastewater and in particular because of the term $(COD_S - 50)^{0.5}$. This expression describes 1/2-order biofilm kinetics of the sulfide formation as discussed in Section 2.2.2.1 and the availability of organic substrate for the bacteria in terms of $COD_S - 50$. The explanation for using only a part of the soluble COD as readily biodegradable is based on the fact that $COD_S$ includes nonbiodegradable and slowly biodegradable organic matter not directly useful as substrate for the sulfate-reducing biomass (SRB) (Tanaka and Hvitved-Jacobsen 1998). Field investigations by Tanaka et al. (2000a, 2000b) showed that the term $(COD_S - 50)$ could be replaced by the most biologically active COD components, $S_S + X_{S1}$, the sum of readily biodegradable and fast hydrolyzable substrate (cf. Section 3.2.6). These two compounds are determined from an analysis of oxygen uptake rate measurements (cf. Section 10.1.3). The temperature coefficient in Equation 4 is low, only 3% per °C. It reflects the long-term finding—at least experienced in a temperate climate—that species of the sulfate-reducing bacteria during a winter period apparently adapt to rather low temperatures (Nielsen et al. 1998). The equation therefore expresses a long-term sulfide production potential depending on the season of the year. For short-term temperature changes, a temperature coefficient as used in Equations 1 and 3 should be recommended. The biofilm surface rate of sulfide formation using Equation 4, is in Figure 6.4, shown for different values of the rate constant, *a*. It is readily seen that the biodegradability of the organic substrate highly affects the sulfide formation rate.

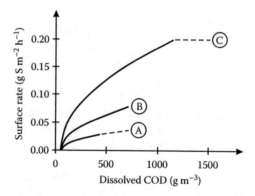

**FIGURE 6.4** Hydrogen sulfide formation rates in pressure mains by Equation 4, Table 6.6, at $T = 20°C$. The biofilm surface rate is shown as curves A, B and C for values of the rate constant, a, equal to 0.0015, 0.003, and 0.006, respectively.

The expressions shown in Table 6.6 include constants that have been found to have good approximate values based on experiments. These values may, of course, be adjusted to account for specific cases. As an example, different flow conditions in continuously and intermittently pumped pressure mains may affect the transfer of substrates and products across the biofilm-water interface, and, thereby, the production of sulfide (Melbourne and Metropolitan Board of Works 1989).

### 6.4.3.2.2 Type II Sulfide Prediction Models for Gravity Sewers

Compared with pressure mains, the sulfur cycle in partly filled gravity sewer pipes is more complex. Emission of hydrogen sulfide into the sewer atmosphere, oxidation of sulfide, and precipitation are three major processes that will reduce its occurrence in the water phase (cf. Figure 4.7). These processes must therefore be taken into account (Pomeroy and Parkhurst 1977; Tchobanoglous 1981; Wilmot et al. 1989). Furthermore, parameters such as pH will play a dominating role in this respect. Expressed in a slightly different way, the formation of sulfide that takes place in the sewer biofilm and causes a corresponding buildup in the water phase is reduced because of emission and oxidation, both processes related to air–water gas transfer. The air–water exchange processes of both hydrogen sulfide and oxygen therefore become central.

Based on several years of intensive studies, Pomeroy and Parkhurst (1977) presented the first reliable empirical model that was formulated according to these facts:

$$r_a = \left(\frac{A}{V}\right)^{-1}\left(\frac{dC_{S(-II)}}{dt}\right) = M' BOD_5 1.07^{T-20} - N(s \cdot u)^{3/8} d_m^{-1} RC_{S(-II)}$$

$$= M' BOD_5 1.07^{T-20} - N(s \cdot u)^{3/8}\left(\frac{P}{b}\right)^{-1} C_{S(-II)}$$

(6.9)

where:

$r_a$ = biofilm area net formation rate of sulfide (g S m$^{-2}$ h$^{-1}$)

$A/V$ = area/volume ratio, i.e., wetted pipe surface area/wet pipe volume (m$^{-1}$)

$C_{S(-II)}$ = sulfide concentration (g S m$^{-3}$)

$t$ = time (h)

$M'$ = empirical rate constant

$BOD_5$ = biochemical oxygen demand (g O$_2$ m$^{-3}$)

$T$ = temperature (°C)

$N$ = empirical sulfide loss coefficient

$s$ = slope (m m$^{-1}$)

$u$ = flow velocity (m s$^{-1}$)

$d_m$ = hydraulic mean depth of the water phase, i.e., the cross-sectional area of the water volume divided by the water surface width (m)

$R$ = hydraulic radius, i.e., the cross-sectional area of the water volume divided by the wetted perimeter (m)

$P$ = wetted pipe-wall perimeter (m)

$b$ = pipe width at the water surface (m)

The first term in Equation 6.9, $M'$, is basically equal to the corresponding constant in Equation 1 in Table 6.6, and thereby expresses an apparent formation rate of sulfide in units of m h$^{-1}$. However, the apparent biofilm area sulfide formation rate is normally found to be lower in a gravity sewer than in a pressure main. The reason is probably an effect of a daily changing flow and water level and the fact that even a small amount of DO in a gravity sewer will reduce the sulfide formation rate. The value of $M'$ is therefore typically lower than shown in Equation 1. At DO concentrations less than about 0.5 g O$_2$ m$^{-3}$, a value of $M'$ equal to 0.32 10$^{-3}$ m h$^{-1}$ was proposed by Pomeroy and Parkhurst (1977).

The second term in Equation 6.9 corresponds to the sinks for sulfide in the water phase that, according to Figure 4.7, are primarily caused by oxidation in the water phase, emission into the sewer atmosphere, and, also to some extent, precipitation. Pomeroy and Parkhurst (1977) propose values for $N$ at two levels, $N = 0.96$ and $N = 0.64$. The first value corresponds to a median buildup of sulfide, whereas the last value is a conservative estimate for prediction of sulfide buildup in a gravity sewer. A relatively high value of $N$ may be applied when focusing on corrosion, whereas odor problems may be best taken into account by a relatively low value. The second term of Equation 6.9 shows that the removal of sulfide from the water phase is considered a 1-order reaction in the sulfide concentration. The term also includes elements related to the reaeration and, thereby, also the emission of hydrogen sulfide (cf. Equations 3, 6, and 7 in Table 4.3).

Equation 6.9 has been the dominating model applied for sulfide prediction in gravity sewers. In a number of studies it has been applied as described but also with values of $M'$ and $N$ adjusted according to local observations. It is furthermore important to note, that Equation 6.9 with the constants proposed by Pomeroy and Parkhurst (1977), may underestimate the formation rate of sulfide because of heavy metal inflows to the sewer network (Morton et al. 1991). Kienow et al. (1982) have produced monographs for a graphical prediction of sulfide built-up.

Although Equation 6.9 is based on intensive investigations, it must—as an empirical model—be applied with caution. It has already been mentioned that sulfide in gravity sewers in general will not appear if the DO concentration exceeds 0.2–0.5 g $O_2$ m$^{-3}$.

Equation 6.9 has been reformulated by Matos (1992) to predict a downstream concentration of sulfide:

$$C_{S(-II),d} = C_1 - (C_{S(-II),u} - C_1)e^{C_2} \tag{6.10}$$

where:

$C_{S(-II),d}$ = downstream concentration of sulfide (g S m$^{-3}$)

$$C_1 = \frac{M'}{N} BOD_5 (s \cdot u)^{-3/8} \frac{P}{b}$$

$C_{S(-II),u}$ = upstream concentration of sulfide (g S m$^{-3}$)

$$C_2 = -\frac{LNs^{3/8}}{3600 d_m u^{0.625}}$$

where $L$ is the sewer length (m).

Other types of empirical gravity sewer models for estimation of the emission rate of $H_2S$ have been developed (Lahav et al. 2004). These authors based the model development on rather few, but central, parameters: the dissolved sulfide concentration, the level of turbulence, pipe flow characteristics (surface area and volume), and pH.

Anaerobic activity in terms of $H_2S$ formation may take place in sediment deposits, particularly in the deeper parts (cf. Section 3.2.8.3). Depending on sewer design and hydraulic conditions, sewer solids can temporarily or more permanently accumulate as sediments in gravity sewer networks. In separate sanitary sewers, it may depend on the daily fluctuations in the flow pattern and in combined sewer networks particularly affected by the varying dry and wet weather flow conditions. At the more or less permanent sewer sediment surfaces, a biofilm will develop that thereby can be characterized as a "thick biofilm."

Focusing on sulfide formation, the sediment is often simply taken into account by considering it covered with a biofilm. The potential for sulfide production in terms of a surface flux may—if a permanent sediment layer exists—exceed what is observed for sewer biofilms, in some cases, being 50% to 100% higher (Schmitt and Seyfried 1992; Bjerre et al. 1998).

## 6.5   SULFIDE-INDUCED CORROSION OF CONCRETE SEWERS

A number of impacts and effects are closely related to the presence of anaerobic conditions in wastewater of sewers. The major phenomena are as follows:

- Health-related effects
- Odor problems
- Concrete and metal corrosion

- Inflow of anaerobic wastewater into treatment plants
- Combined sewer overflows into receiving waters

The first three points are related to the release of volatile substances into the gas phase of the sewer and from there also into the urban atmosphere. These volatile substances are, for example, hydrogen sulfide, volatile sulfur compounds (VSCs) and nitrogenous compounds produced under anaerobic conditions in the wastewater or associated biofilm and sediments (cf. Table 4.1). The first two mentioned points, human health impacts and odor-related effects, are dealt with in Section 4.3 in relation to exchange processes between the water and the gas phase in a sewer. The specific impacts of wastewater inflows onto treatment plant processes and receiving water impacts are not subjects of this text on sewer processes. In the following, concrete corrosion is focused on.

Concrete corrosion is closely related to the formation of hydrogen sulfide and its emission to the sewer atmosphere (cf. Figure 4.7 and Table 6.1). The problems related to anaerobic conditions in sewers including concrete corrosion have been known worldwide for more than 40 to 60 years (Parker 1945a, 1945b, 1951; Fjerdingstad 1969; Thistlethwayte 1972; USEPA 1974). Concrete corrosion in terms of the fundamental chemical and microbial processes was thereby formulated. Although numerous scientific papers and technical reports on concrete corrosion have been published, inappropriate and wrong solutions on sewer design and operation are in this respect still frequently seen. The consequences of anaerobic conditions in sewers are generally well known among both engineers and practitioners. However, the details of the sewer processes behind the phenomena are unfortunately often absent. It is the authors' hope that this book can improve this situation.

Concrete corrosion is a phenomenon with great economic impact. A concrete corrosion rate of, for example, 2 to 3 mm of a pipe surface per year is often observed and may lead to deterioration of a sewer after only a few years of operation. It is therefore important that detailed knowledge is today available for control of problems in existing sewers and for design of new sewer networks with reduced risk for future sulfide attack on the constructions (Vincke et al. 2000; Vollertsen et al. 2008; Jensen et al. 2009b, 2011; Nielsen et al. 2012). The availability of models for prediction of sulfide problems that are formulated on this conceptual basis is invaluable.

### 6.5.1 Concrete Corrosion as a Sewer Process Phenomenon

The sulfur cycle shown in Figure 4.7 outlines the pathways leading to concrete corrosion. It is seen that as long as sulfide remains in the water phase, no harmful effect will occur. In brief, the concrete corrosion problem is caused by hydrogen sulfide that from the gas phase is absorbed in the liquid film that exists at moist concrete surfaces in the sewer system. The concrete surfaces that are mostly corroded are typically close to the anaerobic water phase and at areas directly exposed to the release of $H_2S$ from turbulent water surfaces, for example, at sewer drops and falls. Furthermore, the sewer crown is also reported as being corroded rather fast. At the moist surfaces, oxygen is typically available from the sewer atmosphere, and $H_2S$ is at the concrete surface oxidized to sulfuric acid by microbial reactions (cf. Equation

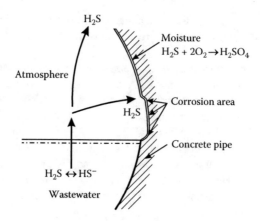

**FIGURE 6.5**  Principle of concrete corrosion in a sewer pipe.

6.11 and Figure 6.5). Sulfuric acid reacts with the alkaline substances of concrete and thereby degrades it.

$$H_2S + 2O_2 \rightarrow H_2SO_4 \tag{6.11}$$

The oxidation of hydrogen sulfide is more complex than expressed by Equation 6.11. Different reaction pathways and sulfur compounds with thiosulfate and elemental sulfur as potential intermediates have been proposed to occur in corroding concrete (Parker 1945a; Islander et al. 1991; Jensen et al. 2009a). The proposal by Jensen et al. (2009a) including two fractions of elemental sulfur, a fast and probably amorphous biodegradable form, $S_{fast}^0$, and a slowly biodegradable form, $S_{slow}^0$, has been experimentally verified (cf. Figure 6.6). The occurrence of a slowly biodegradable fraction may result in delayed corrosion, i.e., corrosion after $H_2S$ has been depleted.

The aerobic bacteria responsible for this oxidation of hydrogen sulfide to sulfuric acid belong to the aerobic and autotrophic *Thiobacillus* family (Parker 1945a; Milde et al. 1983; Sand 1987). Several of these bacteria are active at rather low pH values and can produce solutions of sulfuric acid up to about 7%. *Acidithiobacillus thiooxidans* (previously named *Thiobacillus thiooxidans* and *Thiobacillus concretivorus*)

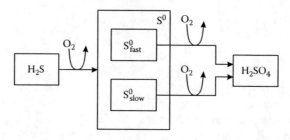

**FIGURE 6.6**  Oxidation of hydrogen sulfide in corroded concrete of a sewer showing different oxidation pathways of elemental sulfur. (From Jensen, H.S. et al., *Water Environ. Res.*, 81(4), 365–373, 2009a; Jensen, H.S. et al., *Environ. Technol.*, 30(12), 1291–1296, 2009b. With permission.)

is considered the dominant species of bacteria on heavily corroded concrete surfaces and active at pH values between about 0.5 and 5. To be active, it requires that other species of the *Thiobacillus* family bring down the pH value. Two of the species, *A. thiooxidans* and *T. neapolitanus*, are, in addition to sulfide, also able to use thiosulfate and elemental sulfur as energy sources.

Sulfuric acid that is generated on the moist sewer surfaces may react with the alkaline cement in the concrete. A simple stoichiometry of this reaction is given by the following expression:

$$H_2SO_4 + \underset{\text{(Cement)}}{CaCO_3} \rightarrow H_2O + CO_2 + \underset{\text{(Gypsum)}}{CaSO_4} \tag{6.12}$$

If the formation rate of sulfuric acid is low, a major part of it will react with the cement and leave a material of loosely bound inert components (e.g., sand and gravel). On the other hand, if the formation rate of sulfuric acid is relatively high, a part of the sulfuric acid will be washed away before reaction and turn up in the wastewater and here react with alkaline components under the formation of sulfate ions. The sulfate ions produced in the liquid film on the sewer walls may, as an associated effect, cause a chemical sulfate attack on the concrete material.

### 6.5.2 PREDICTION OF HYDROGEN SULFIDE-INDUCED CORROSION

#### 6.5.2.1 Traditional Approach of Predicting Concrete Corrosion

Equations 6.11 and 6.12 show that 1 mol of hydrogen sulfide (32 g $H_2S$-S) resulting in the formation of sulfuric acid has a potential for reaction with 1 mol (100 g $CaCO_3$) of cement in the concrete. The corrosion rate of the concrete can, based on this simple mass balance, be empirically expressed as follows (USEPA 1974):

$$r_{corr} = \frac{100}{32} \cdot \frac{F}{A} \tag{6.13}$$

where:

$r_{corr}$ = corrosion rate per unit area of the concrete surface (g concrete m$^{-2}$ h$^{-1}$)

$F$ = rate of hydrogen sulfide absorption in the moisture at the concrete surface (g $H_2S$-S m$^{-2}$ h$^{-1}$)

$A$ = alkalinity of the concrete material in units of g $CaCO_3$ per g concrete material*

Equation 6.13 can be reformulated. If the right side of the equation is divided by the density of the concrete material, estimated to be about $2.4 \times 10^6$ g m$^{-3}$, the area-based corrosion rate can be transferred to an annual corrosion rate in units of depth per time:

$$c = 11.4 \frac{F}{A} \tag{6.14}$$

where $c$ is the annual corrosion rate (mm year$^{-1}$).

---

* A standard test procedure for determination of $CaCO_3$ equivalents can be found in the *Encyclopedia of Chemical Analysis, Interscience Publishers Division*, John Wiley & Sons, New York, vol. 15, p. 230.

It is by Equation 6.14 assumed that all sulfuric acid formed will react with the cement. The following expression is formulated with reduced efficiency of sulfuric acid:

$$c = k11.4\frac{F}{A} \qquad (6.15)$$

where $k$ is the correction factor <1 (–).

In cases of severe concrete corrosion, a corrosion rate of 4–5 mm year$^{-1}$ can be observed (Mori et al. 1991).

For systems where the formation rate of sulfuric acid is low, $k$ is approaching 1. However, at increasing values of $F$, $k$ may decrease to about 0.3–0.4. A combined effect of an incomplete oxidation of sulfide in the corroding matrix and a loss of sulfuric acid through a flow of condensed water back to the bulk water volume may be reasons for the observed reduced efficiency of the corrosion process. Vollertsen et al. (2011) experimentally determined and simulated the results, and confirmed good agreement with these findings (cf. Figure 6.7).

### 6.5.2.2 A Process-Related Approach for Prediction of Concrete Corrosion

Concrete corrosion is, as already shown in this section, a complex phenomenon affected by numerous process-related details and the configuration of the sewer network. Equation 6.15 is a valuable empirical equation for predicting the annual corrosion depth of a sewer pipe. However, this simple approach fails when it comes to the dynamic corrosion phenomena in, for example, larger sewer networks with

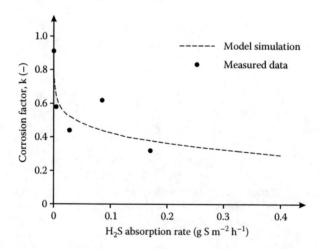

**FIGURE 6.7**   Correction factor, k, for prediction of concrete corrosion rate, *c*, versus rate of hydrogen sulfide absorption at the moist concrete surface in a sewer. (From Vollertsen, J. et al., *12th International Conference on Urban Drainage*, Porto Alegre, Brazil, September 11–16, 2011b. With permission.)

a combination of gravity sewers, pressurized systems, and pumping stations. Furthermore, the short-term changes and impacts that may appear under conditions of, for example, high sulfide formation, are not predictable.

A process-related description of the concrete corrosion rate is by nature complex to formulate but possible to approach based on the knowledge available today. A major problem, however, will appear when the parameters in such detailed descriptions should be selected. The pragmatic solution to this problem is to describe corrosion in terms of one or few governing processes that are feasible to include in a description including other relevant processes related to the sulfur cycle. As shown in Figure 4.7, sulfide formation, emission to the sewer atmosphere, absorption, and oxidation on the sewer walls are major pathways of $H_2S$ to its occurrence in the final corrosion process. It remains for this description to briefly explain the behavior of $H_2S$ in the gas phase and its relation to the two interfaces, the solid concrete surface and the water phase.

The emission of $H_2S$ from the air–water interface is the source for the buildup of hydrogen sulfide in the sewer air (cf. Section 4.4.3). $H_2S$ in the sewer air is absorbed in the liquid film that exists at the concrete surfaces and is then oxidized in the concrete matrix (cf. Figure 6.5). It is important to note that the area absorption rate of $H_2S$ at the concrete surface is high compared with the corresponding typical emission rate from the water phase. When corrosion occurs, the result is a rather low concentration of $H_2S$ in the sewer air compared with the equilibrium concentration. It is therefore in sewers with noncorroding materials such as plastic pipes observed that the hydrogen sulfide concentration in the gas phase is significantly higher than that in concrete sewers with corrosion (Nielsen et al. 2008). As a consequence, it is furthermore important that a high corrosion rate is possible although the concentration of $H_2S$ in the sewer air is low. Emission resulting in odor problems and absorption of $H_2S$ resulting in corrosion are therefore two strongly related—and competing—processes.

Long-term kinetic studies of the corrosion process in pilot-scale concrete pipes have been carried out under sewer conditions (Vollertsen et al. 2008). These studies have shown that a relatively simple conceptual description of the corrosion process mainly depending on the concentration of $H_2S$ in the sewer air is possible. In principle, two types of kinetic expressions are possible: a saturation expression based on Monod kinetics and an exponential expression (cf. Equations 2.21 and 2.18, respectively). The $H_2S$ oxidation rate at a concrete surface based on these two types of expressions are shown in the following (cf. Equations 6.16 and 6.17).

The hydrogen sulfide oxidation rate based on saturation kinetics (Monod kinetics):

$$r_{H_2S} = k_{cc} \frac{p_{H_2S}}{p_{H_2S} + K_{p,H_2S}}$$

(6.16)

where:

$r_{H_2S}$ = area specific oxidation rate of $H_2S$ (to $S^0$ and $H_2SO_4$) at the concrete surface exposed to the sewer atmosphere (g $H_2S$-S m$^{-2}$ h$^{-1}$)

$k_{cc}$ = area specific formation rate of $H_2SO_4$ on concrete surfaces (g $H_2SO_4$–S m$^{-2}$ h$^{-1}$)
$p_{H_2S}$ = partial pressure of $H_2S$ in the gas phase (ppm of $H_2S$)
$K_{p,H_2S}$ = saturation constant for concrete corrosion (ppm of $H_2S$)

In relation to the formulation of Equation 6.16, it should be mentioned that $r_{H2S}$ basically expresses a sorption rate of $H_2S$. However, elemental sulfur, as an intermediate, is expected to be oxidized to sulfuric acid (cf. Figure 6.6).

The hydrogen sulfide oxidation rate at the concrete surface described by an exponential expression ($n$-order reaction):

$$\frac{dp_{H_2S}}{dt} = k_n p_{H_2S}^n \tag{6.17}$$

where:
$t$ = time (h)
$k_n$ = rate constant for oxidation of $H_2S$ (ppm$^{1-n}$ h$^{-1}$)
$n$ = reaction order (–)

The oxidation (adsorption) rate of hydrogen sulfide expressed by Equation 6.17 can, by applying the ideal gas law (Equation 4.1), be converted to area-specific units of the oxidation rate (Nielsen et al. 2008). The reaction order, $n$, was proposed by Vollertsen et al. (2008) to vary between 0.45 and 0.75. Nielsen et al. (2012) found that $n = 0.8$ described the adsorption and oxidation process with good agreement. They furthermore found that the adsorption and oxidation rate of $H_2S$ was affected by the flow conditions in the gas phase. Within the ranges of typical air flows, they showed that a simple power function of Reynolds number, $Re^{0.65}$, in Equation 6.17 can describe this effect.

It can be discussed which expression, Equation 6.16 or 6.17, will best predict concrete corrosion in sewers. Aesoy et al. (2002) concluded that corrosion followed a Monod-type expression (Equation 6.16) with a saturation constant of 2 ppm. However, experiments by Vollertsen et al. (2008) clearly shoved no saturation at $H_2S$ gas concentrations as high as 500 to 1000 ppm and that Equation 6.17 accurately simulated the $H_2S$ oxidation rate.

There are numerous examples from the literature showing that corrosion has seriously—and often fast—deteriorated sewer networks (EWPCA 1982; Aldred and Eagles 1982; ASCE 1989). Although corrosion is difficult to predict, the number of examples and extent of the problems observed have given a comprehensive knowledge of where and when concrete corrosion may exist. A concrete corrosion rate is typically relatively low at both moderate temperatures (<15–20°C) and relatively low sulfide concentrations in the wastewater (<0.5 g S m$^{-3}$). In contrast, the following conditions indicate risk for corrosion:

- The concrete sewer network or pumping station is located downstream of systems with risk for sulfide formation. Such systems are primarily pressure mains, but also gravity sewers with permanent deposits of sewer solids and relatively high temperatures.

- Systems exposed to excessive turbulence of anaerobic wastewater and thereby a potential increased release of hydrogen sulfide. Typical systems with risk for increased turbulence are inlet structures, drops, cascades, sharp bends, and inverted siphons. As an example, changes in the flow regime from a pressure pipe into a gravity sewer may give rise to the release of hydrogen sulfide.

The first stages of concrete corrosion of a sewer pipe wall are typically seen near the daily water surface level and at the sewer crown. Areas near the daily water surface are close to where emission of $H_2S$ takes place, and both types of surface areas are especially exposed to moisture.

## 6.6 METAL CORROSION AND TREATMENT PLANT IMPACTS

In addition to hydrogen sulfide as a malodor and corroding substance, specific impacts also occur in terms of metal corrosion and effects on wastewater treatment processes.

Hydrogen sulfide is a weak acid that reacts with most heavy metals and produces a corresponding metal sulfide with a low solubility in water. With a divalent metal, the total reaction is as follows:

$$H_2S + Me \rightarrow MeS + H_2 \tag{6.18}$$

Metal corrosion has often been observed in pumping stations and sewer structures with electronic equipment.

Inlet structures at treatment plants may increase the turbulence of the wastewater. A corresponding release of hydrogen sulfide and other odorous substances into the air from inflow of anaerobic wastewater is a potential nuisance that may require treatment of the gas released (cf. Section 7.3.7). On the other hand, aeration of anaerobic wastewater may quickly oxidize sulfide and some of the organic odorous substances. However, the N-containing odorous substances are slowly oxidized and are particularly a potential problem whenever they occur in the inflow to the treatment plant (Hwang et al. 1995).

Inflow of wastewater containing sulfide to activated sludge treatment plants may lead to a change in floc structure because of a reduction and precipitation of Fe(III) in the flocs to FeS (Nielsen and Keiding 1998). The change has been observed as a weakening of the floc strength leading to potential disintegration. Release of up to 10% of the total organic matter of the flocs has been observed. Extracellular polymeric substances, colloids, and loosely adhered bacteria have been identified as being released.

Wastewater that has been under anaerobic conditions in a sewer may furthermore lead to inflow of filamentous bacteria, for example, species of *Thiothrix* and *Beggiatoa*, and thereby result in potential reduced settling characteristics of wastewater sludge.

## 6.7 ANAEROBIC MICROBIAL TRANSFORMATIONS IN SEWERS

From a basic point of view—however, still related to sewer systems—aerobic and anaerobic microbial processes was addressed in Chapter 3. The aerobic transformations and a corresponding conceptual formulation of the microbial processes are the main subjects of Chapter 5. Further details and the interactions between anaerobic water phase processes in sewers are in the following focused on, including processes related to both organic carbon and sulfur.

### 6.7.1 ANAEROBIC TRANSFORMATIONS OF ORGANIC MATTER IN SEWERS

The difference between aerobic and anaerobic transformations of organic matter is of fundamental importance when dealing with the conveyance of wastewater.

According to what was dealt with in Chapter 3 and in Section 6.3, the main processes in wastewater under anaerobic conditions are briefly as follows:

- Anaerobic hydrolysis: transformation of hydrolyzable substrate, $X_{Sn}$, to fermentable, readily biodegradable substrate, $S_F$.
- Fermentation: transformation of fermentable substrate, $S_F$, to fermentation products, $S_A$ [volatile fatty acids (VFAs)].
- Methanogenesis: transformation of fermentation products (VFAs) to methane $(CH_4)$.
- Sulfate reduction: formation of hydrogen sulfide.

These anaerobic processes have been investigated under sewer conditions (Tanaka and Hvitved-Jacobsen 1998, 1999, 2001; Hvitved-Jacobsen et al. 1999; Tanaka et al. 2000a, 2000b; Rudelle et al. 2011). The results from experiments in the laboratory, in pilot sewers, and under full-scale conditions have been combined with basic theoretical knowledge on the processes. An overview of the processes in a conceptual form is shown in Figure 6.8.

The concept expressed in Figure 6.8 is described in relatively simple terms. The most important parts are shown with fully drawn lines, whereas the dotted lines are typically less important for the formulation of sewer processes. The processes can be described in further details; however, the major concern is to establish a concept for which components and parameters can be experimentally determined without unrealistic resources for laboratory and field studies. Methods for this determination are subjects of Chapter 10.

One of the details of Figure 6.8 often omitted is the growth of the SRB. Characklis et al. (1990) proposed Equation 6.19 for determination of the stoichiometry for the sulfate reduction by using lactate as the carbon source for energy requirement and growth. This equation can be used to assess the importance of the simplification.

$$CH_3CHOHCOOH + 0.43H_2SO_4 + 0.067NH_3 \rightarrow 0.33CH_{1.4}N_{0.2}O_{0.4}$$
$$\underset{\text{lactic acid}}{} \qquad\qquad\qquad\qquad\qquad\qquad\qquad \underset{\text{biomass}}{}$$

$$+ 0.96CH_3COOH + 0.43H_2S + 0.7CO_2 + 0.94H_2O \quad (6.19)$$

In mass units, the yield constant, $Y$, for the SRB is, according to Equation 6.19:

$$Y = (\text{g biomass})/(\text{g lactic acid}) = 0.083 \text{ g g}^{-1} \quad\quad (6.20)$$

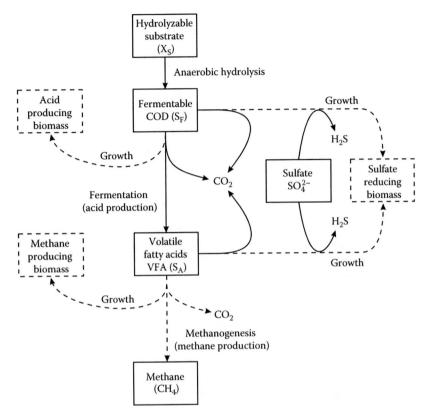

**FIGURE 6.8** Simplified concept for anaerobic transformations of biodegradable organic matter in wastewater and biofilm of a sewer system.

The relatively low yield constant makes it acceptable to omit the growth process for the SRB in the concept.

Figure 6.8 shows that the acid producing biomass via fermentation will compete with the SRB. Hydrolysis resulting in the formation of fermentable substrate, $S_F$, is therefore central, and the fermentation process as well as sulfate reduction are potentially limited by hydrolysis (Rudelle et al. 2011). Fermentation is a rather fast process and after extended anaerobic incubation, the readily biodegradable substrate is mainly converted to VFAs, in particular, acetate and propionate. $S_F$, in general, ethanol and lactate, are good substrates for SRBs but normally not acetate and propionate. Although methanogenesis potentially can proceed in permanent sewer deposits, it is typically not found to be an important process in sewer networks (cf. Figure 3.2). The competition under anaerobic conditions for VFAs is therefore limited, resulting in a buildup of VFAs as a dominating part of $S_S$ (cf. Figure 6.9).

As shown in Figure 6.8, an essential part of the anaerobic sulfur cycle in terms of the sulfate respiration process can be integrated with the anaerobic carbon cycle. As seen from this figure, the readily biodegradable substrate, $S_S$, is subdivided into $S_F$ and $S_A$. Although the SRB consumes both types of organic fractions, there

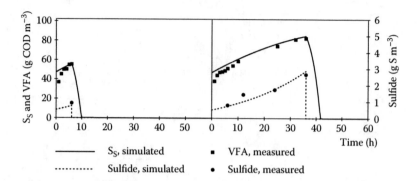

**FIGURE 6.9** Buildup of readily biodegradable substrate ($S_S$), VFA, and sulfide in wastewater under anaerobic conditions followed by depletion under aerobic conditions. Anaerobic periods of two experiments are 6 and 36 h, respectively. (From Rudelle, E. et al., *Water Sci. Technol.*, 66(8), 1728–1734, 2012. With permission.)

exist preferences for specific substances within both groups (Nielsen and Hvitved-Jacobsen 1988). As an example, propionate and ethanol are generally consumed faster than acetate. By integrating the sulfide and the organic carbon cycles, a simple conceptual approach is obtained as a possible substitution of the traditional empirical descriptions as shown in Table 6.6.

The nature as well as the process rates for the transformations of wastewater organic matter under anaerobic conditions are quite different compared with the aerobic transformation rates (cf. Chapter 5). As an example, Figure 6.10 outlines the average mass balance and flows of organic matter for a number of anaerobic experiments performed under laboratory conditions (Tanaka and Hvitved-Jacobsen 1999). The compounds and processes are included according to the concept shown in Figure 6.8. In Figure 6.10, it is seen that an essential part of $S_S$ is found as $S_A$ (VFA) and that no methanogenesis was observed. It is furthermore seen that the production rate of readily biodegradable substrate, $S_S$, by the anaerobic hydrolysis is larger than the removal rate. It is thereby shown that readily biodegradable substrate is not just

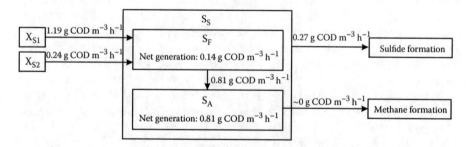

**FIGURE 6.10** Example showing average values of wastewater organic matter transformations under anaerobic conditions. (From Tanaka, N., Hvitved-Jacobsen, T., in: I.B. Joliffe, J. E. Ball (eds.), *Proceedings of the 8th International Conference on Urban Storm Drainage*, Sydney, Australia, August 30–September 3, 1999, pp. 288–296, 1999. With permission.)

preserved but produced under anaerobic conditions. This fact has potential positive implications on the subsequent treatment processes in terms of denitrification and biological phosphorous removal—if $S_S$ is not degraded under aerobic conditions in downstream sewer sections. On the other hand, it is not advisable if wastewater treatment is only required as BOD removal.

The results shown in Figure 6.10 give an error in the overall mass balance. It may be caused by the fact that average values of the experiments are shown but also because $CO_2$ production is not taken into account in the mass balance (cf. Table 3.3).

### 6.7.2 Conceptual Formulations of Central Anaerobic Processes in Sewers

The central anaerobic processes for transformation of organic matter in wastewater and in biofilm are outlined in Figure 6.8. The anaerobic hydrolysis and fermentation are in this respect those processes that primarily determine the quality and the changes in the composition of organic matter. Second, there are two other processes that, to some extent, may contribute to the mass balance of organic matter: growth of the SRB and decay of the heterotrophic biomass mainly produced under aerobic conditions. Referring to Equations 6.19 and 6.20, the relative importance of SRB growth can typically be neglected for the entire mass balance of organic matter. Anaerobic hydrolysis, fermentation, and decay of biomass are in the following described based on conceptual process rate expressions.

In addition to the expressions for transformation of organic matter, an equation for sulfide formation in anaerobic biofilms will be presented.

The mathematical description of the anaerobic processes must include terms that make them active when the concentrations of DO and nitrate approach zero. Two switch terms for both DO and nitrate similar to the one for DO shown in Equation 5.27 are therefore included in the mathematical expressions.

#### 6.7.2.1 Anaerobic Hydrolysis

The rate of anaerobic hydrolysis in the water phase and in the biofilm is in principle formulated similar to Equation 5.31 for anoxic hydrolysis:

$$r_{\mathrm{hydr,ana}} = \eta_{\mathrm{h,ana}} k_{hn} \frac{X_{Sn}/X_{\mathrm{Hw}}}{K_{Xn} + X_{Sn}/X_{\mathrm{Hw}}} \frac{K_O}{K_O + S_O} \frac{K_{NO_3}}{K_{NO_3} + S_{NO_3}} \left( X_{\mathrm{Hw}} + \varepsilon X_{\mathrm{Hf}} \frac{A}{V} \right) \alpha^{(T-20)}$$

$$(6.21)$$

where:

$r_{\mathrm{hydr,ana}}$ = rate of anaerobic hydrolysis (g COD m$^{-3}$ day$^{-1}$)

$\eta_{\mathrm{h,ana}}$ = efficiency constant for anaerobic hydrolysis relative to aerobic hydrolysis (–)

$k_{hn}$ = hydrolysis rate constant, fraction $n$ (day$^{-1}$)

$X_{Sn}$ = hydrolysable substrate, fraction # $n$; $\Sigma n$ typically equals 2 or 3: $n = 1$ (fast degradable), $n = 2$ (slowly degradable); $n = 1$ (fast degradable), $n = 2$ (medium degradable), $n = 3$ (slowly degradable)

$X_{\mathrm{Hw}}$ = heterotrophic biomass concentration in the water phase (g COD m$^{-3}$)

$K_{Xn}$ = saturation constant for hydrolysis of fraction n (g COD g COD$^{-1}$)
$K_{NO_3}$ = saturation constant for nitrate (g NO$_3$–N m$^{-3}$)
$S_{NO_3}$ = nitrate concentration in the water phase (g NO$_3$–N m$^{-3}$)
$\varepsilon$ = relative efficiency constant for hydrolysis of the biofilm biomass (–)
$X_{Hf}$ = heterotrophic biomass in the biofilm (g COD m$^{-2}$)
$K_O$ = saturation constant for DO (g O$_2$ m$^{-3}$)
$S_O$ = DO concentration in bulk water phase (g O$_2$ m$^{-3}$)
$\alpha$ = temperature coefficient (–)
$T$ = temperature (°C)

### 6.7.2.2  Fermentation

The anaerobic fermentation rate in the water phase and in the biofilm is expressed based on Monod kinetics for the fermentable substrate, $S_F$, including switch terms as shown in Equation 6.21:

$$r_{ferm} = q_{ferm} \frac{S_F}{K_{ferm} + S_F} \frac{K_O}{K_O + S_O} \frac{K_{NO_3}}{K_{NO_3} + S_{NO_3}} \left( X_{Hw} + \varepsilon X_{Hf} \frac{A}{V} \right) \alpha^{(T-20)} \qquad (6.22)$$

where:
$r_{ferm}$ = fermentation rate (g COD m$^{-3}$ day$^{-1}$)
$q_{ferm}$ = fermentation rate constant (day$^{-1}$)
$S_F$ = fermentable substrate (g COD m$^{-3}$)
$K_{ferm}$ = saturation constant for fermentation (g COD m$^{-3}$)

### 6.7.2.3  Anaerobic Decay of Heterotrophic Biomass

As described in Section 5.2.2, the endogenous decay of biomass is, under aerobic conditions, expressed as a maintenance energy requirement. When it comes to anaerobic conditions, a corresponding expression is not relevant. Decay or inactivation of the biomass may still be relevant to consider—but, typically, only if the biomass is exposed to long-term anaerobic conditions. The following expression for decay of biomass follows 1-order kinetics in the heterotrophic biomass concentration:

$$r_d = d_{H,ana} \frac{K_O}{K_O + S_O} \frac{K_{NO_3}}{K_{NO_3} + S_{NO_3}} X_{Hw} \alpha^{(T-20)} \qquad (6.23)$$

where:
$r_d$ = anaerobic decay rate of heterotrophic biomass (g COD m$^{-3}$ day$^{-1}$)
$d_{H,ana}$ = decay rate constant for the aerobic heterotrophic biomass under anaerobic conditions (day$^{-1}$)

### 6.7.2.4  Sulfate Reduction

Equation 4 in Table 6.6 is an empirical equation for prediction of the sulfide formation rate (Nielsen et al. 1998). The dissolved COD, COD$_S$, used in the equation is a measure of the wastewater quality relevant for SRBs. However, COD$_S$ does not exist among the COD compounds used in the conceptual formulation of sewer processes

(Table 3.7). A substitute for the term (COD$_S$ – 50) is therefore needed. It is a well-known fact that sulfate-reducing bacteria use readily biodegradable organic matter such as alcohols, lactate, pyruvate, and some aromatic substrates, but generally, not directly, for example, carbohydrates with high or moderate numbers of carbon atoms in the molecule. Based on studies in a pilot pressure sewer and field investigations, Tanaka and Hvitved-Jacobsen (1998) and Tanaka et al. (2000a) found that, among different options, the term ($S_S + X_{S1}$, with $S_S = S_F + S_A$) was an acceptable substitute for (COD$_S$ – 50) (cf. Figure 6.11). Theoretically, experimentally, and from a modeling point of view, there are good reasons for this substitution.

The biofilm sulfide formation rate transferred from a flux to the unit of concentration in the corresponding water phase is shown in Equation 6.24. The formulation of the rate expression follows 1/2-order kinetics as described in Section 2.2.2.1.

$$r_a = a\sqrt{S_F + S_A + X_{S1}} \; \frac{K_O}{K_O + S_O} \frac{K_{NO_3}}{K_{NO_3} + S_{NO_3}} \frac{A}{V} \alpha^{(T-20)} \tag{6.24}$$

where:

$r_a$ = biofilm surface formation rate for sulfide (g S m$^{-3}$ h$^{-1}$)
$a$ = rate constant for sulfide formation (g$^{0.5}$ m$^{-0.5}$ h$^{-1}$)
$A$ = biofilm surface area (m$^2$)
$V$ = water volume (m$^3$)

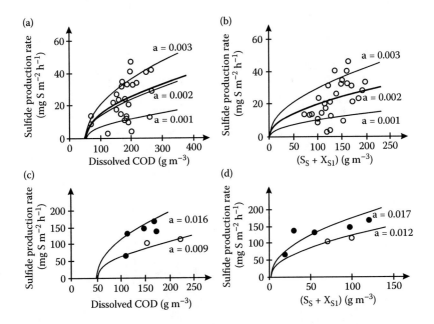

**FIGURE 6.11** Sulfide formation rates in sewers versus wastewater quality characteristics. Results originate from pilot-plant studies (a and b) and field investigations (c and d) in Japan, Kawasaki town (○) and Oga city (●). The a value refers to the coefficient included in Equation 4, Table 6.6. (From Tanaka, N. et al., *Water Environ. Res.*, 72(6), 651–664, 2000a. With permission.)

## 6.8   INTEGRATED AEROBIC–ANAEROBIC CONCEPT FOR MICROBIAL TRANSFORMATIONS

Finally, it is important to highlight the fact that aerobic, anoxic, and anaerobic microbial processes in sewers in varying order may occur sequentially. The conceptual formulations of the sewer processes dealt with in this text must therefore be looked upon as an entity. In other words, the different flows, pathways, and processes for the central substances, i.e., organic carbon, nitrogen, and sulfur, influence in common the quality of a sewer system and must therefore be integrated. The different cycles

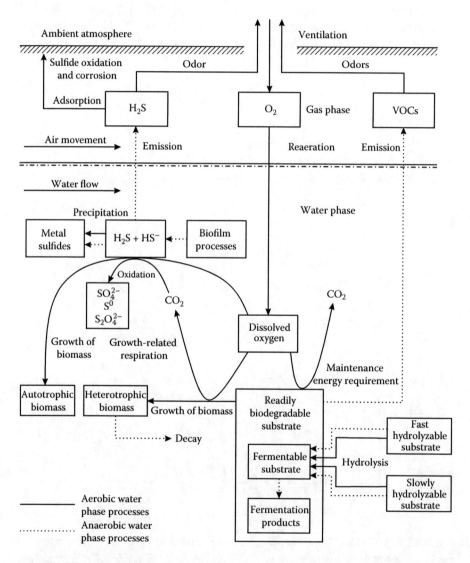

**FIGURE 6.12**   Simplified integrated aerobic–anaerobic concept for microbial transformations of organic carbon and sulfur in sewers.

**TABLE 6.7**

**Overview of Sewer-Related Cycles for Organic Carbon, Nitrogen, and Sulfur Shown in this Text**

| Figure Number | Subject |
|---|---|
| 4.7 | Sulfur cycle |
| 5.5 | Aerobic cycle for organic carbon |
| 5.6 | Aerobic cycle for organic carbon and sulfur |
| 5.17 | Anoxic cycle for organic carbon |
| 6.8 | Anaerobic cycle for organic carbon |

for each of these compounds are in principle those pieces that must be integrated. As an example, the aerobic concept for the integrated organic carbon and sulfur cycles is already shown in Figure 5.6.

It is in continuation of this approach convenient to give an overview of the different cycles of organic carbon, nitrogen, and sulfur that have been presented in this text (cf. Table 6.7).

In brief, it is the different cycles shown in Table 6.7 that in common give an overall description of potentially occurring sewer processes. Furthermore, it is important to note that other substances, for example iron, may also contribute and therefore must be integrated with these processes (cf. Section 2.4).

For practical purposes, it is not feasible to show the "final," fully integrated overview of all relevant cycles in one figure. However, except for cases where nitrate or iron salts are used for sulfide control, it is the integrated aerobic–anaerobic cycle for organic carbon and sulfur that are the most central to consider. This simplified integrated concept is shown in Figure 6.12 as an extension of Figure 5.6.

## REFERENCES

Aesoy, A., S.W. Oesterhus, and G. Bentzen (2002), Controlled treatment with nitrate in sewers to prevent concrete corrosion, *Water Sci. Technol. Water Supply*, 2(4), 137–144.

Aldred, M.I and B. G. Eagles (1982), Hydrogen sulfide corrosion of the Baghdad trunk sewerage system, *Water Pollut. Control*, 81(1), 80–96.

ASCE (American Society of Civil Engineers) and WPCF (Water Pollution Control Federation) (1982), *Gravity Sanitary Sewer Design and Construction*, ASCE manuals and reports on engineering practice No. 60 and WPCF manual of practice No. FD-5, p. 275.

ASCE (American Society of Civil Engineers) (1989), *Sulfide in Wastewater Collection and Treatment Systems*, ASCE manuals and reports on engineering practice No. 69, ASCE, New York, p. 324.

Bechmann, G.-E. (1900), *Notice sur le service des eaux et de l'assainissement de Paris (Exposition Universelle de 1900)*, Librairie Polytechnique Ch. Béranger éditeur, Paris, France, p. 524.

Bertrand-Krajewski, J.-L. (2003), Sewer sediment management: some historical aspects of egg-shaped sewers and flushing tanks, *Water Sci. Technol.*, 47(4), 109–122.

Bjerre, H.L., T. Hvitved-Jacobsen, S. Schlegel, and B. Teichgräber (1998), Biological activity of biofilm and sediment in the Emscher river, *Germany, Water Sci. Technol.*, 37(1), 9–16.

Boon, A.G. (1995), Septicity in sewers: causes, consequences and containment, *Water Sci. Technol.*, 31(7), 237–253.

Boon, A.G. and A.R. Lister (1975), Formation of sulfide in rising main sewers and its prevention by injection of oxygen, *Prog. Water Technol.*, 7, 289–300.

Bowlus, F.D. and A.P. Banta (1932), Control of anaerobic decomposition in sewage transportation, *Water Works Sewerage*, 79(11), 369.

Brüser, T., P.N.L. Lens and H.G. Trüper (2000), The biological sulfur cycle. In: P. Lens and L.H. Pol (eds.), *Environmental Technologies to Treat Sulfur Pollution – Principle and Engineering*, IWA Publishing, London, 47–85.

Characklis, W.G., W.C. Lee, and S. Okabe (1990), Kinetics and stoichiometry of planktonic and biofilm (sessile) sulfate-reducing bacteria, report of Inst. Biological and Chemical Process Analysis, Montana State University, Bozeman, MT.

Davy, W.J. (1950), Influence of velocity on sulfide generation in sewers, *Sewage Ind. Wastes*, 22(9), 1132–1137.

EWPCA (European Water Pollution Control Association) (1982), *Proceedings of EWPCA International State-of-the-art Seminar on Corrosion in Sewage Plants*, Hamburg, January 28–29, 1982.

Fjerdingstad, E. (1969), Bacterial corrosion of concrete in water, *Water Res.*, 3, 21–30.

Hvitved-Jacobsen, T., K. Raunkjær, and P. H. Nielsen (1995), Volatile fatty acids and sulfide in pressure mains, *Water Sci. Technol.*, 31(7), 169–179.

Hvitved-Jacobsen, T., J. Vollertsen, and N. Tanaka (1999), Wastewater quality changes during transport in sewers—an integrated aerobic and anaerobic concept for carbon and sulfur microbial transformations, *Water Sci. Technol.*, 39(2), 242–249.

Hvitved-Jacobsen, T., B. Jütte, P. Halkjær Nielsen, and N.Aa. Jensen (1988), Hydrogen sulphide control in municipal sewers, in: H. H. Hahn and R. Klute (eds.), *Pretreatment in Chemical Water and Wastewater Treatment, Proceedings of the 3rd International Gothenburg Symposium*, Gothenburg, Sweden, June 1–3, 1988, Springer-Verlag, New York, pp. 239–247.

Hwang, Y., T. Matsuo, K. Hanaki, and N. Suzuki (1995), Identification and quantification of sulfur and nitrogen containing compounds in wastewater, *Water Sci. Technol.*, 29(2), 711–718.

Islander, R.L., J.S. Devinny, F. Mansfeld, A. Postyn and S. Hong (1991), Microbial ecology of crown corrosion in sewers, *J. Environ. Eng.*, 117(6), 751–770.

James, C.C. (1917), Sewers and their construction, in: *Drainage Problems of the East* (a revised and enlarged edition of Oriental Drainage), Bennett, Coleman & Company, Ltd., Bombay, India.

Jensen, H.S., A.H. Nielsen, T. Hvitved-Jacobsen, and J. Vollertsen (2009a), Modeling of hydrogen sulfide oxidation in concrete corrosion products from sewer pipes, *Water Environ. Res.*, 81(4), 365–373.

Jensen, H.S., A.H. Nielsen, P.N.L. Lens, T. Hvitved-Jacobsen, and J. Vollertsen (2009b), Hydrogen sulfide removal from corroded concrete: comparison between surface removal rates and biomass activity, *Environ. Technol.*, 30(12), 1291–1296.

Jensen, H.S., P.N.L. Lens, J.L. Nielsen, K. Bester, A.H. Nielsen, T. Hvitved-Jacobsen, and J. Vollertsen (2011), Growth kinetics of hydrogen sulfide oxidizing bacteria in corroded concrete from sewers, *J. Hazard. Mater.*, 189, 685–691.

Kienow, K.E., R.D. Pomeroy, and K.K. Kienow (1982), Prediction of sulfide buildup in sanitary sewers, *J. Environ. Eng. Div., Proc. Am. Soc. Civil Eng. (ASCE)*, 108(EE5), 941–956.

Lahav, O., Y. Lu, U. Shavit, and R.E. Loewenthal (2004), Modeling hydrogen sulfide emission rates in gravity sewage collection systems, *J. Environ. Eng.*, 130(11), 1382–1389.

Matos, J.S. (1992), Aerobiose e septicidade im sistemas de drenagem de águas residuais, Ph.D. thesis, IST, Lisbon Portugal.

Melbourne and Metropolitan Board of Works (1989), Hydrogen sulphide control manual—septicity, corrosion and odour control in sewerage systems, Technological Standing Committee on Hydrogen Sulphide Corrosion in Sewerage Works, vols. 1 and 2.

Meyer, W.J. and G.H. Hall (1979), *Prediction of Sulfide Generation and Corrosion in Concrete Gravity Sewers: A Case Study*, J. B. Gilbert & Associates, A Division of Brown and Caldwell, Sacramento, CA.

Milde, K., W. Sand, W. Wolf, and E. Bock (1983), *Thiobacilli* of the corroded concrete walls of the Hamburg sewer system, *J. Gen. Microbiol.*, 129, 1327–1333.

Miljøstyrelsen (1988), Hydrogen sulfide formation and control in pressure mains, Danish Environmental Protection Agency, project report no. 96, p. 109 (in Danish).

Mori, T., M. Koga, Y. Hikosaka, T. Nonaka, F. Mishina, Y. Sakai, and J. Koizumi (1991), Microbial corrosion of concrete pipes, $H_2S$ production from sediments and determination of corrosion rate, *Water Sci. Technol.*, 23(7–9), 1275–1282.

Morton, R.L., W.A. Yanko, D.W. Graham, and R.G. Arnold (1991), Relationships between metal concentrations and crown corrosion in Los Angeles County sewers, *Res. J. WPCF*, 63(5), 789–798.

Nielsen, P.H. (1991), Sulfur sources for hydrogen sulfide production in biofilm from sewer systems, *Water Sci. Technol.*, 23, 1265–1274.

Nielsen, P.H. and K. Keiding (1998), Disintegration of activated sludge flocs in the presence of sulfide, *Water Res.*, 32(2), 313–320.

Nielsen, P.H. and T. Hvitved-Jacobsen (1988), Effect of sulfate and organic matter on the hydrogen sulfide formation in biofilms of filled sanitary sewers, *J. WPCF*, 60, 627–634.

Nielsen, P.H., K. Raunkjaer, and T. Hvitved-Jacobsen (1998), Sulfide production and wastewater quality in pressure mains, *Water Sci. Technol.*, 37 (1), 97–104.

Nielsen, A.H., J. Vollertsen, H.S. Jensen, T. Wium-Andersen, and T. Hvitved-Jacobsen (2008), Influence of pipe material and surfaces on sulfide related odor and corrosion in sewers, *Water Res.*, 42, 4206–4214.

Nielsen, A.H., T. Hvitved-Jacobsen, and J. Vollertsen (2012), Effect of sewer headspace airflow on hydrogen sulfide removal by corroding concrete surfaces, *Water Environ. Res.*, 84(3), 265–273.

Olmsted, F.H. and H. Hamlin (1900), Converting portions of the Los Angeles outfall sewer into a septic tank, *Eng. News*, 44(19), 317–318.

Parker, C.D. (1945a), The corrosion of concrete 1. The isolation of a species of bacterium associated with the corrosion of concrete exposed to atmospheres containing hydrogen sulphides, *Aust. J. Expt. Biol. Med. Sci.*, 23, 81–90.

Parker, C.D. (1945b), The corrosion of concrete 2. The function of *Thiobacillus concretivorus* (nov. spec.) in the corrosion of concrete exposed to atmospheres containing hydrogen sulphide, *Aust. J. Expt. Biol. Med. Sci.*, 23, 91–98.

Parker, C.D. (1951), Mechanics of corrosion of concrete sewers by hydrogen sulfide, *Sewage Ind. Wastes*, 23, 1477–1485.

Polder, R.B. and T. van Mechelen (1989), Abschätzung des biogenen Schwefelsäureangriffes auf Abwasseranlagen [Assessment of biologically produced sulfuric acid attacks on sewer networks], 2nd International Congress on Construction of Sewer Networks, Hamburg, West Germany, October 23–26, 1989.

Pomeroy, R. (1936), The determination of sulfides in sewage, *J. Sewage Works*, 8(4), 572.

Pomeroy, R.D. and F.D. Bowlus (1946), Progress report on sulphide control research, *J. Sewage Works*, 18(4).

Pomeroy, R. (1959), Generation and control of sulfide in filled pipes, *Sewage Ind. Wastes*, 31(9), 1082–1095.

Pomeroy, R.D. (1970), Sanitary sewer design for hydrogen sulfide control, *Public Works*, 101(10), 93–96 and 130.

Pomeroy, R.D. and J.D. Parkhurst (1977), The forecasting of sulfide buildup rates in sewers, *Prog. Water Technol.*, 9(3), 621–628.

Rudelle, E., J. Vollertsen, T. Hvitved-Jacobsen, and A.H. Nielsen (2011), Anaerobic transformations of organic matter in collection systems, *Water Environ. Res.*, 83(6), 532–540.

Rudelle, E., J. Vollertsen, T. Hvitved-Jacobsen and A.H. Nielsen (2012), Modeling anaerobic organic matter transformations in the wastewater phase of sewer networks, *Water Sci. Technol.*, 66(8), 1728–1734.

Sand, W. (1987), Importance of hydrogen sulfide, thiosulfate and methylmercaptan for growth of thiobacilli during simulation of concrete corrosion, *Appl. Environ. Microbiol.*, 53(7), 1645–1648.

Schmitt, F. and C.F. Seyfried (1992), Sulfate reduction in sewer sediments, *Water Sci. Technol.*, 25(8), 83–90.

Stutterheim, N. and J.H.P. van Aardt (1953), Corrosion of concrete sewers and some remedies, *S. Afr. Ind. Chem.*, 7(10).

Tanaka, N. and T. Hvitved-Jacobsen (1998), Transformations of wastewater organic matter in sewers under changing aerobic/anaerobic conditions, *Water Sci. Technol.*, 37(1), 105–113.

Tanaka, N. and T. Hvitved-Jacobsen (1999), Anaerobic transformations of wastewater organic matter under sewer conditions, in: I. B. Joliffe and J. E. Ball (eds.), *Proceedings of the 8th International Conference on Urban Storm Drainage*, Sydney, Australia, August 30–September 3, 1999, pp. 288–296.

Tanaka, N. and T. Hvitved-Jacobsen (2001), Sulfide production and wastewater quality—investigations in a pilot plant pressure sewer, *Water Sci. Technol.*, 43(5), 129–136.

Tanaka, N., T. Hvitved-Jacobsen, and T. Horie (2000a), Transformations of carbon and sulfur wastewater components under aerobic–anaerobic transient conditions in sewer systems, *Water Environ. Res.*, 72(6), 651–664.

Tanaka, N., T. Hvitved-Jacobsen, T. Ochi, and N. Sato (2000b), Aerobic–anaerobic microbial wastewater transformations and reaeration in an air-injected pressure sewer, *Water Environ. Res.*, 72(6), 665–674.

Tchobanoglous, G. (ed.) (1981), Occurrence, effect and control of the biological transformations in sewers, in: *Wastewater Engineering: Collection and Pumping of Wastewater*, Metcalf and Eddy, Inc., McGraw-Hill, New York, pp. 232–268.

Thistlethwayte, D.K.B. (ed.) (1972), *The Control of Sulfides in Sewerage Systems*, Butterworth, Sydney, Australia, p. 173.

USEPA (1974), *Process Design Manual for Sulfide Control in Sanitary Sewerage Systems*, USEPA 625/1-74-005, Technology Transfer, Washington, DC.

USEPA (1985), *Odor and Corrosion Control in Sanitary Sewerage Systems and Treatment Plants*, USEPA 625/1-85/018, Washington, DC.

Vincke, E., J. Monteny, A. Beeldens, N.D. Belie, L. Taerwe, D. van Gemert, and W.H. Verstraete (2000), Recent developments in research on biogenic sulfuric acid attack of concrete, in: P.N. L. Lens and L. H. Pol (eds.), *Environmental Technologies to Treat Sulfur Pollution—Principles and Engineering*, IWA Publishing, London, pp. 515–541.

Vollertsen, J., A.H. Nielsen, H.S. Jensen, T. Wium-Andersen, and T. Hvitved-Jacobsen (2008), Corrosion of concrete sewers—the kinetics of hydrogen sulfide oxidation, *Sci. Total Environ.*, 394, 162–170.

Vollertsen, J., L. Nielsen, T.D. Blicher, T. Hvitved-Jacobsen, and A.H. Nielsen (2011a), A sewer process model as planning and management tool—hydrogen sulfide simulation at catchment scale, *Water Sci. Technol.*, 64(2), 348–353.

Vollertsen, J., A.H. Nielsen, H.S. Jensen, E.A. Rudelle, and T. Hvitved-Jacobsen (2011b), Modeling the corrosion of concrete sewers, in: *12th International Conference on Urban Drainage*, Porto Alegre, Brazil, September 11–16, 2011, p. 9.

Wilmot, P.O., K. Cadce, J.J. Katinic, and B. V. Kavanagh (1989), Kinetics of sulfide oxidation by dissolved oxygen, *J. Water Pollut. Control Fed.*, 60, 1264–1270.

# 7 Sewer Processes and Mitigation
## *Water and Gas Phase Control Methods*

Adverse effects caused by sewer processes often call for mitigation. Typical examples of these adverse effects are deterioration of the sewer network in terms of concrete corrosion and corrosion of metal installations, toxic effects on sewer personnel, and odor problems in the surrounding environment. Generally, these problems are identified by the presence of specific substances, in particular hydrogen sulfide and volatile organic compounds (VOCs) (cf. Chapter 6). The adverse effects are thereby caused by the occurrence of volatile substances, and it is characteristic that the problems are closely related to the air phase in the sewer or in its adjacent environment.

It is important that the occurrence of hydrogen sulfide and VOCs is basically a result of anaerobic processes. It is likewise important to note that the location where a problem is identified is not necessarily where the generation of the substances takes place. Any specific sewer system should therefore be carefully analyzed and assessed in terms of both generation and transport of problematic substances to select where a technology for mitigation is most conveniently located.

The fact that anaerobic processes are the basic cause of a sewer-related problem correspondingly means that focus is on management of the "anaerobic sewer problem" in some way or another. When identifying these sewer problems directly with the occurrence of anaerobic conditions in the wastewater (and the biofilm), it is also important to realize that aerobic wastewater smells. However, there is a fundamental difference in the odor character between aerobic and anaerobic wastewater (cf. Section 4.3.1).

The ultimate solution to a sewer problem caused by anaerobic conditions is, of course, to build a sewer network where aerobic conditions always occur. However, it is not a "mitigation method," nor is the selection of materials for corrosion-resistant sewer construction (WEF 2007). In this book, mitigation is defined as a technology that—with relatively minor changes of constructions in or close to the sewer system—can be implemented to reduce an identified adverse effect of a sewer process. A technology thereby involves, for example, dosing of a chemical but also implementation of forced ventilation to remove problematic sewer gases.

Adverse effects caused by sewer processes are known worldwide, and the literature is rich in the description of both problems and mitigation methods. Manufacturers of equipment also contribute, often promising general, efficient, and cheap solutions.

This chapter is basically not a manual on how to establish a solution. The objective is to provide the reader with an overview of potential solutions as well as a deeper process insight, typically based on well-established theories. This fundamental basis is found in Chapters 1 through 4. Further details on aerobic, anoxic, and anaerobic process characteristics are discussed in Chapters 5 and 6. This knowledge is central for assessing and selecting mitigation methods among widely different alternatives.

## 7.1 OVERVIEW OF MITIGATION METHODS

In brief, mitigation methods aim at reducing the adverse effects of hydrogen sulfide and VOCs. It is important to note that these substances have different characteristics and that a specific control methodology is not necessarily equally effective in mitigating these different substances. Hydrogen sulfide is, for example, potentially subject to precipitation that is not possible for the molecular types of VOCs. In general, there are more mitigation methods available for managing hydrogen sulfide than for VOCs. As a starting point, it is therefore appropriate to provide a systematic overview of methods that can be used for reducing a hydrogen sulfide problem and then discuss the specific method in terms of VOC reduction.

It is convenient to discuss the use of mitigation methods for hydrogen sulfide based on the sulfur cycle in sewers (Zhang et al. 2008). By doing so, it is possible to identify where in the system and when the control of sulfide is feasible. Figure 4.7 is therefore central in this respect. Figure 7.1 identifies where sulfide control is possible, taking into account that a problem does not occur until hydrogen sulfide appears in the air phase. Figure 7.1 indicates three different points of the sulfur pathways

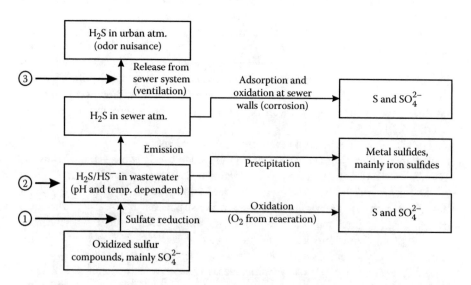

**FIGURE 7.1** Sulfur cycle in sewers with indication of three main points of the cycle where a mitigation method can change the formation, occurrence, or flow of hydrogen sulfide. Point nos. 1, 2, and 3 correspond to the descriptions in Sections 7.1.1 through 7.1.3, respectively.

where a mitigation method has a potential "starting point" by closing, reducing, or changing the generation or the transfer of hydrogen sulfide. These three different ways of tackling the hydrogen sulfide problem are briefly discussed in Sections 7.1.1 through 7.1.3.

The mitigation methods for sulfide, shown in Figure 7.1, are to some extent also active for other volatile sulfur compounds (VSCs). As an example, precipitation of mercaptans with metals will occur parallel with precipitation of sulfides.

### 7.1.1 INHIBITION OR REDUCTION OF SULFIDE FORMATION

Figure 7.2 gives an overview of different control methods for inhibition of sulfide formation. These methods aim at tackling the sulfide problem at the root by either weakening or inhibiting the general biological activity in the sewer or, more specifically, the activity of the sulfate reducing biomass (SRB).

The methods shown in Figure 7.2 will also reduce or eliminate the formation of VOCs. In particular, it is the case when dissolved oxygen (DO) or nitrate is added and the anaerobic formation of VOCs is thereby inhibited (cf. Section 3.2.2).

The nature of the different mitigation methods will be further discussed in the following sections.

### 7.1.2 REDUCTION OF GENERATED SULFIDE

Methods that reduce the generated sulfide typically operate on sulfide occurring in the water phase, often under anaerobic conditions. These methods, which are shown in Figure 7.3, are either biological or chemical in nature and specifically aim at removing sulfide. During the formation of sulfide, VOCs are also normally produced,

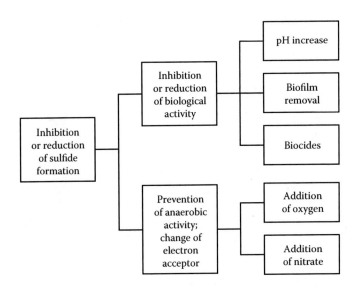

**FIGURE 7.2** Overview of control methods for inhibition of sulfide formation.

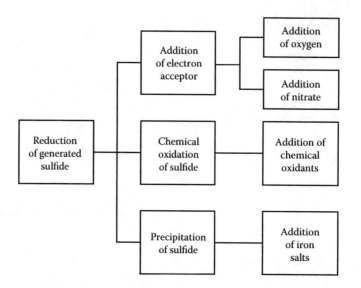

**FIGURE 7.3** Overview of control methods for reduction of generated sulfide. Addition of an electron acceptor mainly results in a biological oxidation of sulfide.

and the mitigation methods generally do not affect these substances. Furthermore, the formation of VOCs continues unchanged. Although some of the methods (addition of electron acceptors) are identical with the methods shown in Figure 7.2, there is basically a "time delay" in the implementation that allows VOCs to be generated under anaerobic conditions. In contrast to the methods shown in Figure 7.2, these controls are, therefore, not feasible (e.g., if an odor problem is dominated by volatile organics).

Figure 7.3 focuses on control methods for sulfide. However, metals can also, as an example, precipitate mercaptans.

These mitigation methods will be further discussed in the following sections.

### 7.1.3 SEWER GAS REDUCTION AND DILUTION

An overview of methods that tend to reduce the impact of hydrogen sulfide by focusing on its potential occurrence in the gas phase is shown in Figure 7.4. These methods are, by nature, "end-of-pipe solutions" for the gas flow. The formation of sulfide and VOCs in the water phase continues unchanged. Such methods are primarily applied by managing the gas phase but are also indirectly changing the composition of the water phase. An example of the latter type is to increase the pH in the water phase and thereby reduce the relative amount of the molecular and volatile form of sulfide, $H_2S$. The methods shown in Figure 7.4, therefore, reduce the concentration of $H_2S$ in the sewer gas phase either by reducing its transfer from the water phase or by applying forced ventilation.

Although a pH increase will reduce the volatility of hydrogen sulfide in the water phase, it is important to note that other VOCs that also exist in ionic and molecular

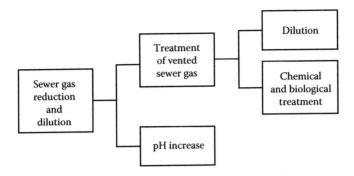

**FIGURE 7.4** Overview of control methods for reduction of hydrogen sulfide in emitted sewer gas.

forms may be affected. As an example, emission of ammonia and amines with high $pK_a$ values will be increased by increasing the pH value and thereby affect the occurrence of these substances in the sewer atmosphere and the odor character hereof (cf. Example 4.8).

## 7.2   SEWER PROCESS CONTROL PROCEDURES

The mitigation methods discussed in Section 7.1, and in this way given a systematic approach, are all more or less process-related controls. There is a need to expand these methods with different means of mitigation that are not directly process related but, from a practical and operational point of view, are important factors to consider (e.g., mechanical methods and management related procedures).

### 7.2.1   GENERAL ASPECTS OF SEWER PROCESS CONTROLS

The need for control of sewer process-related problems has, as discussed in Section 7.1, resulted in the development of numerous strategies and procedures. The effort to avoid sewer gas-related problems should be dealt with during the planning and design phase of the sewer network, although experience shows that it is often not done until the problem appears and requires "cure."

In addition to the systematic approach of sewer process mitigation shown in Section 7.1, control methodologies can be subdivided in terms of the type of control procedures. Such procedures add some kind of practical angle to the more systematic method of process mitigation:

- Design procedures for active control of sulfide problems—procedures typically implemented during the design phase of the sewer network to reduce the conditions for formation of sulfide and VOCs.
- Design procedures for passive control of sulfide problems—procedures typically included during design phase and aim at reducing the effect of sewer gases but not necessarily their formation.

- Operational procedures for control of sewer gas problems—procedures that aim at reducing the formation or related effects by the use of mitigation methods. The procedures are often implemented after a problem has been identified. An overview of these methods is given in Section 7.1.

The three main types of sewer gas control procedures give an overall subdivision, although some overlap between the three groups may exist. The procedures basically aim at reducing sulfide-related problems, although some of them are also relevant for reduction of VOCs. An overview of the methods emphasizing the process aspects is given in the following three subsections. The description of the procedures will not aim at giving information on detailed design and operational principles. A number of handbooks provide further details in this respect (Thistlethwayte 1972; USEPA 1974, 1985; WEF 2007).

### 7.2.1.1 Design and Management Procedures for Active Control of Sewer Gas Problems

These "active" design procedures for sewer systems include the following major tasks:

- Increase of the reaeration of the wastewater
- Reduction of the turbulence of anaerobic wastewater
- Reduction of conditions for sewer solids accumulation
- Reduction of the biofilm thickness

Procedures to observe these tasks tend to reduce the formation of harmful and odorous sewer gases or alternatively affect the occurrence of these substances in the air phase. In the design phase of a sewer network, it is therefore important to consider the implementation of system-dependent conditions that could alleviate a sewer gas problem.

#### 7.2.1.1.1 Reaeration

When designing sewer networks, particularly gravity sewers, reaeration is the major process to be focused on to reduce the formation of sulfide and VOCs (cf. Section 4.4). A number of hydraulic and system characteristics can be managed to increase the reaeration rate and thereby avoid or reduce anaerobic-related problems. The hydraulic mean depth, the hydraulic radius, the wastewater flow velocity, and the slope of the sewer pipe are, in this respect, important factors that are dealt with in Section 4.4. It should furthermore be stressed that it is not necessarily the objective to avoid sulfide formation (in the sewer biofilm), but the concentration of sulfide that occurs in the bulk water phase must remain at a low level. Therefore, the DO concentration in the bulk water phase should not be lower than about $0.2\text{--}0.5 \text{ g } O_2 \text{ m}^{-3}$ and thereby typically sufficiently high to oxidize sulfide—mainly in the upper part of the biofilm—before a considerable amount is emitted to the sewer atmosphere (cf. Section 5.5.2).

#### 7.2.1.1.2 Turbulence

As discussed in Section 6.3.3.6 and generally dealt with in Chapter 4, the turbulence of anaerobic wastewater should be avoided because of the risk of sulfide and VOC problems. The risk of emission is therefore typically high in gravity sewers with a steep slope and at sewer falls and drops (cf. Section 4.4.4). Otherwise, when dealing with aerobic systems, turbulence will enhance the reaeration.

### 7.2.1.1.3  Biofilms and Sewer Solids

The occurrence of both biofilms and sediments in a sewer may enhance the occurrence of anaerobic zones and thereby the risk for formation of both sulfide and VOCs. The formation of these two solid phases in a sewer depends primarily on the hydraulic conditions (cf. Section 1.5).

An increased wastewater flow velocity may lead to a reduction of the biofilm thickness and the occurrence of a permanent sediment layer. At very low velocities in an aerobic gravity sewer, a biofilm thickness may increase to more than 50 mm. However, it may be substantially reduced to approximately 1–5 mm when the flow velocity is increased to 0.5–1 m s$^{-1}$.

In terms of hydraulic conditions, the thickness of the biofilm is determined—although not fully quantified—by the shear stress on the pipe wall caused by the wastewater passing by. The average shear stress around the walled perimeter is:

$$\tau = \rho g s R \qquad (7.1)$$

where:

$\tau$ = average shear stress on the wetted surface (N m$^{-2}$)
$\rho$ = density of the wastewater (kg m$^{-3}$)
$g$ = gravitational acceleration (m s$^{-2}$)
$s$ = slope of the pipe (m m$^{-1}$)
$R$ = hydraulic radius, i.e., the cross-sectional area of the water volume divided by the welled perimeter (m)

A critical shear stress, $\tau_{crit}$ between 1 and 2 N m$^{-2}$, is typically recommended to prevent the formation of a thick biofilm. However, the Melbourne and Metropolitan Boards of Work (1989) proposed $\tau = 3.4$ N m$^{-2}$. In a small diameter pipe (<0.4 m), this value corresponds to a flow velocity about 1 m s$^{-1}$. In a larger pipe with a diameter >1 m, the velocity should exceed 1.2–1.4 m s$^{-1}$ to prevent excessive biofilm growth. It is furthermore important that a biofilm affects the roughness of the sewer pipe.

It is crucial to design a sewer system without permanent solids accumulation. Investigations have shown that the potential for sulfide formation in terms of a surface flux may exceed what occurs in biofilms (Schmitt and Seyfried 1992; Bjerre et al. 1998). Solids deposition and resuspension of deposits in sewers and requirements for self-cleansing conditions are briefly discussed in Sections 1.5.3 and 3.2.8.1. Readers who wish to explore the literature on physical properties and characteristics of sewer solids are referred to these two sections.

### 7.2.1.2  Design Procedures for Passive Control of Sewer Gas Problems

Like the active design procedures, the passive procedures tend to reduce a potential sewer gas problem. However, and in contrast to the active procedures, they do not affect the formation of these gases or their occurrence at a location where they are unwanted. The following two design considerations are particularly relevant for the reduction of corrosion:

1. Selection of corrosion-resistant materials
2. Design of a ventilated sewer system

The concrete material used for sewer construction is typically based on the use of Portland cement. Different types of Portland cement exist, without showing significant differences in the potential concrete corrosion rate. However, an increase in the relative amount of cement increases the alkalinity and thereby the active material for reaction with hydrogen sulfide (cf. Section 6.5.2.1). The corrosion rate—in units of mm year$^{-1}$—is thereby reduced according to the increase in the alkalinity per unit volume of the concrete material (Grennan et al. 1980). The use of high-alkaline materials as aggregates in concrete (e.g., limestone and dolomite) will also decrease the corrosion rate. Corrosion-resistant types of concrete exist, but they are typically relatively expensive. Such types are typically based on supersulfated or aluminous cement or cement with a high content of sulfur. The concrete surface can also be protected by coating or painting (e.g., epoxy resin and coal tar). The impermeability of such coatings is crucial to resist sulfide attack. Furthermore, alternatives for sewer construction exist in terms of, e.g., PVC, ABS, and PE plastics materials. Furthermore, pipes of glazed earthenware have shown good resistance to sulfide corrosion. Examples of corrosion-resistant designs of sewer systems are found in the reports of Kienow and Pomeroy (1978) and ASCE (1989).

Forced ventilation of sewers may not only reduce the hydrogen sulfide concentration in the sewer atmosphere (and VOCs) but also the moisture that is a fundamental requirement for establishment of microbial activity at the sewer walls. It is important that the ventilation be well controlled. Otherwise, odorous problems in the vicinity of the sewer network may occur. In some cases, operational procedures such as treatment of the vented sewer gas are needed (cf. Section 7.3.7).

When extending the understanding of "passive sewer gas control," the number of options becomes legion. For example, the location of the sewer network is, in practice, important to consider. An unsuitable means of handling such a problem may, in terms of malodors, seriously affect certain areas of a city, in particular, those with a high population density. Complaints from the public often constitute the provoking factor for managing odor problems from sewers.

### 7.2.1.3  Operational Procedures for Control of Sewer Gas Problems

Operational procedures for the control of sulfide problems are, in terms of a classification based on the sulfur cycle, outlined in Section 7.1. These different methods have played an important role for mitigation of sewer gas-related problems over the past 50 to 60 years. The methods are being increasingly implemented in existing sewer systems. The reason is that sulfide and VOC problems have not always been expected to occur when designing the network, or it was an acceptable practice to deal with sewer gas problems for the public as well as for managers and workers during the daily operation of the sewer network.

In pressure mains—except for the short ones—sulfide and VOC formation may typically always take place. In such systems, control is therefore generally needed, and appropriate technologies operated by the asset owner should be designed and implemented from the very beginning. In principle, it is not possible to recommend specific mitigation methods. However, it is important to note that not all of them can be used to control VOCs and therefore solve an odor problem completely (cf. Section 7.1). Chemical precipitation by addition of iron salts is probably what is mostly used in practice, although odorous problems caused by VOCs are generally not controlled with mercaptans as a likely exception. The use of an electron acceptor, dissolved oxygen or nitrate, is also frequently applied. Methods that have an oxidizing effect—including the electron acceptors mentioned—will, in addition to sulfide control, also typically reduce malodors caused by VOCs.

There are several ways of classifying the operational procedures for sewer gas mitigation. From a process viewpoint, the classification based on the sulfur cycle is fundamental (cf. Section 7.1). A more management-oriented classification as outlined in Table 7.1, however, is also often used.

Some of the methods outlined in Table 7.1 require the addition of chemicals. In general, it must be considered whether such methods may result in negative effects, e.g., on the successive wastewater treatment processes or in the receiving waters.

Selected and often used mitigation methods for sewer gas control are described in the following sections. Further examples and details of these methods are found in the literature (e.g., USEPA 1974, 1985; Pomeroy et al. 1985; ASCE and WPCF 1982; ASCE 1989; Melbourne and Metropolitan Board of Works 1989; Vincke et al. 2000; Stuetz and Frechen 2001).

## TABLE 7.1
## Operational Methods for Control of Sewer Gases

| General Principles of Mitigation | Examples of Measures |
|---|---|
| Prevention of sulfate reducing conditions (prevention of anaerobic conditions in the water phase) | Addition to the wastewater of: Air Pure oxygen Nitrate |
| Prevention of adverse effects of hydrogen sulfide (methods applied on the water phase) | Chemical precipitation of sulfide: Iron (II and III) sulfate Iron (II and III) chloride |
| Methods aiming at specific effects (reduction of $H_2S$ emission, chemical oxidation and effect on the biological system) | Addition of alkaline substances (increase in pH) Addition of chlorine, hydrogen peroxide, ozone Addition of biocides |
| Mechanical methods | Flushing Use of a ball for detachment of the biofilm |
| Management of the sewer gas | Forced ventilation Chemical or biological treatment of the sewer gas |

## 7.3  SELECTED MEASURES FOR CONTROL OF SEWER GASES

The different control measures for hydrogen sulfide and VOCs are briefly described in Sections 7.1 and 7.2. The fundamental characteristics of these technologies and their overall potentials for treatment are thereby focused on. In the following subsections, selected technologies are further dealt with. As discussed in Sections 7.1 and 7.2, the technologies for control are rather different in their approach and extent to reduce adverse effects. It is therefore also important to note that mitigation methods can be combined and thereby achieve an overall objective, in terms of, e.g., treatment efficiency and cost-effectiveness.

It is crucial to note that a successful use of these technologies requires a basic process insight. The preceding chapters, in this respect, have provided the reader with general and important knowledge and information. This basis is therefore considered well known and only referred to in specific cases.

### 7.3.1  MEASURES AIMED AT PREVENTING ANAEROBIC CONDITIONS OR THE EFFECT HEREOF

The common characteristic of these methods is the addition of an electron acceptor that increases the redox potential. It is thereby possible to suppress anaerobic processes such as sulfate reduction (sulfide formation), anaerobic hydrolysis, and fermentation (production of low molecular organics, VOCs). In practice, these mitigation methods involves the addition of oxygen or nitrate (cf. Figures 7.2 and 7.3). Addition of oxygen is possible in terms of air or as pure oxygen.

Referring to Figures 7.2 and 7.3, it is important to note that there are basically two different ways to apply an electron acceptor: (1) preventing anaerobic conditions in the sewer or (2) oxidizing reduced substances (sulfide and VOCs) that have been produced in, e.g., upstream sewer sections. In practice, the first option is carried out by adding the electron acceptor upstream a section where anaerobic conditions may otherwise occur. The second option involves the addition of oxygen or nitrate downstream a sewer section where anaerobic processes have produced sulfide and VOCs. In general, however, the second method cannot be considered efficient because of an insufficient residence time for reaction. The methods based on the addition of an electron acceptor are applied in systems with rising (pressure) mains as well as in gravity sewer systems.

Both oxygen and nitrate added to wastewater create conditions for a relatively high biological activity. The biological activity is highly temperature dependent, typically resulting in an increase in the transformation rates of about 7% per °C increase in temperature. It is therefore worth remembering that these methods are correspondingly less efficient per unit of electron acceptor and therefore less cost-effective when used in warm climates.

Finally, it is crucial to remember that both hydrogen sulfide and odor problems can be controlled by addition of oxygen or nitrate. It is, however, important to remember that oxidation with nitrate of already produced sulfide is very slow and therefore in practice not feasible. Details concerning the aerobic and anoxic processes are found in Chapter 5.

### 7.3.1.1   Injection of Air

Theoretically, the activity of the SRB requires a redox potential of less than about −200 mV (cf. Section 2.1.3.2 and Boon 1995). Oxygen in injected air will prevent the occurrence of anaerobic conditions in the water phase of the sewer (cf. Figure 5.14 and Gutierrez et al. 2008). To some extent, it will also oxidize already produced sulfide and VOCs. In practice, a DO concentration exceeding 0.5 g $O_2$ m$^{-3}$ is typically sufficient to prevent the occurrence of sulfide in the water phase. By injecting air at an upstream location, the DO concentration in the wastewater establishes an aerobic upper layer in the biofilm, and sulfide produced in the deeper part of the biofilm or in the deposits that otherwise might diffuse into the water phase will be oxidized (cf. Figure 5.14). The oxidation kinetics of sulfide—both chemical and biological oxidation—is described in Section 5.5. Major factors affecting the oxidation rate of sulfide include pH and temperature, but also the presence of sulfide-oxidizing microorganisms and catalysts (e.g., heavy metals).

It is important to note that the amount of oxygen needed to avoid sulfate-reducing conditions is determined by the aerobic respiration rate of the wastewater and the biofilm and not the potential of sulfide production in the sewer. The solubility of DO in water is relatively low under normal atmospheric conditions (8–11 g $O_2$ m$^{-3}$ depending on the temperature) compared with a DO consumption rate as high as 20–40 g $O_2$ m$^{-3}$ h$^{-1}$ (depending on wastewater quality, pipe geometry, and temperature) (cf. Section 5.1). The pressure in a rising main is >1 atm, and the solubility of $O_2$ is correspondingly increased (cf. Equation 4.35). It is typically required that air must be injected at several points of a sewer pipe to ensure aerobic conditions. From a practical point of view, it is expensive and requires the installation of dosing equipment at different locations and the corresponding manpower for operation and maintenance purposes.

Furthermore, the readily biodegradable and fast hydrolyzable fractions of the organic matter are removed and may be more or less depleted (Tanaka et al. 2000). In the case of the requirement for mechanical treatment of the wastewater, it is a positive development. However, the opposite is true if the subsequent treatment requires denitrification and biological phosphorous removal.

A large number of technical systems and methods for air injection exist. Boon (1995) proposed a method that reduces the requirement for oxygen in a rising main. In this system, air is injected continuously at the discharge end of the pipe, where the wastewater is recirculated in a relatively short length of the sewer line close to its end until sulfide is sufficiently depleted. By oxidation at a downstream point, less oxygen is used for organic carbon oxidation. However, release of hydrogen sulfide to the air is a potential risk.

When injecting air in pressurized sewer systems, it must be taken into account that the air volume of oxygen only amounts to about 20%. The inert $N_2$ gas—to some extent, mixed with the incoming air—may accumulate at the top part of a pressure sewer. In order to maintain the hydraulic capacity of the pressure main, the installation of an automatic vent system to remove this gas may be required. The pumping performance of an air-injected force main can be assessed using a model for the energy loss.

### 7.3.1.2   Injection of Pure Oxygen

The injection of pure oxygen overcomes some of the problems by using air injection but also creates problems. The solubility of pure oxygen in water, when compared with the oxygen in air increases with a factor of about 5, i.e., it is about 45–50 g $O_2$ m$^{-3}$

at a pressure of 1 atm. In a pressure main with higher pressure, this value is increased even more (cf. Equation 4.35). Furthermore, the amount of inert gas to be managed is reduced to about nil. A disadvantage is that pure oxygen results in a DO nonlimited biomass activity and therefore also both high consumption rate of the oxygen added and a reduction of readily biodegradable organic matter. Furthermore, pure oxygen must be available at the injection point, kept in containers, or generated on-site.

It should also be noted that the high solubility of oxygen in water relies on contact with pure oxygen in the air phase. The use of pure oxygen is therefore in practice limited to pressure sewers, in particular, those with a relatively high pressure.

### 7.3.1.3 Addition of Nitrate

The addition of nitrate will establish anoxic conditions when DO is depleted and will, thereby, increase the redox potential and suppress sulfate reduction and production of noxious VOCs. Details of the anoxic processes in the bulk water phase and in the biofilm are discussed in Section 5.6. Supplementary literature is referred to in this section. In addition to preventing the occurrence of anaerobic conditions, autotrophic anoxic oxidation of sulfide may also occur (cf. Section 5.6.2.2). The rate of oxidizing already produced sulfide by nitrate is, however, low and therefore not feasible in practice as an end-of-pipe solution.

The microbial activity of wastewater is typically lower under anoxic conditions than at the corresponding aerobic state (Abdul-Talib et al. 2002). From a practical point of view, it is important, because a low nitrate uptake rate (NUR) compared with the oxygen uptake rate in units of electron equivalents means a corresponding reduced transformation rate of the most biodegradable fractions of organic matter (cf. Figure 5.16). As discussed in Section 7.3.1.1, this aspect is relevant in terms of preservation of readily biodegradable substrate for use in wastewater treatment. Furthermore, a relatively low NUR value has operational and economic advantages because of a reduced demand for nitrate to maintain anoxic conditions. Nitrate is, particularly at high temperature, a relatively expensive chemical to use for sulfide control.

Nitrate can be added as different salts, e.g., $Ca(NO_3)_2$, well known as a fertilizer product. The rate of nitrate addition to the wastewater must be controlled to avoid the inexpedient inflow of nitrate to the wastewater treatment plant located downstream (Bentzen et al. 1995; Einarsen et al. 2000).

The effectiveness of nitrate addition to control sulfide was investigated in a sewer network in the Lake Balaton catchment, Hungary (Jobbágy et al. 1994). In this study, Figure 7.5 shows the effect of nitrate addition on the $H_2S$ concentration in the headspace of a manhole located downstream of the sewer section where nitrate was added.

### 7.3.2 Chemical Precipitation of Sulfide

Referring to Figure 7.3, chemical precipitation aims at preventing the adverse effects of the generated sulfide. Anaerobic conditions still prevail in the sewer, and the formation of sulfide and VOCs is basically not affected. Odor problems caused by VOCs are in general not mitigated by this methodology. However, mercaptans (thiols) may,

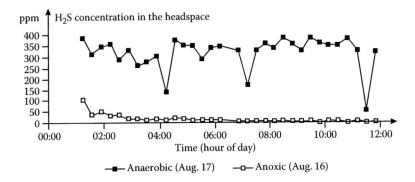

**FIGURE 7.5** H$_2$S concentration in headspace of a sewer manhole with and without addition of nitrate at an upstream located station.

to some extent, be precipitated, and the emission of amines is reduced at low pH values (cf. Section 7.1.3). As seen in Figure 7.1, precipitation of sulfide reduces the soluble sulfide concentration in the water phase and thereby the emission of H$_2$S to the sewer atmosphere. The negative effects of sulfide are therefore avoided by precipitation and inactivating it as a solid.

The most common metal salts used for precipitation of sulfide are ferrous iron, iron(II), and ferric iron, iron(III), as either sulfate or chloride. Basic aspects and characteristics of the iron chemistry relevant for sulfide control are described in Section 2.5.2. Although the kinetics of sulfide and a mixture of iron(II) and iron(III) are complex, the use of iron salts for sulfide control is, in practice, rather efficient, simple, and relatively cheap. This mitigation method is therefore applied worldwide (Thistlethwayte 1972; USEPA 1985; Hvitved-Jacobsen et al. 1988; Jameel 1989).

Knowledge on the iron chemistry related to sulfide control—as discussed in Section 2.5.2—is important when optimizing precipitation with iron salts. In the following, additional aspects important for chemical sulfide precipitation in practice are outlined.

As described in Section 2.5.2, three major aspects are principally determining the combined efficiency of iron(II) as soluble Fe$^{2+}$ and iron(III) as Fe$^{3+}$, namely, the integrated reaction kinetics of these species, the pH-dependent precipitation of FeS, and the fact that the governing processes are either biological or chemical of nature. In practice, this means that although Fe$^{3+}$—as shown by Equations 2.63 and 2.64 in terms of stoichiometry—is more efficient than Fe$^{2+}$, the relatively low reaction rate of Equation 2.64 may reduce this advantage. Only when adding iron salts upstream a sewer section where anaerobic conditions occur, is the residence time typically sufficient to benefit from the rather slow autotrophic microbial process where ferric iron oxidizes sulfide to elemental sulfur. Another important aspect is that FeS does not precipitate efficiently at pH values less than about 8. Efficient sulfide control by using iron salts may therefore require an increase of pH by adding a base, e.g., Ca(OH)$_2$.

When estimating the amount of iron needed for chemical precipitation, it is often assumed that $Fe^{3+}$ is reduced to $Fe^{2+}$ by chemoheterotrophic bacteria (cf. Section 3.1.3), without any additional effect on sulfide removal. Pragmatically, and depending on the pH value, the requirement of iron to precipitate sulfide can therefore be determined by the stoichiometry of the following equation:

$$Fe^{2+} + HS^- \rightarrow FeS + H^+ \tag{7.2}$$

Sulfide is thereby removed from the water phase as a precipitate, and emission to the sewer atmosphere is hindered. When adding iron salts at an upstream location, the amount needed is therefore determined by the formation rate of sulfide in the sewer section located downstream this point. It is anticipated that the addition of iron salts has no effect on the formation rate of sulfide.

The stoichiometry expressed by Equation 7.2 shows that addition of an iron salt requires 1.75 g Fe per g of $(H_2S + HS^-)$-S. In other words, the stoichiometric determined ratio is g Fe/g S = 1.75. In practice, however, the iron requirement for sulfide control is attributed to side reactions higher than those expressed by the stoichiometry. Table 7.2 outlines estimated iron/sulfur ratios when adding iron to an upstream station in a pressure main. Compared with the theoretical stoichiometric Fe/S ratio, there may be further need for addition of iron salts to control sulfide. In addition to the pH-dependent efficiency of FeS precipitation, such requirement for iron salt is caused by a potential reaction of iron to form precipitates and complexes with, e.g., phosphorous compounds, organic compounds, and alkaline substances. The chelating agents EDTA (ethylenediaminetetraacetic acid) and NTA (nitrilotriacetic acid), which are widely used in detergents and therefore found in wastewater, can form complexes with iron. Furthermore, iron can form complexes with other types of organic matter.

The interactions between iron(II) and iron(III) are described as complex and uncertain. Experiments show that some governing processes for sulfide removal by addition of iron salts are slow and apparently dependent on rather unknown conditions in the anaerobic environment of the sewer. It is therefore likely that the residence time may play a role for the effectiveness of sulfide precipitation. The values shown in Table 7.2 are therefore rough estimates that should be assessed in each specific case.

When adding iron salts at a downstream end of a sewer section, iron(II) salts should be recommended because of an insufficient residence time for $Fe^{3+}$ to oxidize

---

**TABLE 7.2**

**Estimated Iron to Sulfur Ratios Depending on Ionic Form of Iron and pH when Adding Iron Salts to an Upstream Point of a Pressure Main**

| Ionic Form of Iron | pH = 7.0 (g Fe/g S) | pH = 7.5 (g Fe/g S) | pH = 8.0 (g Fe/g S) |
|---|---|---|---|
| 100% $Fe^{2+}$ | 4.4 | 2.5 | 1.8 |
| 100% $Fe^{3+}$ | 4.4 | 2.5 | 1.8 |
| 50% $Fe^{2+}$ and 50% $Fe^{3+}$ | 3.5 | 2.0 | 1.4 |

$HS^-$ to $S^0$ and form $Fe^{2+}$. The subsequent precipitation of sulfide according to Equation 7.2 is thereby suppressed.

Sulfate is needed as an electron acceptor for the sulfate reduction process, but normally, it is available in nonlimiting concentrations in wastewater. In practice, this means a concentration of sulfate of more than about 5–10 gS m$^{-3}$. If this is not the case, addition of iron sulfate may affect the potential amount of sulfate produced. The anaerobic hydrolysis of organic matter and fermentation processes still proceed, and the formation of odorous organic compounds (VOCs) is therefore not affected by the addition of iron salts.

Iron salts are normally added to wastewater in a dissolved form and mainly as iron(II) with some contents of iron(III). $Fe^{3+}$ is hydrated (typically with six water molecules associated with iron) and reacts as an acid:

$$Fe(H_2O)_X^{3+} \rightarrow Fe(H_2O)_{X-1}(OH)^{2+} + H^+ \tag{7.3}$$

The hydrated $Fe^{3+}$ is a relatively strong acid with $pK_a = 2.2$. Noncorrosive materials should therefore be used for the dosing equipment. The addition of iron salts reduces the alkalinity of the wastewater, and the pH value is typically affected (cf. Section 2.4.2). As shown in Table 7.2, a decrease in pH will reduce the efficiency of iron addition for the precipitation of sulfide. It is furthermore important to note that an effect of a low alkalinity of wastewater is a potential reduction of the nitrification rate in the subsequent treatment process.

The precipitated FeS typically remains in the wastewater phase as small, suspended particles that make the color black. Iron salts are normally found in wastewater in small—but not necessarily insignificant—amounts. Therefore, a small amount of the sulfide produced in a sewer will typically be removed by precipitation of the iron that naturally occurs.

Although a number of precautions have been mentioned for the control of sulfide by precipitation as FeS, this methodology is often considered not only acceptable but also cheap and efficient. Iron salts added in the sewer can be reused in the treatment process. Under aerobic conditions in the wastewater treatment plant, the amorphous FeS is typically rapidly oxidized, and the iron can be released for the chemical removal of phosphate.

### 7.3.3 Chemical Oxidation of Sulfide

A number of chemicals can be used to control sulfide and malodors, typically chlorine compounds, hydrogen peroxide, ozone, and permanganate. The primary effect of these chemicals is the oxidation of sulfide to elemental sulfur or sulfate, but VOCs are also, to some extent, oxidized. Some of the chemicals will, in different ways, affect the general biological activity in the sewer. The chemicals have varying effectiveness depending on the characteristics of the chemical and the amount added.

Dosing equipment is, in principle, simple for most chemicals. However, it should be noted that these chemicals are oxidants, and that materials for the equipment must be selected based on this fact. Furthermore, precaution must always be taken to protect operators against the potential toxicity of the chemicals.

The following four subsections outline the main characteristics of the most commonly used chemicals. Further details of these control measures are found in several reports (e.g., WEF 2004).

### 7.3.3.1 Chlorine Compounds

Chlorine ($Cl_2$) is added directly to the wastewater as a gas or as hypochlorite ($OCl^-$) (often as NaOCl) producing $Cl_2$. The sulfide-controlling effect of chlorine compounds is based on the oxidation of sulfide to elemental sulfur, $S^0$, or sulfate, $SO_4^{2-}$, with $S^0$ as the main product at pH > 7:

$$HS^- + Cl_2 + OH^- \rightarrow 2Cl^- + S^0 + H_2O \qquad (7.4)$$

Chlorine also has, to some extent, a poisoning effect on the biological system. Chlorine is nonspecific, and a large number of side reactions with organic wastewater components producing chlorinated organics may occur.

In Equation 7.4, it is seen that the stoichiometric $Cl_2/S$ ratio in mass units is 2.2, whereas oxidation to sulfate requires a ratio of 8.8. The side reactions of chloride furthermore require chlorine, and a $Cl_2/S$ ratio of 10–15 is typically needed for sulfide control.

In contrast to hypochlorite, chlorine gas has serious poisoning effects on humans, and use of safety equipment is needed when it is added as a gas. Because of the formation of chlorinated organic matter, e.g., chlorinated VOCs, chlorine is also a potentially environmentally problematic chemical that cannot always be recommended for use as a control agent. Furthermore, addition of chlorine compounds in a sewer may also result in adverse effects on the biological processes in a wastewater treatment plant.

### 7.3.3.2 Hydrogen Peroxide

Hydrogen peroxide ($H_2O_2$) oxidizes sulfide to elemental sulfur ($S^0$) or sulfate ($SO_4^{2-}$), with $S^0$ as the main product at a pH of less than about 8.5. Under such conditions typical for wastewater, the redox reaction is:

$$HS^- + H^+ + H_2O_2 \rightarrow S^0 + 2H_2O \qquad (7.5)$$

The stoichiometric $H_2O_2/S$ mass ratio is therefore 1. Although hydrogen peroxide compared with chlorine is rather specific in its ability to oxidize sulfide, this ratio should typically be increased to about 2 for practical purposes.

Hydrogen peroxide can, to some extent, degrade the sulfate-reducing bacteria in the biofilm. Although $H_2O_2$ is relatively expensive, it may be economical in its use. A concentration of 5–10 g $H_2O_2$ $m^{-3}$ is often sufficient to suppress the occurrence of sulfide.

### 7.3.3.3 Ozone

Ozone is a very strong and unstable oxidant that will disproportionate to molecular and elemental oxygen:

$$O_3 \rightarrow O_2 + O^0 \qquad (7.6)$$

Elemental (atomic) oxygen reacts with almost all reduced substances in wastewater including sulfide, VOCs, and microorganisms. It is effective, however, and relatively expensive to use.

#### 7.3.3.4 Permanganate

The oxidation characteristics of permanganate mostly used as potassium permanganate ($KMnO_4$) are complex. It oxidizes sulfide to a mixture of elemental sulfur, sulfate, and other oxidation products of sulfur. Permanganate is thereby reduced to manganese dioxide ($MnO_2$), seen as brown fluffy flocs.

### 7.3.4 ALKALINE SUBSTANCES INCREASING pH

Alkaline substances such as sodium hydroxide (NaOH), calcium hydroxide (lime), [$Ca(OH)_2$], and magnesium hydroxide [$Mg(OH)_2$] increase pH. An increase in pH increases the amount of HS$^-$ relative to the amount of the volatile, molecular form of sulfide, $H_2S$ (cf. Figure 2.7). According to this figure and Equation 2.7, Table 7.3 shows a considerable pH-dependent distribution between $H_2S$ and HS$^-$ for values relevant for wastewater (cf. Section 2.1.3.3).

Addition of alkaline chemicals can be done in different ways, resulting in different types of effects on the occurrence of sewer gases, in particular, hydrogen sulfide:

- As shown in Table 7.3, an increase in pH increases the relative amount of sulfide that remains in the water phase. The emission of hydrogen sulfide is thereby depressed. Continuous addition of an alkaline chemical for adjustment of the pH value to >8.5 will considerably reduce the potential for emission of $H_2S$.
- Periodic "shock dosing" of an alkaline chemical to about 12 for 0.5–1 h affects the microorganisms in the biofilm and typically results in sloughing. In case of "shock addition," the effect on sulfide formation is only temporary until a new biofilm has been reestablished, typically after few days to about 2 weeks depending on, e.g., the temperature.

Any negative effects of alkaline dosing on the subsequent wastewater treatment processes should be carefully considered.

The amount of an alkaline chemical needed to increase pH depends on the alkalinity and buffer system of the wastewater (cf. Section 2.4).

**TABLE 7.3**
**Distribution of Sulfide Species at Selected pH Values at 20°C**

| pH | 7.0 | 7.5 | 8.0 | 8.5 | 9.0 |
|---|---|---|---|---|---|
| $H_2S$ (%) | 51 | 25 | 9 | 3 | 1 |
| HS$^-$ (%) | 49 | 75 | 91 | 97 | 99 |

## 7.3.5   Addition of Biocides

Different types of biocides can be used to inhibit the biological system in a sewer. As an example, anthraquinone will affect the respiration system of the sulfate-reducing biomass in biofilms and sediments and thereby the formation of sulfide. Anthraquinone is an aromatic organic compound with several applications in industry and medicine (e.g., as a precursor for dyestuffs and as an agent for bleaching pulp for papermaking). The fermentation, however, is not affected by addition to wastewater, and the formation of VOCs will therefore continue.

Addition of specific enzymes and microbial species has been proposed to control the formation of sewer gases. It remains uncertain whether or not such products are effective.

## 7.3.6   Mechanical Methods

Sewer biofilms and deposits can be partially removed by flushing or use of a ball, often named a "cleaning pig" or a "sewer pig." Through the removal of the biofilm, the occurrence of sulfide and VOCs are reduced. Mechanical cleaning is possible in both pressure mains and gravity sewers and should typically be repeated regularly depending on the growth characteristics of the biofilm. Mechanical cleaning is a technology that can reduce the requirement for other types of control measures.

## 7.3.7   Treatment and Management of Vented Sewer Gas

Inlet structures to treatment plants, pumping stations, and manholes are typical locations where sewer gases can be emitted to the atmosphere. To avoid odor nuisance, the gas flow out of these structures may need to be managed and undergo treatment. In general, gas treatment is a well-known technology. Among other industries, gas treatment is widely applied in the mining and coal gas industries to remove hydrogen sulfide from gases (Herrygers et al. 2000).

A variety of methodologies exist for the treatment of foul air. These different methods can, in terms of processes, be subdivided as follows:

- Absorption of a gas in a liquid stream (wet or dry scrubbing)
- Catalytic combustion (incineration)
- Activated carbon adsorption
- Chemical oxidation
- Biological oxidation

These treatment methods rely in different ways on chemical, physicochemical (gas adsorption and absorption), or biological processes for treatment of a gas stream. The number of ways to combine processes and reactor design is legion and dependent on the manufacturer of the equipment. In the following, examples of sewer gas treatment systems will be outlined.

It is crucial to note that treatment of sewer gases requires that equipment for forced ventilation be installed. Systems with adsorption, e.g., activated carbon units and biofilters, may need regular media replacement, although some of these are effective over several years.

In addition to treatment of the sewer gas, it is important that dilution as a management technology can reduce the concentration of odorous substances to a level where they do not cause a nuisance.

### 7.3.7.1  Wet and Dry Scrubbing

Wet scrubbing typically occurs as a countercurrent process where a compound in a gas stream is absorbed by a liquid stream in a packed bed of a tower (cf. Figure 7.6). In the dry scrubber, the gas stream is in contact with the solution or a suspension in a spray process. A scrubbing system (chemical scrubbing) is based on a chemical reaction between a gas compound (e.g., $H_2S$) and a reactant in the liquid stream. By applying the countercurrent principle, the advantage in terms of adsorption is that the low concentration gas—at the point where it leaves the reactor as treated gas—is in contact with the liquid in its highest concentration. Several design principles (e.g., single-stage towers, multistage systems, and horizontal reactor types) can be used for scrubbing (Lagas 2000). Figure 7.6 shows the principle of wet scrubbing.

As an example, $H_2S$ in the sewer gas stream will, as an acidic gas, react with an alkaline liquid stream (e.g., sodium hydroxide). At pH 10–11, $H_2S$ is absorbed as $HS^-$, whereas at pH >12 it is absorbed as $S^{2-}$ (cf. Figure 2.7 and Table 7.3). The reaction occurs efficiently within a packed contact tower. Excess scrubbing liquid is typically returned to the waste stream of the sewer or directly to a wastewater treatment plant. Alternatively, scrubbing with oxidative solutions, e.g., ozone, is also possible.

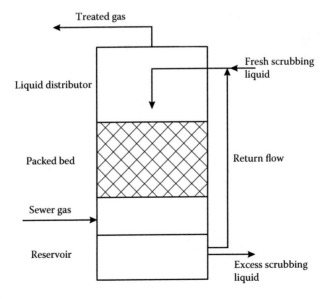

**FIGURE 7.6**  Principle of a wet scrubber with a packed contact tower.

A different example of $H_2S$ gas absorption is shown in Figure 7.7 (Gholami et al. 2009). In this scrubber system, $H_2S$ is oxidized to elemental sulfur ($S^0$) by iron(III) (cf. Section 2.4.2):

$$H_2S + 2Fe^{3+} \rightarrow S^0 + 2Fe^{2+} + 2H^+ \qquad (7.7)$$

Elemental sulfur is removed in a separator, and $Fe^{2+}$ is in a liquid stream oxidized to iron(III):

$$2Fe^{2+} + 2H^+ + 0.5O_2 \rightarrow 2Fe^{3+} + H_2O \qquad (7.8)$$

In total, the oxidation of $H_2S$ occurs by oxygen in the air. By this principle, iron is circulated and hydrogen sulfide is removed as elemental sulfur (cf. Figure 7.7).

### 7.3.7.2 Biological Treatment of Vented Sewer Gas

Biological treatment of sewer gas involves both transport and transformations. Typically, the sewer gas is first absorbed or adsorbed in a liquid suspension or a biofilm, respectively. Second, the gas compounds are degraded and typically oxidized by microorganisms. The type of reactor used for biological treatment is normally either a scrubber type or a filter (Herrygers et al. 2000).

The microorganisms involved in the treatment processes of sulfide are named sulfide oxidizing bacteria (SOB), i.e., mainly aerobic, chemoautotrophic bacteria that oxidize $H_2S$ to either elemental sulfur ($S^0$) or sulfate ($SO_4^{2-}$) (cf. Section 3.1.3). Furthermore, it is also likely that other volatile and malodorous sulfur compounds (VSCs) are oxidized. The bacteria are typically inoculated in the biologically active medium.

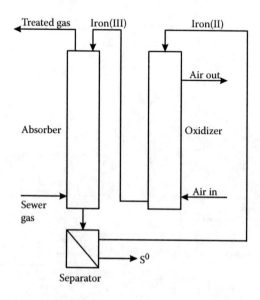

**FIGURE 7.7** Principle of $H_2S$ removal from a gas stream by oxidation to elemental sulfur.

Biofilters – also named biotrickling filters – are widely used by passing the polluted gas through a filter medium. Efficient and simple biofilters may use packing materials such as iron ore (hydrated hematite, i.e., iron(III) oxide) or peat. If not inoculated, an aerobic biofilm will, under moist conditions, develop at the surface of such materials. The biofilm will normally include complex, mixed populations of microorganisms and is therefore capable of degrading (oxidizing) sulfide as well as VOCs, including VSCs. Often, a rather short residence time (e.g., 30–60 s) is typically needed to treat the polluted sewer gas.

A bioscrubber is a scrubber controlled by biochemical processes. In a bioscrubber, the polluted gas is absorbed in a liquid and, in a next separate reactor step, oxidized (e.g., in an activated sludge process).

### 7.3.7.3  Activated Carbon Adsorption

In several industries, activated carbon adsorption is a well-known technology used to remove pollutants from both liquid streams and air. The activated carbon can be differently impregnated (e.g., with alkaline substances) and thereby improve the adsorption capacity, not just for hydrogen sulfide but also for VOCs. These systems may need regular regeneration and, less frequently, also media replacement. Activated carbon units are therefore best suited where rather low-concentrated flows should be treated.

### 7.3.7.4  Forced Ventilation and Dilution

The occurrence of odorous substances in the air is a nuisance for the public, and public complaints are often the direct cause for municipalities to manage sewer gases. In reality, it is therefore important that dilution of sewer gases to a level where malodors are no longer a concern may reduce the need for treatment—e.g., when the concentration of odorous substances is rather low. Treatment of the vented sewer air is, however, typically required.

Ventilation of sewers is discussed in Section 4.3.4. It is important that only forced ventilation—and not natural ventilation—is feasible to manage odor problems in the vicinity and that stacks for emission of the sewer air typically be installed. In addition to dilution in the surrounding atmosphere, forced ventilation will reduce the hydrogen sulfide concentration in the sewer and typically also the moisture content. Both phenomena may have a positive effect on the overall concrete corrosion of pipes, manholes, and pumping stations.

### 7.3.8  Evolving Mitigation Methods

It is a never-ending process to find the ideal mitigation method—effective under changing conditions, economic in terms of both equipment and operation, without environmental risks, and simple to operate. There are, of course, gradual improvements in the existing mitigation methods described in Section 7.3. However, there is also potential for new concepts, and even to some extent, a shift in paradigm.

It is outside the range of this book to discuss potentials and details for new concepts in terms of technologies and procedures for avoiding adverse effects of sewer gases. However, it is always an interesting point that process developments by the integration of wastewater transport in sewer networks and end-of-pipe wastewater treatment might lead to overall improvements. From this standpoint, wastewater sludge has potential for oxidation of reduced compounds such as sulfide and VOCs from the sewer. A technology applying this principle is known by diffusion of odorous air into activated sludge basins at wastewater treatment plants (Bowker 1999). In general, targeted biological wastewater treatment methods are of interest to investigate.

Slow release of oxygen from solid-phase peroxides and use of microbial fuel cells are potential future technologies for sulfide control in sewers (Zhang et al. 2008). Furthermore, catalytic oxidation of reduced waste compounds is generally well known but might be developed to a technology feasible for sewers.

When assessing new types of control technologies, it is important not just to focus on the reduction of adverse effects caused by sulfide and VOCs but also on the potential negative consequences. As an example, numerous chemicals inhibit SOB, but may also result in improper impacts on wastewater treatment processes and the adjacent environment.

## 7.4   FINAL COMMENTS

This section aims at focusing on aspects that are generally important to consider when selecting a technology for sewer gas control. The highlighted points should be considered guidelines and generally expressed recommendations. They are important to consider in the design phase of a sewer network as well as when a technology is being selected for an existing sewer network.

It appears from the preceding sections of this chapter that numerous mitigation methods exist for control of sulfide problems and malodors caused by VOCs. The effectiveness of these control methods has been described. It is, however, not an easy task to select the "most appropriate" technology in an actual case. Basically, there are both advantages and disadvantages in all methods. In particular, it is important to note that several methods are useful for reducing problems related to sulfide, but are not expedient for odor problems caused by VOCs.

The cost of mitigating sulfide and VOC-related problems in sewers vary considerably. Recommendation of specific methods only based on cost is not the best way to make a comparison. Cost-effectiveness as a more relevant parameter should be determined in each specific case. If we are just limiting "cost" to the price of chemicals used, a typical range is $5–20 per kg sulfur removed.

The mitigation methods dealt with in this chapter are—in reference to the preceding chapters—expressed in quantitative terms and therefore, by nature, appropriate to include in sewer process modeling (cf. Chapters 8 and 9). By doing so, it is possible to assess the different mitigation methods in an actual case, and among these alternatives select the most appropriate technology. It is furthermore important that solutions to sewer gas problems can be designed at different levels. It is therefore possible to assess solutions at catchment scale as well as for specific sewer pipe sections. It is the objective of this chapter to point out which technologies are available. The selection of the most appropriate methodology in an actual case—including, e.g., the location of the technology—is a result of an assessment based on modeling the current "problem" (cf. Chapter 9).

Finally, it is important to emphasize that mitigation should not solely be considered a "cure" for problems that occur in existing sewer networks. Although different types of actions can—and should—be taken to avoid sewer gas problems in the design phase of a sewer network, they are not always possible to avoid. The mitigation methods that are dealt with in this chapter are all controls that are relevant to consider when designing both new sewer systems and upgrading existing networks. An understanding of the processes in the sewer combined with the corresponding knowledge on interacting processes of mitigation is, in this respect, invaluable.

## REFERENCES

Abdul-Talib, S., T. Hvitved-Jacobsen, J. Vollertsen, and Z. Ujang (2002), Anoxic transformations of wastewater organic matter in sewers—process kinetics, model concept and wastewater treatment potential, *Water Sci. Technol.*, 45(3), 53–60.

ASCE (American Society of Civil Engineers) and WPCF (Water Pollution Control Federation) (1982), *Gravity Sanitary Sewer Design and Construction, ASCE Manuals and Reports on Engineering Practice No. 60 or WPCF Manual of Practice No. FD-5*, ASCE, New York, p. 275.

ASCE (American Society of Civil Engineers) (1989), *Sulfide in Wastewater Collection and Treatment Systems, ASCE Manuals and Reports on Engineering Practice No. 69*, ASCE, New York, p. 324.

Bentzen, G., A.T. Smith, D. Bennett, N.J. Webster, F. Reinholt, E. Sletholt, and J. Hobson (1995), Controlled dosing of nitrate for prevention of $H_2S$ in a sewer network and the effects on the subsequent treatment processes, *Water Sci. Technol.*, 31(7), 293–302.

Bjerre, H.L., T. Hvitved-Jacobsen, S. Schlegel, and B. Teichgräber (1998), Biological activity of biofilm and sediment in the Emscher river, Germany, *Water Sci. Technol.*, 37(1), 9–16.

Boon, A.G. (1995), Septicity in sewers: causes, consequences and containment, *Water Sci. Technol.*, 31(7), 237–253.

Bowker, R.P.G. (1999), Activated sludge diffusion; clearing the air on an overlooked odor control technique, *Water Environ. Technol.*, 11(2), 30–35.

Einarsen, A.M., A. Aesoey, A.-I. Rasmussen, S. Bungum, and M. Sveberg (2000), Biological prevention and removal of hydrogen sulphide in sludge at Lillehammer Wastewater Treatment Plant, *Water Sci. Technol.*, 41(6), 175–182.

Gholami, Z., M.T. Angaji, F. Gholami, and S.A.R. Alavi (2009), Reactive absorption of hydrogen sulfide in aqueous ferric sulfate solution, *World Acad. Sci., Eng. Technol.*, 49, 208–210.

Grennan, J.M., J. Simpson, and C.D. Parker (1980), Influence of cement composition on the resistance of asbestos cement sewer pipes to $H_2S$ corrosion, *Corros. Australas.*, 5(1), 4–5.

Gutierrez, O., J. Mohanakrishnan, K.R. Sharma, R.L. Meyer, J. Keller and Z. Yuan (2008), Evaluation of oxygen injection as a means of controlling sulfide production in a sewer system, *Water Res.*, 42(17), 4549–4561.

Herrygers, V., H. van Langenhove, and E. Smet (2000), Biological treatment of gases polluted by volatile sulfur compounds, in: P.N.L. Lens and L.H. Pol (eds.), *Environmental Technologies to Treat Sulfur Pollution—Principles and Engineering*, IWA Publishing, London, pp. 281–304.

Hvitved-Jacobsen, T., B. Jütte, P. Halkjær Nielsen, and N.A. Jensen (1988), Hydrogen sulphide control in municipal sewers, in: H.H. Hahn and R. Klute (eds.), *Pretreatment* in *Chemical Water and Wastewater Treatment, Proceedings of the 3rd International Gothenburg Symposium*, Gothenburg, Sweden, June 1–3, 1988, Springer-Verlag, New York, pp. 239–247.

Jameel, P. (1989), The use of ferrous chloride to control dissolved sulfides in interceptor sewers, *J. Water Pollut. Control Fed.*, 61(2), 230–236.

Jobbágy, A., I. Szántó, G.I. Varga, and J. Simon (1994), Sewer system odour control in the Lake Balaton area, *Water Sci. Technol.*, 30(1), 195–204.

Kienow, K.K. and R.D. Pomeroy (1978), Corrosion resistant design of sanitary sewer pipe, ASCE (American Society of Civil Engineers) Convention and Exposition, Chicago, IL, October 16–20, 1978, p. 25.

Lagas, J.A. (2000), Survey of $H_2S$ and $SO_2$ removal processes, in: P.N.L. Lens and L.H. Pol (eds.), *Environmental Technologies to Treat Sulfur Pollution—Principles and Engineering*, IWA Publishing, London, pp. 237–264.

Melbourne and Metropolitan Board of Works (1989), *Hydrogen Sulphide Control Manual— Septicity, Corrosion and Odour Control in Sewerage Systems*, vols. 1 and 2, Technological Standing Committee on Hydrogen Sulphide Corrosion in Sewerage Works, Melbourne, Australia.

Pomeroy, R.D., J.D. Parkhurst, J. Livingston, and H.H. Bailey (1985), Sulfide occurrence and control in sewage collection systems, Technical Report, US Environmental Protection Agency, USEPA 600/X-85-052, Cincinnati, OH.

Schmitt, F. and C.F. Seyfried (1992), Sulfate reduction in sewer sediments, *Water Sci. Technol.*, 25(8), 83–90.

Stuetz, R. and F.-B. Frechen (2001) (eds.), *Odours in Wastewater Treatment—Measurement, Modelling and Control*, IWA Publishing, London, p. 437.

Tanaka, N., T. Hvitved-Jacobsen, T. Ochi, and N. Sato (2000), Aerobic-anaerobic microbial wastewater transformations and reaeration in an air-injected pressure sewer, *Water Environ. Res.*, 72(6), 665–674.

Thistlethwaite, D.K.B. (ed.) (1972), *The Control of Sulfides in Sewerage Systems*, Butterworth, Sydney, Australia, p. 173.

USEPA (1974), *Process Design Manual for Sulfide Control in Sanitary Sewerage Systems*, USEPA 625/1-74-005, Technology Transfer, Washington, DC.

USEPA (1985), *Odor and Corrosion Control in Sanitary Sewerage Systems and Treatment Plants*, USEPA 625/1-85/018, Washington, DC.

Vincke, E., J. Monteny, A. Beeldens, N.D. Belie, L. Taerwe, D. van Gemert, and W.H. Verstraete (2000), Recent developments in research on biogenic sulfuric acid attack of concrete, in: P.N.L. Lens and L.H. Pol (eds.), *Environmental Technologies to treat Sulfur Pollution—Principles and Engineering*, IWA Publishing, London, pp. 515–541.

WEF (2004), *Control of Odors and Emissions from Wastewater Treatment Plants*, WEF (Water Environment Federation) Manual of Practice No. 25, WEF, Alexandria, VA, USA, p. 537.

WEF (2007), *Gravity Sanitary Sewer Design and Construction*, WEF (Water Environment Federation) Manual of Practice No. FD-5, WEF, Reston, VA, USA, p. 422.

Zhang, L., P.D. Schryver, B.D. Gusseme, W.D. Muynck, N. Boon, and W. Verstraete (2008), Chemical and biological technologies for hydrogen sulfide emission control in sewer systems: a review, *Water Res.*, 42, 1–12.

# 8 Sewer Process Modeling
## *Concepts and Quality Assessment*

First of all, it is important to realize that, among engineers and practitioners, modeling related to sewers is normally understood as computer modeling aimed at analyzing the hydraulic performance. This type of modeling is often related to an analysis of stormwater runoff flows from urban and road surfaces. The objective of modeling related to this chapter is different. The focal point is an analysis of the overall chemical and microbiological performance of the sewer network using sewer process models. Here, the main point is an assessment of the impact of in-sewer chemical and biological processes in terms of, for example, hydrogen sulfide occurrence, odor nuisance, and concrete corrosion. As discussed in the preceding chapters, sewer processes do not proceed without potential negative impacts. In particular, hydrogen sulfide formation and anaerobic degradation of wastewater organic matter producing volatile odorous substances (VOCs) are often a major concern.

As an introduction to this chapter, it should be mentioned that sewer processes are highly dynamic and subject to considerable variability in time and place. Details in terms of the temporal dynamics at a specific location in a sewer are in general not predictable with sufficient accuracy. This fact has been observed in several field investigations and is well known from practice. Because of a high variability, it is consequently out of our reach to model such quality changes. However, what is possible based on solid knowledge on sewer processes—and also focused on in details in this book—is to predict an average performance. Models designed to predict this average behavior may also include modeling of an average daily variability of a phenomenon. Stochastic models that are designed to simulate sewer processes by varying central model parameters within their realistic ranges are examples of what is possible. Sewer process models are like any type of model "equipped" with limitations defined by the reality that they are intended to simulate. Sewer process models are, however, valuable tools in the hands of those who—based on solid process knowledge—are able to define the conditions under which they work and furthermore assess the outcome critically!

It should be clearly stressed that use of sewer process models for analysis and for selection of control strategies and methods is without sense if not based on an understanding of the nature of the governing chemical and microbiological processes. It is therefore the basic and overall objective of this book to serve as a link between this understanding and the use of models as tools for quantification of solutions to problems in real situations.

Relatively simple empirical equations and models for description of specific phenomena in sewers have been developed and presented in the preceding chapters. In continuation of these formulations, it is the objective of this chapter to expand these calculation tools. The models that are dealt with will be expressed in terms of a "framework" that can be used for simulation of conceptually expressed sewer processes. These conceptual descriptions of the aerobic, anoxic, and anaerobic sewer processes and related processes for transport and exchange of substances across interfaces are in details dealt with particularly in Chapters 4 through 6. The general characteristics of models that can be used in this respect are the core objective of this chapter. The specific details of a sewer process model will be the main subject of Chapter 9.

## 8.1   TYPES OF PROCESS MODELS

This book was developed based on the scientific and in practice observed fact that a cause–effect relationship exists for physical, chemical, and biological processes in sewers. This relationship can consequently be formulated in mathematical terms. Under this concept, it is possible to formulate models in various ways:

- With different objectives (odor and corrosion, planning at catchment scale, pipe section analysis, etc.)
- At different levels of details (daytime and yearly variability, average and stochastic performance, etc.)
- For different types of sewer networks (pressure mains, gravity sewers, sewer drops, etc.)
- At different scale of sewer network size (pipe-scale for analysis of specific details and catchment-scale for performance analysis of large sewer networks)

These different ways of approaching process modeling in sewers call for a rather "flexible" type of model. Under these constraints, it is a central requirement that the model is designed to simulate the outcome of the dynamics for the chemical and microbial processes and the exchange of components across the different boundaries (cf. Figure 8.1). Furthermore, it is crucial that a sewer process model—still depending on the objective—includes all relevant phases (in particular, water, air, and biofilm) and the corresponding governing processes for transfers of substances across the different interfaces.

Sewer process models include as central elements mass transport for central wastewater constituents and their rate of formation and transformation. In this way, the following factors become the core elements of a process model:

- Stoichiometric characteristics for the transformations of the governing substances
- Kinetic characteristics of these transformations
- Transport characteristics of the substances, i.e., a specification of where and when they are occurring in the sewer network

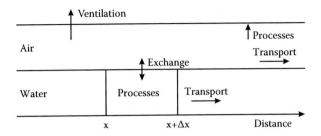

**FIGURE 8.1** Principle of a sewer segment (pipe) with major elements related to process modeling (cf. Figures 1.6 and 6.12 for additional information).

The criteria for selection of the governing processes are:

- Processes affecting the formation and occurrence of substances with negative effects, i.e., processes that are important for impacts such as malodor and corrosion
- Processes controlling the occurrence of these substances and thereby reducing the negative impacts

These different characteristics constitute the basis from where a model should be developed, typically with a mass balance as a central characteristic. Unsteady state flow conditions are basically important if the objective is to simulate the hydraulics of wet weather conditions. It is, however, important to stress that impacts of sewer processes typically occur during "normal" dry weather conditions—although they are affected by varying flows. It is therefore often acceptable to consider steady-state flow as a first approximation.

Two types of models, the deterministic type and the stochastic type, are the most relevant to consider for the description of the dynamics of sewer processes and the corresponding impacts and controls. These two types of models are therefore given particular attention in Sections 8.1.3 and 8.1.4, respectively. The empirical model types are not a subject of this chapter. However, these empirical models are generally important and have been represented in the preceding chapters. Therefore, as an introduction, it is also useful to describe the overall characteristics of this type (cf. Section 8.1.2).

Before these different types of models are described, it is appropriate to define and discuss the basic and central requirements for establishing and applying a model that simulates the real occurring sewer processes in a specific case.

### 8.1.1 Model Validation, Calibration, and Verification

The first step in establishing a model is to consider its validity in terms of a validation.

#### 8.1.1.1 Validation

Validation of a model means a process to ensure that sound scientific information, logically and sufficiently, has been included in the model. Basically, it is expected that the developer of the model has taken into account aspects that are the most central for prediction. The user of the model should consider if the validity of the model expression is observed in the actual case. The purpose of validation is not just to ensure that sound deterministic information is included but also potentially that a stochastic description follows the nature of the phenomena and processes.

In the preceding chapters, focus has been placed on establishing mathematically formulated expressions for relevant sewer processes. Furthermore, it has been crucial to discuss the prerequisites and limitations of these expressions. The different pieces for model validation are thereby present.

Because of the pronounced variability of boarder conditions and process parameters, in time as well as in place, rather extensive data sets that observe the intended use of the model are typically needed. Such data are normally divided into two groups and applied for the two procedures: calibration and verification.

### 8.1.1.2   Calibration

The objective of this procedure is to determine the specific values of the model parameters based on empirically determined data that characterize the system and processes in question. The procedure represents a goodness of fit and the model parameters are adjusted until a certain "optimal" level of accuracy of the model output is obtained. The procedure results in values of model parameters, which are defined as calibrated.

### 8.1.1.3   Verification

The model with the calibrated model parameters is tested on the part of the data set that was not used for calibration. The level of accuracy should thereby be assessed under conditions where the model will be applied.

The extent of the data set used for calibration and validation will depend on the actual situation. It is clear that complex models with several parameters require more information than simple models to comply with the possible conditions under which the model should operate.

There are different approaches for how calibration should be understood and performed. The ideal approach is that a parameter should express a specific system characteristic, for example, that a growth rate parameter exclusively expresses the true growth characteristics under given conditions. The real situation, however, is that different system characteristics via empirically determined data influence the value of a calibrated parameter. Calibration thereby becomes complex. A model may for example require a large amount of data for its calibration, and more than one set of parameters might observe a given criterion in terms of model performance. Such details of overparameterized models are discussed in Section 8.1.3. The stochastic approach dealt with in Sections 8.1.4 is a possible answer to this problem.

It is outside the scope of this chapter to discuss how calibration and validation can be performed in practice. Several basic texts on mathematical modeling and statistics deal with this aspect.

### 8.1.2   Empirical Models

The formulation of an empirical model is in principle based on results originating from systematically performed experiments or monitoring programs. Such data are organized according to the objective. Rather crude knowledge of a phenomenon is typically included in empirical models. Such models can thereby predict a cause–effect phenomenon, but only corresponding to conditions that were originally observed when the model was developed, i.e., in similar systems and under similar constraints.

The degree of theoretical knowledge of a phenomenon that is modeled may vary considerably from one type of model to another. The very simple empirical models are black-box models, e.g., input–output models, which predict a phenomenon without knowledge on the governing processes. Other model types include some basic theoretical knowledge on a phenomenon, e.g., the so-called gray-box models.

Empirical models are of particular importance when rather crude knowledge on the performance of a system exists. Such models are also useful when limited information in terms of data is available. However, it should be stressed once again that empirical models can only be used within their area of definition. This means that the system or phenomenon modeled or predicted must be within the range at which the model was originally developed. Typically, the constants and parameters of an empirical model do not reflect any general understanding of underlying processes and phenomena. Such parameters must consequently be calibrated on-site and the calibration must be validated to obtain reliable results.

### 8.1.3 DETERMINISTIC MODELS

A deterministic model expresses the performance of a system in mathematical terms via a number of physical, chemical, or biological characteristics and processes. The mathematical formulation of the processes in a system and their interactions are often the core elements of the deterministic description. A deterministic model is thereby in principle based on theoretically well-accepted and relevant scientific knowledge that is applied for simulation of the performance of a system in process terms.

It is a main characteristic of a deterministic model that one set of input parameters and initial conditions has one and only one outcome, i.e., when these conditions of a system are stated, the model result is constant.

In general, deterministic models, and particularly those for simulation of quality aspects, include several parameters expressing, for example, stoichiometric and kinetic characteristics of the processes. Deterministic models are strong tools but they require careful calibration and verification, often based on an extensive set of measured data. If such site-specific data are not available, the use of a noncalibrated deterministic model may give misleading results. It is therefore crucial to understand that a complex deterministic model requires corresponding extensive data but also that a user must have fundamental knowledge of the underlying processes. If these requirements are not observed, simpler model tools should typically be recommended.

The process related in-sewer phenomena are by nature complex and typically show high variability—not necessarily to be understood as high uncertainty. The fact that corresponding deterministic models therefore also turn out to be rather complex, often results in a structure that is overparameterized. A large number of parameters for expressing a given phenomenon are likely to be required to maintain a high complexity within a mathematical framework that is not correspondingly complexly expressed. It is therefore always a point to obtain an appropriate balance between the complexity of a model structure and the corresponding number of model parameters. At a pragmatic level, it is often needed to accept an overparameterized model in order to obtain an output that reflects the reality in terms of a "correct" response under varying external conditions. However, a large number of parameters may result in a difficult calibration procedure, and an extensive amount of data is often needed for the calibration process. Furthermore, one must be aware that good model fits to the verifying data are obtained, not just good fits to calibration data.

### 8.1.4 STOCHASTIC MODELS

Stochastic models represent an extension of the deterministic models by including model parameters with a known or expected statistic distribution. By using, for example, Monte Carlo simulation techniques, each single model sequence is repeated a large number of times—often several thousands. For each repetition of the Monte Carlo procedure, a random number generator is used for drawing a parameter value from the statistic distribution. The outcome of the modeling is analyzed statistically with the distribution of the output parameters being the model result. This result expresses the variability caused by the variability of the input parameters. As a consequence, stochastic models are well suited for sensitivity analysis.

As an example, a deterministic model simulating the transformations of organic matter in a sewer can be applied as a stochastic model if the input parameters are expressed in terms of a statistic distribution. A prerequisite is that the variability of the model parameters, for example, those included in the biomass growth rate and the rate of oxygen transfer across the air–water interface, is known. Ideally, this variability should be known from relevant field investigations and determined a representative number of times.

The ideal requirements for using stochastic models in sewer networks hardly exist and basically is just as an approximation. Some model parameters, for example, the sulfide concentration in the water phase, may under certain conditions vary with a time resolution of less than 1 min. A pragmatic approach is to apply a simple and expected parameter distribution based on a limited number of investigations. It is often acceptable to apply such an approach; however, any prerequisite used must be taken into account when assessing the model output.

## 8.2   DETERMINISTIC SEWER PROCESS MODEL APPROACH

Several fundamental requirements and constraints must be observed when establishing a concept of understanding and a corresponding sewer process model. As the starting point, the processes must be theoretically understood before they can be formulated mathematically in a conceptual manner. This work must in principle be continuously repeated, and corrections must be made by iterative procedures influenced by results from relevant experiments and observations. Not until a point has been reached where an acceptable agreement between a concept and its experimental verification has been obtained, can a concept be defined and successfully formulated. Mass balances are, in this respect, a basic engineering requirement. A relatively simple concept including central but also a limited number of compounds and parameters to be experimentally determined is generally preferred. The overall procedure in terms of validation, calibration, and verification is outlined in Section 8.1.1.

In addition to the kinetics of the sewer processes, the stoichiometry of the transformations of the compounds is crucial for the mass balance. As an example related to organic matter in the water phase, the stoichiometry of the biomass–substrate relationships is determined by the heterotrophic biomass yield constant, $Y_{Hw}$, in units of g COD g COD$^{-1}$ (cf. Section 5.3.2.1). As shown in Figure 5.6, the yield constant is a central factor related to the consumption of both readily biodegradable substrate, $S_S$, and dissolved oxygen (DO), $S_O$, for the production of the heterotrophic biomass, $X_{Hw}$.

The concept for modeling of in-sewer transformations of substances follows the principle used in activated sludge modeling (Henze et al. 1987, 1995, 2000). The similarities between activated sludge modeling and sewer process modeling concern the principle of the dynamic descriptions and the computation procedures. The details of which processes and substances are relevant and their formulations are definitely not comparable.

## 8.2.1   Principle of a Sewer Process Model

The principle of a sewer process model is outlined in the following discussion. For simplicity reasons, it exemplifies only processes in the water phase. The biofilm and the gas phase are both excluded. The example concerns the aerobic growth of heterotrophic biomass, $X_{Hw}$, based on the utilization of readily biodegradable organic matter, $S_S$, produced by hydrolysis of fast hydrolyzable substrate, $X_{S1}$ (cf. Figure 8.2). The example in this simplistic version is not a realistic picture of which processes will proceed in a sewer. It is solely selected to outline the principle of sewer process modeling. When comparing Figure 8.2 with, for example, Figure 6.12, it is readily seen that the present example concerns just a small segment of what potentially may occur in a sewer.

Figure 8.2 outlines in this example the conceptual approach to be transferred into a quantitative description in model terms. Based on this description, the following three elements constitute in general what quantifies the process dynamics:

1. Kinetic description of the processes
2. Stoichiometric formulations
3. Mass balances

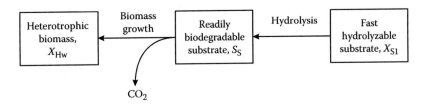

**FIGURE 8.2**   Example showing water phase DO nonlimited aerobic growth of heterotrophic biomass based on hydrolysis. The figure shows a selected segment of Figure 5.6.

As outlined in Figure 8.2, only two aerobic water phase processes, heterotrophic biomass growth and hydrolysis, are considered.

As a first step, it is realized that three fractions of organic matter are relevant for the concept expressed in Figure 8.2:

$X_{Hw}$ = heterotrophic biomass concentration in the water phase (g COD m$^{-3}$)

$S_S$ = readily biodegradable substrate (g O$_2$ m$^{-3}$)

$X_{S1}$ = fast hydrolyzable substrate (g O$_2$ m$^{-3}$)

Figure 8.2 furthermore shows that two processes are relevant: aerobic growth of heterotrophic biomass and hydrolysis of fast hydrolyzable substrate.

According to the sewer process concept shown in Figure 8.2, points 1–3 are in the following further exemplified:

*Point #1*

The kinetic description of the processes:

The kinetics of these processes is described in Equations 5.1 and 5.4, respectively. In this specific case, these two expressions are formulated for water phase processes in a simplified way corresponding to DO nonlimited growth at constant temperature. The actual simplified expressions for the growth rate and the rate of hydrolysis are shown in Equations 8.1 and 8.2, respectively:

$$r_{grw} = \mu_{H,O_2} \frac{S_S}{K_{Sw} + S_S} X_{Hw} \tag{8.1}$$

where:

$r_{grw}$ = growth rate of heterotrophic biomass in suspension (g COD m$^{-3}$ day$^{-1}$)

$\mu_{H,O_2}$ = maximum specific growth rate (day$^{-1}$)

$K_{Sw}$ = saturation constant for readily biodegradable substrate (g O$_2$ m$^{-3}$)

$$r_{hydr} = k_{h1} \frac{X_{S1}/X_{Hw}}{K_{X1} + X_{S1}/X_{Hw}} X_{Hw} \tag{8.2}$$

where:

$r_{hydr}$ = rate of hydrolysis (g COD m$^{-3}$ day$^{-1}$)

$k_{h1}$ = hydrolysis rate constant for the fast hydrolyzable substrate (day$^{-1}$)

$K_{X1}$ = saturation constant for hydrolysis of the fast hydrolyzable substrate (g COD g COD$^{-1}$)

The kinetic description related to the concept shown in Figure 8.2 is, by the formulations of Equations 8.1 and 8.2, hereby completed.

*Point #2*

The stoichiometric formulation:

Before the mass balances in this example are dealt with, it should be recalled that readily biodegradable organic matter—in addition to its utilization for biomass growth—is degraded for energy purposes (cf. Figures 2.3, 5.6, and 8.2). This aspect is in terms of the mass balance for carbon accounted for by introducing the biomass yield constant, $Y_{Hw}$, as a stoichiometric parameter. Before the mass balance is dealt with, it is crucial to note that $Y_{Hw}$ is defined relative to $S_S$ directly utilized for biomass growth:

$$Y_{Hw} = -\frac{dX_{Hw}}{dS_S} \tag{8.3}$$

It is furthermore important to note that 1 unit of $S_S$ is produced when 1 unit of $X_{S1}$ is hydrolyzed. This means that hydrolysis does not affect the oxidation level of the substance (cf. Sections 2.1.4.1 and 3.2.3).

*Point #3*

The mass balances:

Based on the two rate expressions, Equations 8.1 and 8.2 (i.e., the kinetic descriptions), Equation 8.3 (i.e., the stoichiometry), and the characteristics of the hydrolysis, the mass balances for the three organic carbon fractions can be formulated. These three mass balances are based on the concept shown in Figure 8.2 and expressed as differential equations in Equations 8.4 through 8.6:

$$\frac{\partial X_{Hw}}{\partial t} = r_{grw} \tag{8.4}$$

$$\frac{\partial S_S}{\partial t} = -\frac{1}{Y_{Hw}} r_{grw} + r_{hydr} \tag{8.5}$$

$$\frac{\partial X_{S1}}{\partial t} = -r_{hydr} \tag{8.6}$$

The set of coupled differential equations, Equations 8.4 through 8.6, constitute as the mass balances for the compounds in common the very central part of the sewer process model.

It is important to recall that the model is based on the concept shown in Figure 8.2 and that a sewer process model therefore always refers to the concept from which it is formulated. A conceptual formulation of the sewer processes and a corresponding sewer process model are therefore always interrelated and must be considered identical, although differently expressed.

It is the actual problem that defines which concept is relevant. This concept is expressed by mass balances including the relevant substances and process rate expressions. As an example, the aerobic–anaerobic concept shown in Figure 6.12 for organic carbon and sulfur will definitely result in more complex expressed differential equations than in this example (cf. Chapter 9). Depending on the objective—and thereby the selection of relevant processes and components—the number of processes is, for example, 15 to 25, and the number of mass balances is in the order of 5 to 15.

Under such more complex conditions, it is often convenient to give an overview of the sewer process model different from mass balances expressed as coupled differential equations (cf. Equations 8.4 through 8.6). Referring to this example and the concept shown in Figure 8.2, an overview with the relevant compounds and processes is shown in Table 8.1. Traditionally, and as, for example, done in the activated sludge model, this overview is named a "process matrix." It is, however, important to stress that it is not in mathematical terms a matrix.

The process matrix shown in Table 8.1 is a systematic arrangement of the kinetics and stoichiometry for sewer processes. A row in the matrix represents a process and each column refers to the mass balances for the compounds in the model. By including coefficients in the process matrix, Table 8.1 shows that the mass balances of the three compounds, Equations 8.4 through 8.6, can be deducted from the columns by multiplication with the relevant process rates, which are shown at the right side of the matrix.

Nonlinear, coupled differential equations are in practice solved by numerical methods, for example, a finite difference method using an Euler approach. Equations 8.4 through 8.6 are therefore transformed to difference equations:

$$\Delta X_{Hw} = r_{grw}\, \Delta t \tag{8.7}$$

$$\Delta S_S = \left( -\frac{1}{Y_{Hw}} r_{grw} + r_{hydr} \right) \Delta t \tag{8.8}$$

$$\Delta X_{S1} = -r_{hydr}\, \Delta t \tag{8.9}$$

---

**TABLE 8.1**

**Sewer Process Model Concept as Expressed by Equations 8.4 through 8.6 in a Process Matrix Formulation**

| Process | $X_{Hw}$ | $S_S$ | $X_{S1}$ | Process Rate |
|---|---|---|---|---|
| Growth of heterotrophic biomass | 1 | $-1/Y_{Hw}$ | | Equation 8.1 |
| Hydrolysis | | 1 | $-1$ | Equation 8.2 |

---

These increments determine the change in the values of the three organic fractions from time $t$ to $t + \Delta t$, typically with $\Delta t$ in the order of a few seconds. If $X$ represents a compound concentration, the computation procedure can briefly be expressed as:

$$X_{t+\Delta t} = X_t + \Delta X \qquad (8.10)$$

## 8.2.2 THE PRINCIPLE OF A SOLUTION TO A SEWER PROCESS MODEL

In general, the following four points should be considered when a sewer process concept is transferred to a model, i.e., when seeking a specific solution to a sewer process problem in a given sewer network:

1. The sewer network construction details

     First of all, the boundaries of the sewer network system in question must be defined. These boundaries define corresponding inflows and outflows to the system. Within these boundaries, it must be decided which details are important to include in the model. Dimensions of gravity sewer sections, pressurized systems, and pumping stations should typically be known. Furthermore, information on construction materials is important when dealing with, for example, concrete corrosion.

2. Flow conditions

     In general, flow conditions include information on all types of mass movements in the water phase and the overlying air phase including interfacial transfer of mass and ventilation (cf. Figure 4.5). In practice, this means that important inflows and outflows of water (wastewater) at the boundaries are known, and that a hydraulic model for the system describes the water flows. In particular when dealing with an air phase process for concrete corrosion and odor nuisance, the movement in the air phase and ventilation are crucial. Depending on the conditions, such descriptions can be simple or complex. For example, the different flow rates of water and the overlying gas phase is crucial for where emitted compounds will occur in the gas phase.

3. Initial conditions for the relevant compounds

     Referring to Equation 8.10, the initial concentration values for the compounds in the model must be known.

4. Kinetic and stoichiometric parameters

     Process-related parameters quantify the magnitude of the rate of transformation for the governing chemical and microbial processes. It is important to realize that the values of these parameters are strongly related to the formulation of these processes and typically varying in time and place. The determination of these parameters is discussed in detail in Chapter 10. It is furthermore important to note that some parameters can be determined explicitly, i.e., based on specific designed experiments, whereas others must be determined based on model calibration.

A solution to the sewer process model dealt with in Section 8.2.1 is exemplified in the following. The exemplification follows the four points mentioned, but, for illustration purposes, this was carried out in a very simple version.

### Example 8.1: Solution to a Sewer Process Model

The concept for the sewer process model is in this example based on what is shown in Figure 8.2 and the corresponding formulations shown in Table 8.1 and Equations 8.7 through 8.9. The example therefore only refers to water phase processes, and the biofilm and the sewer atmosphere are both excluded. The example concerns the very simple objective to show the variations of the three fractions of organic matter during 12 h of transport in the sewer.

The sewer network is a simple gravity sewer pipe where the prerequisites for applying Equations 8.7 through 8.9 are observed, for example, that the DO concentration is not limiting the aerobic transformations.

The flow velocity in the pipe is determined by the pipe dimensions and the friction. In this example, the velocity is considered constant and equal to 0.5 m s$^{-1}$, i.e., corresponding to 1.8 km/h.

The initial concentrations at the upstream point of the pipe is based on a total COD concentration of 500 g m$^{-3}$ and a distribution of the organic matter fractions corresponding to what is shown in Figure 3.10: $X_{Hw} = 50$ g m$^{-3}$, $S_S = 30$ g m$^{-3}$, and $X_{Hw} = 70$ g m$^{-3}$.

The values selected for the kinetic parameters in Equations 8.1 and 8.2 are not discussed in this example but follow what might be "typically" observed (cf. Chapter 10). The yield constant is the stoichiometric parameter.

Kinetic parameters: $\mu_{H,O_2} = 6.0$ day$^{-1}$, $K_{Sw} = 2.0$ g O$_2$ m$^{-3}$, $k_{h1} = 5.0$ day$^{-1}$, and $K_{X1} = 2.0$ g COD g COD$^{-1}$

Stoichiometric parameter: $Y_{Hw} = 0.6$ g COD g COD$^{-1}$

With these parameter values, the two rate expressions, Equations 8.1 and 8.2, are formulated as follows:

$$r_{grw} = 6.0 \frac{S_S}{2.0 + S_S} X_{Hw}$$

$$r_{hydr} = 5.0 \frac{X_{S1}/X_{Hw}}{2.0 + X_{S1}/X_{Hw}} X_{S1}$$

These rate expressions are included in Equations 8.7 through 8.9.

With the selected initial concentration values for the three organic matter fractions, the initial process rates for biomass growth and hydrolysis are $r_{grw,init} = 11.72$ g COD m$^{-3}$ h$^{-1}$ and $r_{hydr,init} = 6.00$ g COD m$^{-3}$ h$^{-1}$, respectively.

In this example, there are no abrupt changes in the three organic matter fractions, and a relatively large value of $\Delta t$ can be selected, for example, $\Delta t = 1$ min. The combined modeling result applying Equations 8.7 through 8.10 on the conditions explained in this example is shown in Figure 8.3.

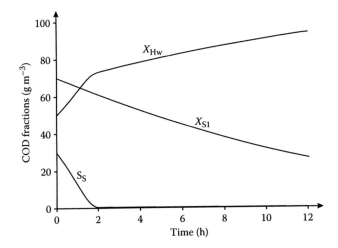

**FIGURE 8.3**  Modeling of varying organic matter fractions in wastewater during its transport in a gravity sewer. Simulations follow the conditions explained in Example 8.1.

Compared with the processes taking place in real sewers, this example—as previously noted—does not show transformations of specific relevance. In Figure 8.3, however, it is noteworthy that when $S_S$ is depleted, the growth rate of $X_{Hw}$ is reduced to a level determined by the rate of hydrolysis producing $S_S$ and by the saturation constants.

## 8.3  ADDITIONAL MODELING APPROACHES

In addition to the deterministic sewer process model as described in Section 8.2, the stochastic model version is particularly relevant when dealing with large sewer networks, i.e., when applied for analysis at catchment scale. The principle of this type of modeling approach is further described in the following.

### 8.3.1  Modeling at Catchment Scale

The existing versions of sewer process models are based on the deterministic model approach. The reason is that a conceptual understanding and description of the underlying chemical and microbial processes are considered the solid starting point for any type of sewer process modeling.

The deterministic model is in general terms discussed in Section 8.1.3, and in a simplified version outlined and exemplified in Section 8.2. For rather small systems, such as single pipe sections, the deterministic model provides a consistent description. However, when it comes to large sewer networks with, for example, several hundred kilometers of pipes, the stochastic modeling approach becomes extremely useful (cf. Section 8.1.4). The stochastic model applied at what can be characterized as catchment scale can provide an overview of where, when, and how often process-related problems are likely to occur in the sewer network.

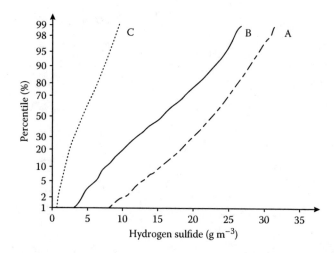

**FIGURE 8.4** Example showing the result of a stochastic modeling of hydrogen sulfide in three points of a sewer.

The idea behind the catchment scale modeling of sewer processes is that it provides information on where the so-called "hot-spots" are located, i.e., areas of the sewer network where specific problems are particularly expected. It is thereby possible to identify which type of processes and substances should be controlled. The use of stochastic models at catchment scale means that variability of inputs as well as process parameters is possible, whereas computation with a deterministic model only provides the outcome with fixed values of all kinetic and stoichiometric parameters. A stochastic model is in principle an extension of a deterministic model by selecting values of these parameters according to their expected variability in practice. The variability is mathematically described in terms of a distribution, and a parameter value is drawn from this distribution by means of a random number generator. This procedure is repeated based on the distributions of all relevant parameters, for example, several thousand times. The results of all single computations with the stochastic model are finally statistically analyzed.

Figure 8.4 shows an example of a stochastic simulation with the WATS (Wastewater Aerobic/Anaerobic Transformation in Sewers) sewer process model. The example illustrates the distributions of the sulfide concentration in three points—A, B, and C—of a gravity sewer located downstream a pressure pipe section in a major city located in the United Arab Emirates. The simulations have been repeated 1000 times. Increased number of simulations would result in more smooth curves but basically not different results.

## REFERENCES

Henze, M., C.P.L. Grady Jr., W. Gujer, G.v.R. Marais, and T. Matsuo (1987), *Activated Sludge Model No. 1*, *Scientific and Technical Report No. 1*, IAWPRC (International Association on Water Pollution Research and Control), London.

Henze, M., W. Gujer, T. Mino, T. Matsuo, M.C. Wentzel, and G.v.R. Marais (1995), *Activated Sludge Model No. 2, Scientific and Technical Report No. 3,* IAWQ (International Association on Water Quality), London, p. 32.

Henze, M., W. Gujer, T. Mino, and M.v. Loosdrecht (2000), *Activated Sludge Models ASM1, ASM2, ASM2d and ASM3, Scientific and Technical Report No. 9,* IWA (International Water Association), London, p. 121.

# 9 WATS

## *A Sewer Process Model for Water, Biofilm, and Gas Phase Transformations*

Chapters 1 through 8 and Chapter 10 include information needed for the prediction and simulation of sewer processes, their impacts, and their control. This chapter is a kind of a "melting pot" for this information by mixing it together in what is termed a model. This model—or more correctly defined as a model concept—is named WATS, an abbreviation of Wastewater Aerobic/anaerobic Transformations in Sewers.

WATS is in traditional terms not a computer model that is generally available for the user and that can be applied for any type of sewer catchment, any objective, and with preset and reliable parameter values. The reality behind such a wish is not realistic because of the extreme complexity of the underlying aspects and problems. The variability in, for example, the objectives of modeling, catchment characteristics, and climate conditions is in practice legion and becomes therefore a major obstacle for a general model formulation. At least this is the case under limited resource constraints and when one has ambitions to produce reliable modeling results. Furthermore, the knowledge on sewer processes is still progressing and should continuously be included in WATS. Last but not least, a sewer process model for simulation of complex interactions requires a corresponding extensive knowledge on the specific sewer-related chemistry, biology, and physics in order not to produce surrogate results.

The solution to these problems is to develop WATS as a flexible and conceptual model for simulation of sewer processes and thereby a predictive tool in terms of which problems may arise. This principle relies on collecting and integrating those segments of model-formulated details that in an actual case describes the relevant processes, the related in-sewer impacts, and their control. The WATS model, expressed as mass balances via relevant coupled differential equations, is thereby ready for a numerical computation and simulation according to an actual objective. The procedure exemplified in Section 8.2 is, in this respect, a simple version of how to apply WATS.

## 9.1 WATS MODEL: AN OVERVIEW

The WATS model is according to these general expressed principles described in the following. The first and central point is to relate the objectives and selection of the corresponding mass balances. These mass balances are in a simple version visualized and exemplified in Table 8.1.

The first version of the WATS model concept was published in 1998 (Hvitved-Jacobsen et al. 1998). The aerobic microbial transformations of organic carbon in a gravity sewer were in this version focused on. Since then, numerous other wastewater compounds, different types of sewer networks, varying process conditions, and sewer process phenomena have been included (cf., e.g., Vollertsen and Hvitved-Jacobsen 1998; Abdul-Talib et al. 2002; Vollertsen et al. 2005, 2008; Nielsen et al. 2008; Vollertsen et al. 2011a, 2011b). As may appear from this list of references, it is worth noting that the development of WATS is a continuing process, based on extensive laboratory investigations, pilot-scale studies, and full-scale experiments. In this way, the advances in the understanding of sewer processes can be expressed conceptually in mathematical terms and included in the WATS model. The WATS model is thereby under continuous development and improvement by including more and more relevant aspects of the in-sewer microbial and chemical world. WATS is therefore a very flexible sewer process model that in principle can be "designed" to any intended use if the underlying understanding of the sewer processes is available.

The WATS model is, for each specific use and according to the principle exemplified in Section 8.2, closely related to a selected concept for sewer processes. The very close relation between a solid conceptual understanding of sewer processes and the formulation of WATS in model terms is a core characteristic.

WATS is described according to these fundamental characteristics. In practice, this means that it is the actual objective that determines which specific model concept is selected and then transformed into WATS model terms. The differential equations derived from this concept express the relevant mass balances of the compounds based on process rate equations. The number of relevant objectives is in principle legion, but in practice often related to the in-sewer problems, i.e., an analysis of these problems and potentially including corresponding process control and management strategies:

- Concrete corrosion caused by hydrogen sulfide
- Impacts on human health of hydrogen sulfide in the sewer atmosphere
- Odor nuisance caused by hydrogen sulfide and volatile organic compounds (VOCs) being emitted and vented from the sewer
- Hydrogen sulfide and VOC controls
- Analysis of wastewater quality at inflows to wastewater treatment plants

These sewer process aspects show that organic carbon substances and sulfur compounds constitute the central wastewater constituents to deal with. In this context, it is important that dissolved oxygen (DO) and nitrate determine whether aerobic, anoxic, or anaerobic processes dominate the transformations. Furthermore, pH must be included in the WATS model as an important parameter and also, for example, processes related to iron salts for control of sulfide. These briefly expressed aspects clearly show that the understanding and quantification of the in-sewer processes as dealt with in the preceding chapters play a central role and a background for the formulation of WATS.

It is therefore convenient to recall the following overview of conceptual sewer process descriptions shown as figures in the preceding chapters:

1. The sulfur cycle (Figure 4.7)
2. Aerobic, heterotrophic transformations of organic matter (Figure 5.5)
3. Aerobic transformations of organic carbon and sulfur (Figure 5.6)
4. Anoxic, heterotrophic transformations of organic matter (Figure 5.17)
5. Aerobic and anaerobic transformations of organic carbon and sulfur (Figure 6.12)

The WATS model, with a focus on descriptions of the different integrated elements, is discussed in Section 9.2 and expressed as process matrices (cf. Table 8.1).

## 9.2   PROCESS ELEMENTS OF WATS MODEL

It appears from the five figures referred to in Section 9.1 that when summarizing the central elements, the backbone of the WATS process concept is the carbon cycle including the related aerobic, anoxic, or anaerobic transformations. As a consequence, the aerobic, heterotrophic organic carbon model concept, expressed in a process matrix formulation, will be outlined as the first WATS model approach.

In general, the nomenclature follows the definitions explained in the preceding chapters. An overview of the nomenclature is given in Appendix A. This is done in order to give an easily understandable and direct overview of the WATS model, the details of which will not be repeated in this chapter. Further details and comments related to the general characteristics of the process matrix formulations are given in Section 8.2.1 and depicted in Table 8.1.

### 9.2.1   Process Matrix for Aerobic, Heterotrophic Organic Matter Transformations

Table 9.1 summarizes the aerobic, heterotrophic transformations of organic matter in a sewer. It should be noted that the formulations in Table 9.1 follow the concept shown in Figure 5.5. The rate expressions for organic matter are discussed in detail in Section 5.3, and the reaeration is dealt with in Section 4.4.

Referring to Table 8.1, the mass balances of the five substances derived from Table 9.1 follow the principle shown in relation to Table 8.1. As an example, the mass balance of $S_S$ as a differential equation is expressed as follows:

$$\frac{\partial S_S}{\partial t} = -\frac{1}{Y_{Hw}} r_{grw} - \frac{1}{Y_{Hf}} r_{grf} - r_{maint} + r_{hydr,1} + r_{hydr,2} \tag{9.1}$$

**TABLE 9.1**

**WATS Sewer Process Model for Aerobic, Heterotrophic Transformation of Organic Matter in Wastewater Expressed as a Process Matrix**

| Process | $X_{Hw}$ | $S_S{}^a$ | $X_{S1}$ | $X_{S2}$ | $-S_O$ | Process Rate |
|---|---|---|---|---|---|---|
| Growth of biomass in bulk water phase | 1 | $-1/Y_{Hw}$ | | | $(1 - Y_{Hw})/Y_{Hw}$ | Equation 5.1 |
| Growth of biomass in biofilm | 1 | $-1/Y_{Hf}$ | | | $(1 - Y_{Hf})/Y_{Hf}$ | Equation 5.3 |
| Maintenance energy requirement | $(-1)^b$ | $-1$ | | | 1 | Equation 5.2 |
| Hydrolysis, fast | | 1 | $-1$ | | | Equation 5.4, $n = 1$ |
| Hydrolysis, slow | | 1 | | $-1$ | | Equation 5.4, $n = 2$ |
| Reaeration | | | | | $-1$ | Equation 4.39 and Equation 6 in Table 4.3$^c$ |

$^a$ $S_S = S_F + S_A$.

$^b$ If $S_S$ is not sufficiently available to support the biomass maintenance energy requirement.

$^c$ Unit used: day$^{-1}$.

It is worth noting that the aerobic, heterotrophic transformations are central when simulating sewer processes. A main reason is that the DO mass balance is determined by the processes in Table 9.1 and that it therefore also determines when, for example, anaerobic conditions occur.

### 9.2.2 PROCESS MATRIX FOR ANOXIC, HETEROTROPHIC TRANSFORMATIONS

The matrix shown in Table 9.1 for the aerobic, heterotrophic processes can be extended to include transformations under anoxic conditions (cf. the concept shown in Figure 5.17). The expressions for anoxic, heterotrophic growth in the water phase, maintenance energy requirement, biofilm growth, and hydrolysis are shown in Equations 5.28 through 5.31, respectively. In addition to the substances in the aerobic matrix, nitrate and elemental nitrogen must be added. It should be noted that nitrite according to previous considerations is not included as an intermediate (cf. Section 5.6.2.1).

Before the anoxic matrix is shown, it is important to recall that the concentration unit of organic matter [chemical oxygen demand (COD)] is g $O_2$ m$^{-3}$, i.e., the units of all substances in the aerobic process matrix—including DO as the electron acceptor—are identical, but with DO having an opposite sign. When nitrate occurs as electron acceptor, it should be noted that when balancing the stoichiometry of redox reactions, all units are in principle electron equivalents (e-eq) (cf. Section 2.1.4). According to Example 5.6, the stoichiometric conversion factor between organic matter as electron donor and nitrate as electron acceptor is:

$$R_{O_2,N} = 2.86 \text{ g } O_2 \text{ (gNO}_3 - \text{N)}^{-1} \tag{9.2}$$

## TABLE 9.2
## Extract from WATS Sewer Process Model for Anoxic, Heterotrophic Transformation of Organic Matter in Wastewater

| Process | $S_{NO_3}$ | $N_2$ | Process Rate |
|---|---|---|---|
| Growth of biomass in bulk water phase | $-\dfrac{1-Y_{Hw,NO_3}}{2.86Y_{Hw,NO_3}}$ | $\dfrac{1-Y_{Hw,NO_3}}{2.86Y_{Hw,NO_3}}$ | Equation 5.28 |
| Growth of biomass in biofilm | $-\dfrac{1-Y_{Hf,NO_3}}{2.86Y_{Hf,NO_3}}$ | $\dfrac{1-Y_{Hf,NO_3}}{2.86Y_{Hf,NO_3}}$ | Equation 5.30 |
| Maintenance energy requirement | $-\dfrac{1}{2.86}$ | $\dfrac{1}{2.86}$ | Equation 5.29 |
| Hydrolysis, fast | | | Equation 5.31, $n = 1$ |
| Hydrolysis, slow | | | Equation 5.31, $n = 2$ |

The extract of the anoxic process matrix in Table 9.2 shows the processes related to nitrate and elemental nitrogen. The influence on the organic matter fractions is, for simplicity reasons, not included but follows in principle what appears from the aerobic process matrix (cf. Table 9.1).

Except for situations where nitrate is added for anoxic control of hydrogen sulfide and anaerobic produced odorous VOCs—and therefore occur in relatively high concentrations compared with typical wastewater—the anoxic processes as shown in Table 9.2 are generally not active.

### 9.2.3 PROCESS MATRIX FOR ANAEROBIC, HETEROTROPHIC TRANSFORMATIONS

The concept for anaerobic transformations of biodegradable organic matter is shown in Figure 6.12. Growth of anaerobic, heterotrophic bacteria can normally be neglected and is therefore not included in the anaerobic process matrix. However, in specific situations it may be relevant to include growth of the sulfate-reducing biomass (SRB). This process is not included in the presented version of the WATS model.

The activities of the SRBs in terms of hydrogen sulfide production, emission of H$_2$S, and corresponding effects such as corrosion are explained in Section 9.2.4. In the extract of the anaerobic process matrix in Table 9.3, the anaerobic transformations of organic matter are emphasized. In this respect, the following processes are important:

- Anaerobic decay of $X_{Hw}$ produced under aerobic conditions (cf. Equation 6.23)
- Anaerobic hydrolysis (cf. Equation 6.21)
- Fermentation in the water phase (cf. Equation 6.22)

Methanogenesis (methane production) is typically not observed in sewer networks, although it is a potentially occurring process in the permanent layers of deposits. Methane production is therefore not included in the process matrix (Table 9.3).

**TABLE 9.3**

**Extract from WATS Sewer Process Model for Anaerobic, Heterotrophic Transformation of Organic Matter in Wastewater**

| Process | $X_{Hw}$ | $S_F$ | $S_A$ | $X_{S1}$ | $X_{S2}$ | Process Rate |
|---|---|---|---|---|---|---|
| Decay of biomass, $X_{Hw}$ | −1 | | | | 1 | Equation 6.23 |
| Hydrolysis, fast | | 1 | | −1 | | Equation 6.21, $n = 1$ |
| Hydrolysis, slow | | 1 | | | −1 | Equation 6.21, $n = 2$ |
| Fermentation in the water phase | | −1 | 1 | | | Equation 6.22 |

In particular, because of fermentation, the readily biodegradable organic matter, $S_S$, is subdivided into fermentable COD, $S_F$, and fermentation products, VFAs (volatile fatty acids) or $S_A$.

The two process matrices, Table 9.1 and 9.3, for aerobic and anaerobic transformations of organic matter can be integrated (cf. Figure 6.12). Although the active biomass is different for aerobic and anaerobic processes, the same fractions of organic substrates—readily biodegradable substrate and hydrolyzable substrate—are relevant in both cases. The fractions of substrate and their formation and utilization constitute, in addition to the DO mass balance, a natural link between the aerobic and anaerobic heterotrophic processes. Changing aerobic and anaerobic conditions in sewers are frequently occurring and therefore crucial to formulate in model terms.

### 9.2.4 PROCESS MATRIX FOR THE SULFUR CYCLE

The sulfur cycle in a sewer is complex for several reasons (cf. Figures 4.7 and 6.12):

- The sulfur cycle concerns both anaerobic and aerobic processes.
- The number and characteristics of relevant chemical and biological processes are large.
- Several sulfur-containing substances occur in the cycle.
- The sulfur cycle includes several phases: the biofilm, the water phase, the sewer atmosphere, and surfaces exposed to the sewer atmosphere.

Because of the complexity, it is often—depending on the objective—convenient to simplify the WATS model in the case of sulfur-related transformations. As an example, precipitation with iron salt may be excluded if not added for hydrogen sulfide control. It is therefore important to stress that different WATS model formulations are relevant to consider. In the version depicted in Table 9.4, the following sulfur-containing compounds are included:

- $S(-II) = C_{H_2S} + C_{HS^-}$ in the water phase (g S m$^{-3}$)
- $S_{SO_4}$ = concentration of sulfate in the water phase (g S m$^{-3}$)
- MeS, precipitated metal sulfide expressed in units corresponding to a water phase concentration (g S m$^{-3}$)

## TABLE 9.4

## Extract from WATS Sewer Process Model for Formulation of Sulfur Cycle

| Process | $S(-II)$ | $S_{SO_4}$ | MeS | $p_{H_2S}$ | $S_{H_2SO_4}$ | $S_O{}^a$ | $d_{corr}$ | Process Rate |
|---|---|---|---|---|---|---|---|---|
| $H_2S$ formation | 1 | -1 | | | | | | Equation 6.24 |
| Chemical oxidation of sulphide | -1 | 1 | | | | $-\dfrac{1}{R_{Cwc}}$ | | Equations 5.13 and 5.14 |
| Biological oxidation of sulphide | -1 | 1 | | | | $-\dfrac{1}{R_{Cwb}}$ | | Equations 5.13 and 5.15 |
| Oxidation of sulfide in biofilm | -1 | 1 | | | | $-\dfrac{1}{R_{Cfb}}$ | | Equation 5.16 |
| Precipitation of sulfide by metals | -1 | | 1 | | | | | b |
| $H_2S$ emission | -1 | | | $\dfrac{RT}{32}\dfrac{V_w}{V_g}$ | | | | Equation 4.40 and Equation 6 (Table 4.3) |
| $H_2S$ oxidation at concrete surfaces | | | c | -1 | $\dfrac{32}{RT}$ | | d | e |

a  The values of the reaction coefficients, $R_C$, are discussed in Section 5.5. The unit used here is g S $gO_2^{-1}$.

b  The basic characteristics of applying iron salts for control of sulfide and the corresponding practical aspects are in details dealt with in Sections 2.4.2 and 7.3.2, respectively. The iron/sulfur ratio is important as a pragmatic way of estimating the consumption of iron for sulfide removal (cf. Table 7.2). Here it should briefly be mentioned that removal of sulfide by precipitation depends on the efficiency of the metal, Fe(II) and Fe(III), to form stable metal sulfides in wastewater. The pH value and the redox potential, i.e., the distribution between Fe(II) and Fe(III), are crucial. Furthermore, the formation of iron complexes with different types of wastewater compounds reduces its efficiency for sulfide precipitation. It is also important that iron is a normal occurring metal in wastewater and available for partial precipitation of sulfide.

c  $\dfrac{32(1-\partial_{corr})}{RT}$.

d  $\dfrac{100\partial_{corr}}{\sigma_{conc}ART}\dfrac{V_g}{A_c}$.

e  As discussed in Section 6.5.2.2, the oxidation of $H_2S$ on concrete surfaces can be described based on either Monod kinetics or by an exponential expression (cf. Equations 6.16 and 6.17, respectively). When applying Equation 6.16 in WATS, this equation must be multiplied with the concrete surface area, $A_c$, to sewer gas volume, $V_g$, ratio to transform the unit to a volumetric basis. It is furthermore from the coefficients in the process matrix for both $H_2S$ emission and oxidation at concrete surfaces seen that transformation from water phase (w) to gas phase (g) concentration unit is included by applying the ideal gas law (cf. Equation 4.1). From practice, it is known that only a fraction, $\partial_{corr}$, of the produced sulfuric acid results in corrosion; the rest may be flushed back to the water phase. The value of $\partial_{corr}$ varies but is as low as about 0.5. The alkalinity, $A$, of the concrete material is expressed in g $CaCO_3$ per g concrete material (the molar weight of $CaCO_3$ is 100 g mol$^{-1}$). The value of $A$ depends on the concrete material used in the constructions. However, typically $A$ is in the order of 0.2 g $CaCO_3$ per g concrete material. The specific mass of the concrete material, $\sigma_{conc}$, is expressed in g m$^{-3}$ and included to transform the loss of mass to a corrosion depth.

- $p_{H_2S}$ = partial pressure of $H_2S$ in the sewer atmosphere (ppm)
- $S_{H_2SO_4}$ = sulfuric acid concentration in the water film of moist sewer surfaces (g S m$^{-3}$)

As discussed in Section 5.5, the end products of biological sulfide oxidation are primarily elemental sulfur ($S^0$) and sulfate ($SO_4^{2-}$). In order to simplify the description of the presented WATS model version, the two oxidized sulfur compounds, $S^0$ and $SO_4^{2-}$, are considered just one fraction and equal to an equivalent amount of sulfate. It is in several cases an acceptable assumption, and if the relative amount of elemental sulfur is not well known, the calculated amount of DO consumed becomes incorrect. However, it is a problematic assumption to neglect $S^0$ if the different types of aerobic oxidation of sulfide in the water phase are important processes.

The oxidation process of $H_2S$ at the concrete surfaces is furthermore simplified in the WATS model shown in Table 9.4. In this version, it is not taken into account that the oxidation of $H_2S$ to sulfuric acid at the sewer walls may proceed via the two intermediates, fast biodegradable elemental sulfur, $S_{fast}^0$, and slowly biodegradable elemental sulfur, $S_{slow}^0$ (cf. Section 6.5.1 and Figure 6.6). These two fractions are, however, included in the WATS model version reported by Vollertsen et al. (2011b).

The rate of concrete corrosion is in the WATS model calculated as a surface corrosion depth, $d_{corr}$, expressed in m day$^{-1}$.

In relation to the formulations of the sulfur cycle in Table 9.4, it should be noted that release of $H_2S$ out of the sewer and into the urban atmosphere is not taken into account. The release (ventilation) is highly dependent on the conditions of the local sewer network and therefore not possible to describe in general terms (cf. Section 4.3.4.1). If ventilation is an important phenomenon, the WATS model can, in the gas phase, be extended with a sink term.

## 9.2.5  ACID–BASE CHARACTERISTICS AND WATS MODELING

The pH value of wastewater is typically within the interval of 7–8.5. Several process rates are pH-dependent, and this dependency is within this pH interval often crucial for their potential impacts. Equations 5.15 shows how a pH dependent process rate mathematically can be formulated. The emission of $H_2S$ and the water phase oxidation of sulfide are important examples (cf. Sections 4.1.3 and 5.5.1.2, respectively). Furthermore, the pH value is crucial for the efficiency of sulfide control applying iron salts (cf. Sections 2.4.2 and 7.3.2). Expressed in general terms, the pH value in wastewater is determined by the actual acid–base distribution and the chemical and biological reactions of acid–base compounds. The pH value and its resistance to its change, i.e., the acid–base buffer capacity, are therefore important factors to consider.

Referring to Sections 2.4 and 4.5, the importance of pH and the buffer characteristics of typical wastewater compositions can briefly be summarized as follows:

- The carbonate system dominates pH and pH changes (the buffer capacity) within the pH range of 5.5–7.5.
- VFAs have, in addition to the carbonate system, an impact at low pH values. It may, for example, occur downstream along pressure mains where the

fermentation process can result in formation of rather high concentrations of VFAs ($S_A$).

- Ammonia and amines affect pH and the buffer capacity at pH values higher than about 7–7.5, and these substances may constitute a dominating system at pH higher than about 8–8.5. In addition to the carbonate system, they must be taken into account, for example, in cases where sulfide is being controlled by addition of bases.

Simulation of pH and pH changes with the WATS model requires basically extensive information on the acid–base systems (cf. Example 4.7). Such details are not always available, and a more pragmatic approach is needed. The use of pH values based on monitoring in the actual sewer network is a first and often acceptable solution. In specific cases, for example, when adding bases and thereby increasing the pH value, a calculated acid–base buffer capacity will give valuable information on which amount of bases is needed to reach a preset pH interval.

## 9.3 WATER AND GAS PHASE TRANSPORT IN SEWERS

Problems related to microbial and chemical processes in sewers occur typically during dry weather flows. The WATS model is therefore normally implemented to simulate such conditions. It is furthermore important to realize that sewer processes occur at a timescale that makes it irrelevant to simulate flow under nonuniform or unsteady state conditions. The hydraulics of the water flow in WATS is therefore described rather simply, typically as steady and uniform flow, applying, for example, the Manning formula or the Colebrook–White equation (cf. Equations 1.7 and 1.8, respectively).

The composition of the wastewater entering the sewer is assumed to be in a quasi-steady state. Therefore, the dispersion of wastewater substances is typically not considered, and plug flow can be assumed when simulating wastewater transformations.

The WATS model is designed to include both water phase and gas phase processes. The water–gas and the gas–solid exchange processes are consequently important factors to take into account. The model is therefore suited to simulate, for example, corrosion and VOC impacts where the process relations between the water phase and the gas phase play a central role. It is in this respect particularly important for the water–gas exchange processes in gravity sewers that the movement of the water phase and the gas phase are different (cf. Section 4.3.4.2). A proportional impact of the flowing water drag onto the movement of the gas phase is a first and simple estimate for determining the velocity of movement in the gas phase. The different transport velocities of the water phase and the overlying gas phase must be considered in the modeling process. The emission of, for example, $H_2S$ from a moving water phase, will therefore take place in a gas phase with a composition that is typically different from what was the case in the previous step of the modeling procedure (cf. Figure 8.1).

## 9.4 SEWER NETWORK DATA AND MODEL PARAMETERS

The layout of the WATS model and the requirement of data to generate the input to WATS will be dealt with in this section. These data concern the physical characteristics of the sewer network, the hydraulic description of WATS, and the parameters needed to describe process and exchange rates for the substances in the sewer. General and, to some extent, also specific comments concerning the determination of process parameters for the WATS model will be focused on. Further details are discussed in Chapter 10, with emphasis on specific methods for determination of model parameters.

### 9.4.1 SEWER NETWORK DATA AND FLOWS

It is not possible to give specific and generally valid recommendations to the extent of network data and water flow inputs. It is from a pragmatic point of view important to realize that widely varying information on the details of a sewer network is available in practice.

First, it should be mentioned that consistent geometric information on the sewer network is needed for the WATS model. In principle, this means that the system must be described in three dimensions with its nodes and manholes connected by pipes. Basic data such as length and diameters of the individual pipes and pipe slopes are thereby defined.

This description of the sewer network means that information on, for example, pumping stations, pressurized pipes, and gravity sewer sections is available. Network data are thereby at a level that sufficiently defines the system and makes it possible to simulate flow inputs and resulting flows in the system. In general, information on the sewer layout and geometry is equivalent to what is needed for a hydrodynamic model. The backbone of WATS is thereby a mass transport model developed for the water phase and the gas phase. On this basis, the relevant sewer processes in these two phases including interfacial transport processes for the corresponding substances form the WATS model. When using the WATS model, and particularly as a simulation model for assessing corrosion and odor problems, it is important to stress that the water phase, the gas phase and the exchange processes are included (cf. Section 9.3).

### 9.4.2 WASTEWATER COMPOSITION

An example of a composition of COD fractions in wastewater is shown in Figure 3.10. It is, however, important that a composition is highly site specific and therefore must be determined in each case. The data in Table 9.5 are presented to demonstrate a typical relative distribution of wastewater compounds.

### 9.4.3 WATS PROCESS MODEL PARAMETERS

In general, there are four different ways to determine the parameters for the WATS process model:

1. Direct monitoring using sensors in sewer network
2. Sampling in the sewer network followed by chemical or biological analysis
3. Experiments specifically designed to determine WATS model parameters
4. Determination of parameter values via model calibration and verification

**TABLE 9.5**

**Characteristic Values of COD Components and Dissolved Oxygen at an Upstream Location of an Intercepting Sewer (cf. Figure 3.10)**

| Component | Description | Characteristic Value | Unit |
|---|---|---|---|
| $X_{Hw}$ | Heterotrophic active biomass in the water phase | 20–100 | g COD m$^{-3}$ |
| $X_{Hf}$ | Heterotrophic active biomass in the biofilm | ~10 | g COD m$^{-2}$ |
| $X_{S1}$ | Hydrolyzable substrate, fast biodegradable | 50–100 | g COD m$^{-3}$ |
| $X_{S2}$ | Hydrolyzable substrate, slowly biodegradable[a] | 300–450 | g COD m$^{-3}$ |
| $S_F$ | Fermentable substrate | 0–40 | g COD m$^{-3}$ |
| $S_A$ | Fermentation products (i.e., VFAs) | 0–20 | g COD m$^{-3}$ |
| $S_S$ | Readily biodegradable substrate ($S_F + S_A$) | 0–40 | g COD m$^{-3}$ |
| $S_O$ | Dissolved oxygen | 0–4 | g O$_2$ m$^{-3}$ |
| COD | Total COD | About 600 | g COD m$^{-3}$ |
| $S_{H_2S}$ | Total sulfide | 0–5 | g S m$^{-3}$ |

*Note:* Wastewater includes particulate (*X*) and soluble (*S*) fractions.

[a] Includes very slowly biodegradable and inert organic matter.

These different options are further dealt with in Chapter 10. All options are relevant to consider depending on the needs and the local conditions and possibilities.

WATS process model parameters are in principle not constant but are subject to variability in time and place. Although this statement is correct and always important to remember, it is from a pragmatic point of view likewise important to take note that the variability may take place within a rather narrow interval. Furthermore, some parameters, for example, reaction orders for rate expressions, are based on intensive studies and for WATS model uses therefore considered known. If an essential part of the parameters were not approximately constant, any type of a realistic modeling would be absurd.

Although it is not generally possible to provide the user of the WATS model with recommended values, it is for some parameter types possible to give estimates. Examples of such estimates are shown in Table 9.6. It should be noted that the list is far from complete and that it is just shown to give an impression of the magnitude and the variability of different parameters. Further details and examples are given in Chapters 2 through 8 in relation to the description of specific sewer processes.

## 9.5 SPECIFIC MODELING CHARACTERISTICS

This section concerns some overall considerations of the WATS model layout and the modeling process.

### 9.5.1 PROCESS CONTENTS OF WATS MODEL

It has previously been referred to that the WATS model in each specific case is designed according to the objective. Basically, this means that the layout of the

**TABLE 9.6**

**Selected Examples of WATS Process Model Parameters**

| Symbol | Definition (unit) | Typical Value |
|---|---|---|
| $\alpha_w$ | Temperature coefficient for heterotrophic, aerobic water phase processes (–) | 1.07 |
| $\alpha_f$ | Temperature coefficient for aerobic biofilm processes (–) | 1.05 |
| $\alpha_r$ | Temperature coefficient for reaeration (–) | 1.024 |
| $\alpha_{Sf}$ | Temperature coefficient for sulfide formation in the biofilm (–) | 1.03 |
| $\mu_{Hw,O_2}$ | Maximum specific aerobic growth rate for heterotrophic biomass in the water phase (day$^{-1}$) | 4–8 |
| $\mu_{Hw,NO_3}$ | Maximum specific anoxic growth rate for heterotrophic biomass in the water phase (day$^{-1}$) | 2–6 |
| $\varepsilon_f$ | Relative efficiency constant for hydrolysis of the biofilm biomass (–) | 0.1–0.2 |
| $k_{h1}$ | Hydrolysis rate constant, fraction 1 (fast) (day$^{-1}$) | 5 |
| $k_{h2}$ | Hydrolysis rate constant, fraction 2 (slow) (day$^{-1}$) | 0.5 |
| $K_O$ | Saturation constant for DO (g O$_2$ m$^{-3}$) | 0.01–0.5 |
| $K_{NO_3}$ | Saturation constant for nitrate (g N m$^{-3}$) | 0.5–1.0 |
| $K_{Sw}$ | Saturation constant for readily biodegradable substrate in the water phase (g COD m$^{-3}$) | 0.5–2.0 |
| $K_{X1}$ | Saturation constant for hydrolysis, fraction 1 (fast) (g COD (g COD)$^{-1}$) | 1.5 |
| $K_{X2}$ | Saturation constant for hydrolysis, fraction 2 (slow) (g COD (g COD)$^{-1}$) | 0.5 |
| $q_{m,O_2}$ | Maintenance energy requirement rate constant for aerobic respiration in the water phase (day$^{-1}$) | 0.5–1.0 |
| $Y_{Hw,O_2}$ | Yield constant for aerobic growth of heterotrophic biomass in the water phase (g COD (g COD)$^{-1}$) | 0.50–0.60 |
| $Y_{Hw,NO_3}$ | Yield constant for anoxic growth of heterotrophic biomass in the water phase (g COD (g COD)$^{-1}$) | 0.30–0.40 |
| $Y_{Hf,O_2}$ | Yield constant for aerobic growth of heterotrophic biomass in the biofilm (g COD (g COD)$^{-1}$) | 0.50–0.60 |

actual sewer network is included in WATS and that site-specific conditions such as flows and process parameters have been determined and selected. Concerning the processes for WATS modeling as shown in Tables 9.1 through 9.4, the anoxic processes in Table 9.3 are just in an active mode when nitrate is added for anoxic sulfide control. The aerobic, anaerobic, and sulfur cycles shown in the other three tables are, however, typically always included in WATS. The reason is that anaerobically produced sulfide and VOCs are relevant for most sewer networks where process-related problems occur. In this respect, it is important to note that the aerobic, heterotrophic processes—including reaeration and DO consuming processes as described in Table 9.1—determine when and where DO is depleted and anaerobic conditions may occur.

## 9.5.2 WATS MODELING PROCEDURES

For the planning and execution processes, it is relevant to know which main working procedures and steps are related to WATS modeling in practice. These procedures focus on which information is required and which working steps are typical when applying the WATS sewer process model. Although specific details depend on the objective and local constraints, there are general aspects to consider. The following main activities exemplify what might be typical steps for catchment scale modeling and corresponding pipe scale analyses, for example, when sulfide problems are relevant to assess:

- Collection of data on geometry and layout of the sewer network.
- Collection of data for wastewater flow, flow variation, and quality inputs, for example, wastewater characteristics, pH, and temperature.
- Design of the WATS model, i.e., selection of relevant processes and parameters, and test runs with the noncalibrated model.
- Identification of "hot spots" based on the test runs.
- Design of a monitoring program to be implemented in the sewer network, typically both water phase and gas phase measurements—for example, measurements of DO, organic matter fractions, sulfur compounds, and pH.
- Collection and analysis of data from the monitoring program.
- Calibration and verification of the WATS model.
- Use of the WATS model at catchment scale for analysis according to the objective—for example, analysis of transformation and variability of concentration levels for selected substances and their potential control.
- Analysis of selected pipe sections, for example, "hot spots". The analysis includes, for example, occurrence in time and place of problematic substances and their control.

The ultimate objective of modeling is the results that constitute the basis for an efficient and optimal management of the sewer network. In particular, it is important that an analysis of the performance of a sewer system can provide the owner with information on potential structural changes in the network and implementation of equipment for process controls, for example, for dosing of chemicals. Several examples are given in Section 9.6.

## 9.6   EXAMPLES OF WATS MODELING RESULTS

Different types of programming languages can be used for numerical modeling with WATS—in principle, any computer application program from a simple spreadsheet to a more complex linear programming language. The WATS model has been programmed in Delphi Pascal, and it thereby exists as a tool for integrated simulation of water and gas transport, water–gas–solid exchange processes, and chemical and biological processes in all phases of a sewer. The WATS model can thereby be used for sewer process modeling at catchment scale—small or large—as well as for analysis of selected pipe systems. In this version, WATS includes all types of pipes and network structures, and it is capable of holding an unlimited number of pipes and nodes. Network data and flow inputs to network nodes at the boundaries are typically exported to WATS from an existing sewer network database. The inputs to all network nodes can be routed through the sewer by the WATS model, typically

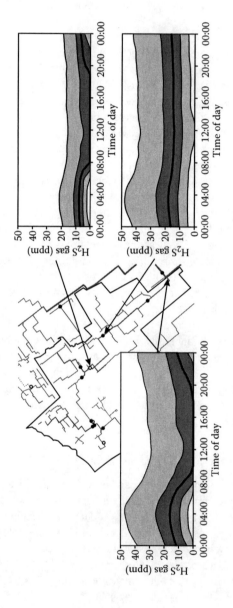

**FIGURE 9.1**  Stochastic simulations with the WATS model showing diurnal variation of H₂S gas phase concentrations at three points of a subcatchment. The centerline is the 50 percentile. Light gray areas and dark gray areas represent 5–95 and 25–75 percentiles, respectively. (From Vollertsen, J. et al. *Water Environ. Res.*, 80(2), 118–126, 2011. With permission.)

applying stationary hydraulics. The modeling results include, for example, assessment and impacts of sewer processes such as odor and corrosion, different types of control measures, and strategic management of sewer networks.

Results based on WATS modeling are in several cases presented in this book. The following examples are selected to give a brief impression of which simulation results are possible. They are not particularly presented to illustrate a specific phenomenon. Further details and examples on WATS modeling results can be found in several studies (e.g., Vollertsen et al. 2005, 2008, 2011; Nielsen et al. 2008).

The characteristics of sewer catchment scale modeling and analysis are described in Section 8.3.1. The results shown in Figure 9.1 refer to the asset owner's objective to formulate quantitative service levels and to ensure that the sewer system is being operated within these limits. The stochastic simulation procedure with WATS at catchment scale is an appropriate way of analyzing such management strategies.

Results from an analysis of a single gravity sewer pipe are illustrated in Figure 9.2. The simulations show both water phase and gas phase concentrations of sulfide/$H_2S$ and methyl mercaptan ($CH_3SH$). Methyl mercaptan—also named methylthiol—is an example of volatile odorous compound produced under anaerobic conditions in a sewer (cf. Tables 4.1 and 4.4).

The simulations in Figure 9.2 show that the maximum gas phase concentration of $H_2S$ occurs closer to the force main than the maximum gas phase concentration

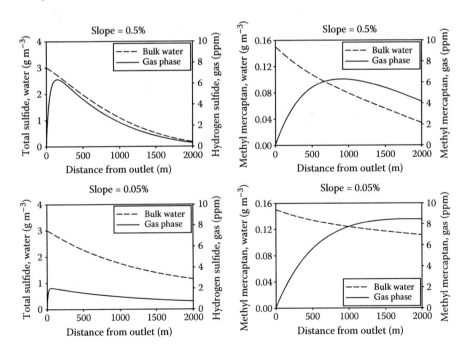

**FIGURE 9.2** WATS model simulation results of water and gas phase concentrations of sulfide/$H_2S$ and methyl mercaptan ($CH_3SH$) in a gravity sewer pipe receiving anaerobic wastewater from a force main. (From Vollertsen, J. et al. *Water Environ. Res.*, 80(2), 118–126, 2008. With permission.).

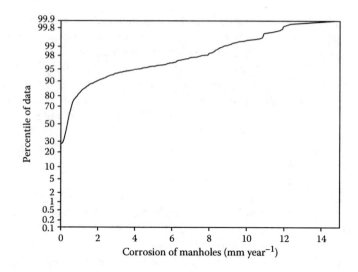

**FIGURE 9.3** Corrosion rate statistics for a total of 2600 manholes in a sewer catchment area of about 3000 ha located in United Arab Emirates. Number of inhabitants in the catchment is approximately 250,000.

of $CH_3SH$. The higher Henry's law constant for hydrogen sulfide and its potential oxidation at the moist concrete walls of the sewer are two possible reasons for this behavior. The simulations show that assessment of odor problems cannot solely be done based on the occurrence of hydrogen sulfide.

Figure 9.3 illustrates a WATS model result in terms of statistics on concrete corrosion in manholes of a relatively large sewer network. It can be seen from this figure that almost 30% of the manholes are not subject to corrosion, whereas the corrosion rate is >5 mm year$^{-1}$ in about 5% of the manholes.

# REFERENCES

Abdul-Talib, S., T. Hvitved-Jacobsen, J. Vollertsen, and Z. Ujang (2002), Anoxic transformations of wastewater organic matter in sewers–process kinetics, model concept and wastewater treatment potential, *Water Sci. Technol.*, 45(3), 53–60.

Hvitved-Jacobsen, T., J. Vollertsen, and P.H. Nielsen (1998) A process and model concept for microbial wastewater transformations in gravity sewers. *Water Sci. Technol.*, 37(1), 233–241.

Nielsen, A.H., J. Vollertsen, H.S. Jensen, H.I. Madsen, and T. Hvitved-Jacobsen (2008), Aerobic and anaerobic transformations of sulfide in a sewer system—field study and model simulations, *Water Environ. Res.*, 80(1), 16–25.

Vollertsen, J. and T. Hvitved-Jacobsen (1998), Aerobic microbial transformations of resuspended sediments in combined sewers—a conceptual model, *Water Sci. Technol.*, 37(1), 69–76.

Vollertsen, J., T. Hvitved-Jacobsen, and A.H. Nielsen (2005), Stochastic modeling of chemical oxygen demand transformations in gravity sewers. *Water Environ. Res.*, 77(2), 331–339.

Vollertsen, J., A.H. Nielsen, H.S. Jensen, and T. Hvitved-Jacobsen (2008), Modeling the formation and fate of odorous substances in collection systems, *Water Environ. Res.*, 80(2), 118–126.

Vollertsen, J., L. Nielsen, T.D. Blicher, T. Hvitved-Jacobsen, and A.H. Nielsen (2011a), A sewer process model as planning and management tool—hydrogen sulfide simulation at catchment scale, *Water Sci. Technol.*, 64(2), 348–353.

Vollertsen, J., A.H. Nielsen, H.S. Jensen, E.A. Rudelle, and T. Hvitved-Jacobsen (2011b), Modeling the corrosion of concrete sewers, *12th International Conference on Urban Drainage*, Porto Alegre, Brazil, September 11–16, 2011, p. 9.

# 10 Methods for Sewer Process Studies and Model Calibration

Experimental investigations and studies on microbial and chemical sewer processes are needed for two major reasons. First of all, basic studies on the nature of these processes are crucial for formulation of, for example, process rates and stoichiometric characteristics. The preceding chapters are rich in examples in this respect. Second, site-specific investigations in sewer networks are required to provide information on central parameters for sewer process models, for example, the WATS (Wastewater Aerobic/anaerobic Transformations in Sewers) model (cf. Chapter 9). This chapter focuses on two main subjects:

1. Methodologies of good practice in the study of chemical and microbial processes in wastewater collection systems. The information on such processes is provided by investigations, measurements, and analyses performed at bench scale, pilot scale, and field scale.
2. Determination of parameters to be used for calibration and verification of sewer process models.

These two main objectives of this chapter are integrated: sampling, bench scale studies, pilot scale and field measurements, and different types of analyses are generally needed to determine wastewater characteristics, including kinetic and stoichiometric parameters.

Different procedures for sewer process studies will be dealt with, primarily those directly related to the determination of process-relevant characteristics. Procedures and measurements of, for example, sewer hydraulics and solids transport characteristics, will not be covered in this chapter. Although information from such measurements is relevant for sewer process models, relevant literature is generally available for this purpose. Hydraulic measurements in sewers are discussed in several publications (e.g., ASCE 1983; Bertrand-Krajewski et al. 2000). An overview of the physical processes in sewers is given by Ashley and Verbanck (1998) and Ashley et al. (2004).

This chapter will not include all relevant methods but representative examples will be focused on.

## 10.1  METHODS FOR BENCH SCALE, PILOT SCALE, AND FULL SCALE STUDIES

### 10.1.1  GENERAL METHODOLOGY FOR SEWER PROCESS STUDIES

A sound theoretical basis, reliable and relevant methodologies for investigations, and a conceptual description of sewer processes constitute a fundamental triangle for process studies. This framework forms the basis for engineering tools that make it possible to design and manage sewer systems with a process dimension.

These methodologies have their origin in bench scale, pilot plant, and field studies. An overall and central objective for these experimental methodologies often involves the determination of model parameters. General aspects including examples of such types of methodologies will therefore be addressed. Experiments must be carefully considered according to the objectives set. It is furthermore a general and basic requirement for an experimental study that the mass balances for the compounds that are focused on are observed.

#### 10.1.1.1  Bench Scale Analysis and Studies

Laboratory methods for chemical analyses form basically the backbone of any type of experiment for the study of sewer processes. Such methods are available worldwide. The so-called "Standard Methods" (APHA–AWWA–WEF 2005), is in general representative of these methods.

Process studies are often either batch or flow reactor experiments. A general and important aspect of laboratory studies is that they can be performed under controlled conditions. Reactor experiments are therefore typically elaborated to fulfill the objective of a basic study on sewer processes. In addition to the experimental setup, a carefully considered program for sampling, handling, and analysis is required. A laboratory reactor experiment needs a lot of planning—and also typically, experience—to be successful.

An example of a biofilm reactor setup shown in Figure 10.1 demonstrates how an experiment can be performed under controlled conditions (Raunkjaer et al. 1997). The objective of the study is to determine the substrate (here, acetate) and dissolved oxygen (DO) surface removal rates of aerobic biofilms grown on wastewater as substrate. Careful experimental planning and handling are needed when studying substrate and DO transformations under limiting conditions. A great number of specific details must therefore be dealt with.

#### 10.1.1.2  Pilot Plant Studies

Pilot and laboratory studies are generally selected when controlling the conditions of the different factors that have an impact on the processes. The advantage of a pilot study compared with a laboratory experiment is the scale that is closer to the real system. Such "scale factors" can, for example, be a more realistic volume/area ratio and a more relevant flow regime. Drawbacks include the high demands of resources and extensive manpower requirement for manufacturing and operation. Furthermore, pilot studies may often be difficult to run and require specific practical skills.

Pilot sewer studies are often carried out in systems operating with recirculation. Specific care must be taken in systems where water–gas exchange processes form a part of the mass balance. Critical points are pumps and bends that may change

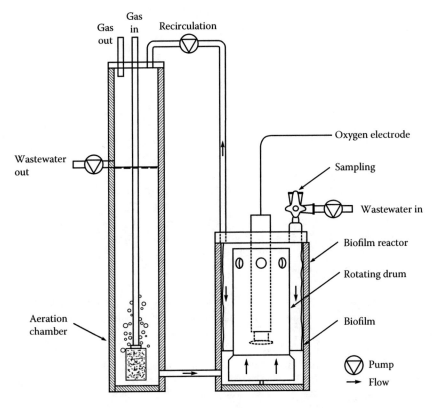

**FIGURE 10.1** Example of a reactor setup to study wastewater biofilm processes under controlled laboratory conditions.

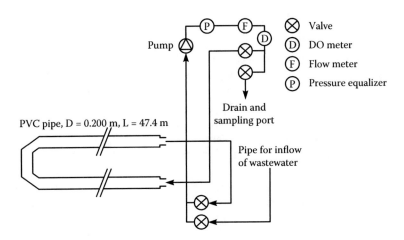

**FIGURE 10.2** Outline of a pilot plant sewer. Studies can be performed under both gravity sewer and pressure sewer conditions.

the flow regime, air–water exchange processes, biofilm development, and particle structure. Figure 10.2 shows a sketch of a pilot sewer used for sewer process studies (Tanaka and Hvitved-Jacobsen 2000).

Pilot scale studies can be implemented in a way that is almost identical with full-scale studies, but still being carried out under strictly controlled conditions. A pilot-scale sewer reactor observing such characteristics is described by Vollertsen et al. (2008) and Nielsen et al. (2008). This experimental setup is equipped for the investigation of concrete corrosion with sections of concrete sewer pipes and a possibility for continuously receiving wastewater from a sewer network.

### 10.1.1.3 Field Experiments and Monitoring

Field experiments and monitoring are important for two major reasons:

1. Assessment of process relations and mathematical expressions that are developed under bench scale or pilot plant conditions.
2. Monitoring and data collections for use in calibration and verification of sewer process models.

Field investigations concern typically water phase and gas phase measurements but may also include the biofilm or samples from, for example, corroding concrete surfaces.

It is in the real sewer systems that the benefits of a proper engineering of the processes must be proven. However, the sewer is not an optimal system for detailed studies of sewer processes because of the difficulties encountered when establishing controlled conditions. Field investigations are, however, needed to determine reasonable values for those compounds and parameters that are normally not feasible to measure explicitly by laboratory or pilot-plant studies.

Field experiments are often carried out by sampling and monitoring at different locations in a sewer network. The transformations in the water phase can, for example, be studied by following the course of a tracer added at an upstream location. Substances such as rhodamine, radiotracers, and salts may typically be selected for that purpose.

It is generally preferred to carry out field experiments in long sewer lines that make it possible to measure relatively large differences in the wastewater quality. Furthermore, a sewer line without tributary sewers, infiltration, and exfiltration must be preferred because of a less complex sampling program that otherwise requires identification of all important inputs and outputs to the system.

The selection of appropriate locations in a sewer for monitoring and collection of data for model calibration and verification are generally determined by the so-called "hot spots" in a sewer catchment, i.e., points where critical process-related impacts are expected to occur. Such monitoring programs require, in general, extensive resources in terms of equipment and manpower. In practice, it is crucial to realize that limited resources are typically available for such purposes and that an "optimal" outcome of a program is often identical with a "minimum" of data needed for model calibration and verification. Monitoring with sensors and use of automatic equipment for data collection and transfer are therefore—if possible—typically selected under such conditions. Monitoring of pH in the water phase and $H_2S$ in the gas phase in four to eight different locations of a sewer catchment for a period of 2 to 4 weeks is a very simple example of a monitoring program focusing on assessment of sulfide problems. In order to minimize cost, it is of course possible to move equipment from one location to another.

## 10.1.2 Sampling, Monitoring, and Handling Procedures

In all types of experimental work, it is crucial that sampling, handling, and analysis are carefully considered and that basic experimental techniques are observed. These three procedures must furthermore conform with the overall objective of the study.

Sewer process studies require that samples and sensor measurements are representative for the system—or a part of a system—and furthermore result in central information needed. Sampling and monitoring in a gravity sewer shortly downstream a force main is an example of an appropriate location for assessment of sulfide problems. Locations for sampling and monitoring are typically manholes and pumping stations. Furthermore, diurnal and seasonal changes must be observed. Sampling and monitoring should therefore be repeated to investigate the importance of these variations. Sampling is often carried out automatically with web-based transfer of data and with a corresponding monitoring of the flow to allow for the determination of a mass transport.

Handling by transport from the sampling site to a laboratory must ideally be done without changing the sample, i.e., without changes in the parameters that will be analyzed. The potential process rates and the time needed for transport must be considered. Process rates are typically lower under anaerobic conditions than under aerobic conditions. If possible, a wastewater sample should be transported under anaerobic conditions.

Automatic sampling, sensor measurements, and web-based transfer of data are all procedures for minimizing manpower resources in sewer process studies. It is, however, important to realize that equipment and technical systems require careful supervision to ensure that they operate stably and in accordance with the objective. As an example, sensor measurement of $H_2S$ in a sewer atmosphere is an easy and frequently used method for sulfide assessment. It is, however, important that these sensors are not continuously exposed to sulfide but frequently are being placed in clean air.

## 10.1.3 Oxygen Uptake Rate Measurements of Bulk Water

As discussed in Section 3.2.6 and illustrated in Figure 3.11, the OUR (oxygen uptake rate)-versus-time curves for wastewater plays a central role in the assessment of the biodegradability of organic matter in terms of its fractionation. As also dealt with in Chapters 8 and 9, this fractionation is crucial for modeling wastewater transformations in sewers. Bulk water OUR measurements are carried out as bench scale experiments.

In brief, OUR is an activity-related quantitative measure of the influence of the aerobic biomass on the relationship between the electron donor (organic substrate) and the electron acceptor (DO). It is thereby a measure of the "flow of electrons" through the entire process system under aerobic conditions (cf. Figure 2.3). The OUR-versus-time relationship of wastewater samples from sewers serves as the backbone for analysis of the microbial system. It is crucial for characterization of the suspended wastewater phase in terms of both COD (chemical oxygen demand) compounds and the corresponding kinetic and stoichiometric parameters of in-sewer processes.

A methodology for OUR measurements in wastewater systems was originally developed for the characterization of activated sludge in terms of COD compounds and process parameters (Ekama and Marais 1978; Dold et al. 1980). The procedure included OUR measurements of the activated sludge under substrate-limited and substrate-nonlimited growth conditions, typically carried out by discontinuous addition of wastewater. Further development of respirometry principles and techniques has taken place, for example, motivated for the control of the activated sludge processes (Spanjers et al. 1998).

The OUR measurement principle developed for nonseeded wastewater (wastewater without addition of biomass and as it occurs in sewer networks) differs from what is typically relevant for activated sludge characterization. OUR-versus-time measurements for wastewater in sewers are typically investigated in 1 to 2 days and in a batch reactor. This procedure will lead to a quantification of the heterotrophic processes and more generally allow operators to determine COD fractions of different biodegradability. An OUR experiment is typically carried out at constant temperature (e.g., 20°C) and under DO nonlimiting growth conditions for the biomass in the wastewater sample. The DO concentration may vary from 8 to 6 g $O_2$ m$^{-3}$ for the determination of a single OUR value. Figure 10.3 shows a schematic outline of the relation between such DO concentration versus time measurements and the calculation of the OUR value at time $t_n$.

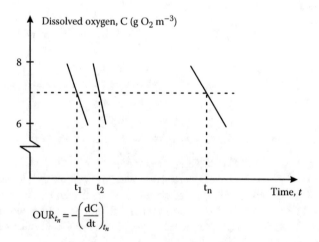

$$OUR_{t_n} = -\left(\frac{dC}{dt}\right)_{t_n}$$

**FIGURE 10.3**  Schematic outline of DO concentration versus time measurements for determination of OUR values.

The aerobic, heterotrophic in-sewer microbial processes are in a process matrix form shown in Table 9.1. When omitting the reaeration and the growth of the biofilm biomass in this description, the remaining processes proceed interactively in the water phase under the DO nonlimited growth conditions established in the OUR experiments. Furthermore, the processes take place at a constant temperature. The relevant processes and the formulation of rate expressions are shown in Table 10.1.

A characteristic example of an OUR experiment of a wastewater sample with readily biodegradable substrates and hydrolyzable substrates is shown in Figure 10.4. The figure shows that the original amount of readily biodegradable substrate is depleted within the first 4 h of the experiment, while increasing the biomass

**TABLE 10.1**

**Matrix Formulation of Aerobic Microbial Transformations of Organic Matter in an OUR Batch Experiment with Nonseeded Wastewater**

|  | $S_S$ | $X_{S1}$ | $X_{S2}$ | $X_{Hw}$ | $-S_O$ | Process Rate |
|---|---|---|---|---|---|---|
| Growth of biomass in bulk water phase | $-1/Y_{Hw}$ |  |  | 1 | $(1 - Y_{Hw})/Y_{Hw}$ | Equation a |
| Maintenance energy requirement | $-1$ |  |  | $-1^a$ | 1 | Equation b |
| Hydrolysis, fraction 1 (fast) | 1 | $-1$ |  |  |  | Equation c, $n = 1$ |
| Hydrolysis, fraction 2 (slow) | 1 |  | $-1$ |  |  | Equation c, $n = 2$ |

The formulation shown includes two fractions of hydrolyzable substrate.

$^a$ If $S_S$ is not present in sufficient concentration, $X_{Hw}$ is used for endogenous respiration.

(a) $\mu_H \dfrac{S_S}{K_{Sw} + S_S} X_{Hw}$.

(b) $q_m X_{Hw}$.

(c) $k_{hn} \dfrac{X_{Sn}/X_{Hw}}{K_{Xn} + X_{Sn}/X_{Hw}} X_{Hw}$.

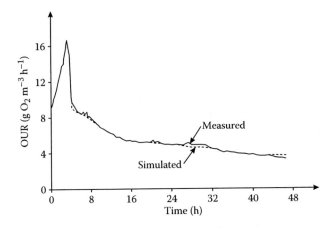

**FIGURE 10.4**   Measured and simulated OUR values of a wastewater sample.

concentration and, consequently, the respiration rate (the OUR value) from about 9 to 17 g $O_2$ m$^{-3}$ h$^{-1}$. After this period, readily biodegradable substrate is only available as a result of production from hydrolysis that is immediately consumed because of the growth-related respiration rate and the maintenance energy requirement rate of the biomass. At about 15 h after the experiment started, the fast hydrolyzable fraction ($n = 1$) is depleted. After this time, only hydrolysis of the slowly hydrolyzable fraction ($n = 2$) remains a source for production of readily biodegradable substrate.

This interpretation of the OUR-versus-time variability for a wastewater sample is, in terms of a quantitative description in model terms, formulated in Table 10.1 according to Table 9.1. The example in Figure 10.4 shows what is often observed, that a good agreement between measured and simulated OUR values can be obtained by the rather simple description of the microbial processes.

In general, it can be concluded that an OUR experiment reflects the different activity levels that the heterotrophic biomass is exposed to, and that the outcome depends on the availability and quality of the substrate (cf. Section 3.2.6). The principle outlined is the experimental basis for the transfer of the concept for microbial transformations of wastewater into experimentally determined values for parameters that via modeling (e.g., with the WATS model) can be applied for design and management of sewer systems.

Different types of equipment, depending on the resources available and the number of measurements required, can be used for determination of an OUR-versus-time curve. A simple and manually operated equipment was used by Bjerre et al. (1995). A relatively inexpensive apparatus, simple to operate automatically, was designed by

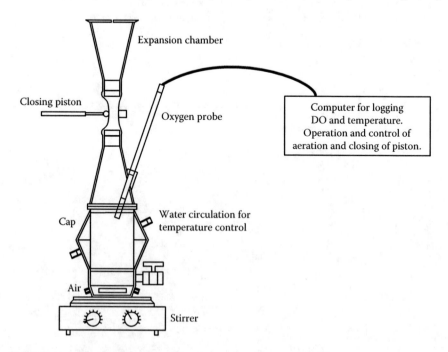

**FIGURE 10.5**  Automatically operated type of equipment for OUR measurements.

Tanaka and Hvitved-Jacobsen (1998). However, this type of equipment may introduce a minor error at low OUR values because of a potential release of oxygen into a headspace of nitrogen gas in the reactor.

An advanced, automatically operated type of equipment was designed and used by Vollertsen and Hvitved-Jacobsen (1999) (cf. Figure 10.5). The equipment shown in this figure is produced in stainless steel with a 2.2-L reactor volume. It operates at a DO concentration between about 6 and 8 g $O_2$ m$^{-3}$ and with the temperature of the wastewater kept constant at 20°C by circulating water through a cap around the reactor. When the DO concentration is below about 6 g $O_2$ m$^{-3}$, an aeration cycle starts by blowing compressed air into the reactor and by opening the piston to the expansion chamber. After the end of the aeration period, the piston is closed with a preset time delay to ensure that the air bubbles are removed from the reactor. Data on DO measurements are automatically sampled, and OUR values are calculated for each period corresponding to a reduction of the DO concentration from about 8 to 6 g $O_2$ m$^{-3}$.

### 10.1.4 MEASUREMENTS IN SEWER NETWORKS

In addition to sampling in a sewer followed by analysis of specific substances or use of a sample for further laboratory or pilot-scale experiments, a number of direct or indirect measurements in the sewer itself are often feasible. Such measurements can be carried out in the water phase, the sewer biofilm, and the gas phase. Measurements related to sewer process investigations include DO, reaeration, biofilm characterization, odor, and hydrogen sulfide in the gas phase. Sensor measurement of $H_2S$ in the gas phase is briefly discussed in Section 10.1.2.

#### 10.1.4.1 DO Measurements

Sensor measurements of DO in sewers require special attention because of the risk for clogging the DO probe with gross solids. Figure 10.6 shows an arrangement to

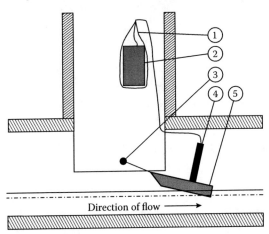

**FIGURE 10.6** Exemplification of DO measurement in a manhole of a gravity sewer. A bag (1) contains a waterproof box (2) with electrical equipment, data logger, and batteries for a DO meter (4). The float (5) is anchored by a steel rod (3).

avoid this problem (Gudjonsson et al. 2002). The DO meter is placed in a small float of polyurethane that keeps the probe submerged and allows the gross solids to pass without being retained at the sensor.

A reliable measurement of DO in a wastewater system requires that the surface of the sensor be regularly cleaned to avoid development of a biofilm that otherwise will consume oxygen and disturb the measurement.

### 10.1.4.2 Measurement of Reaeration

Air–water mass transfer, i.e., reaeration, is dealt with in Section 4.4. Determination of reaeration relies on the measurement of the air–water oxygen transfer coefficient. Measurement of this coefficient—the reaeration coefficient—in gravity sewer pipes basically follows the methods that have been developed for and applied in flowing waters such as rivers.

Methods for determination of the oxygen transfer coefficient in flowing waters can be divided into two groups: direct and indirect methods. Indirect methods are generally based on the principle of a mass balance for DO. Direct methods of the oxygen transfer coefficient make use of inert substances with a constant air–water mass transfer coefficient compared with that of oxygen. An overview of direct and indirect methods for determination of the reaeration is given by Jensen and Hvitved-Jacobsen (1991).

An indirect method to determine reaeration in gravity sewers was developed by Parkhurst and Pomeroy (1972). They conducted measurements in gravity sewers where the biofilm was removed mechanically followed by a shock load with caustic soda. During the measurement stage, the biological activity was suppressed in the water phase by a chemical substance. Measurement of upstream and downstream DO concentrations in the sewer determined the reaeration by using a simple DO mass balance.

Jensen and Hvitved-Jacobsen (1991) developed a direct method for the determination of the air–water oxygen transfer coefficient in gravity sewers. This method is based on the use of the radioisotope krypton-85 ($^{85}$Kr) for the air–water mass transfer and tritium ($^3$H) for dispersion followed by a dual counting technique with a liquid scintillation counter (Tsivoglou et al. 1965, 1968; Tsivoglou and Neal 1976). A constant ratio between the air–water mass transfer coefficients for DO and krypton-85 makes it possible to determine reaeration using a direct method. Sulfur hexafluoride ($SF_6$) is another example of an inert substance that has been used as a tracer for reaeration measurements in sewers (Huisman et al. 1999). The use of propane is furthermore an option as a tracer gas (Genereux and Hemond 1992).

Determination of reaeration at sewer falls and drops is briefly dealt with in Section 4.4.4.

### 10.1.4.3 *In Situ* Measurement of Biofilm Respiration

Measurement of biofilm activity can be carried out based on laboratory reactor experiments or with a technique combining sampling of biofilms grown in a sewer followed by measurements in the laboratory (Raunkjaer et al. 1997; Bjerre et al. 1998). Huisman et al. (1999) developed a sewer *in situ* biofilm respiration chamber. It includes a DO sensor and a chamber that can be pressed onto the sewer wall. It is designed to achieve an even and unidirectional flow distribution over the entire measurement area. Pure oxygen is injected for oxygenation.

#### 10.1.4.4  Gas Phase Movement and Ventilation in Gravity Sewers

Movement in the gas phase of a sewer and ventilation of gases is important when focusing on, for example, hydrogen sulfide and odorous substances (cf. Section 4.3.4).

A direct and rather simple method for determination of the horizontal movement of the gas phase and the ventilation rate in a gravity sewer is based on injection of a gas (Madsen et al. 2006). The gas is injected at a manhole and changes the composition of the sewer atmosphere (cf. Figure 10.7). Different gases can be used, for example, nitrogen and oxygen. Although oxygen is not an inert gas, it was found to give the best response relative to the amount injected, and the loss of oxygen was found to be negligible. Pulses of oxygen are injected to the sewer atmosphere at an injection station and monitoring takes place at an upstream and downstream station, respectively. The oxygen gas acts as a tracer, allowing the gas velocity, dispersion coefficient, and ventilation rate to be calculated. The distances between the three stations may vary but are typically a few hundred meters.

### 10.1.5  Odor Measurements

The basic properties and characteristics of odors and odor assessment are dealt with in Section 4.3.1. Further details are, in this respect and concerning odor measurements, found in several reports (e.g., Fenner and Stuetz 2000; Sneath and Clarkson 2000; Stuetz and Frechen 2001; WEF 2004). Numerous methods have traditionally

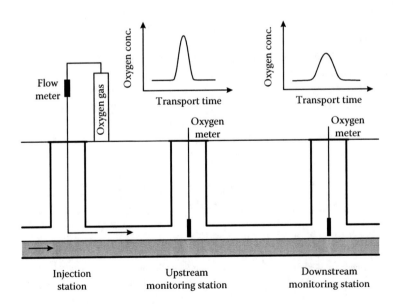

**FIGURE 10.7**  Principle of oxygen injection and monitoring for determination of horizontal gas movement in a gravity sewer.

been used for sampling and measurement of odors. This section will focus on dynamic olfactometry applying the human nose as a sensor. In the United States and in Europe, the need for well-defined procedures for odor measurement has led to the development of standards (ASTM 1991 and CEN 2003, respectively).

Odors must be detected at very low concentrations, and sampling, handling, and analysis of odors must therefore be carefully carried out to be reliable. Well-defined procedures and high-quality equipment is consequently needed. Procedures for odor measurements specifically related to sewer networks do not exist. The following discussion outlines the general fundamentals of sampling and measurement of dynamic olfactometry as also applied at wastewater treatment plants.

Samples of odorous air for olfactometric analysis are usually collected into a sample bag and transported to a laboratory for analysis. A sample bag can be placed in a container, and the bag is gradually filled with odorous air when air is removed from the gap between the container and the bag using a vacuum pump.

The laboratory where the measurement takes place must be free from odor and is typically air-conditioned with air filtration. The odor sample is placed in an olfactometer, which is basically a device for dilution of the sample. Typically, the meter has two outlet ports: diluted odorous air flows from one, and clean odor-free air flows from the other. In dynamic olfactometry, panel members assess the two ports of the olfactometer. The assessors indicate from which of the ports the diluted sample is flowing. The measurement starts with a dilution that is large enough to make the odor concentration exceed the panelists' threshold. This concentration is normally increased by a factor of 2 in each successive presentation. Only when the correct port is chosen and the assessor is certain that the choice is correct and not just a guess, is the response considered a true value.

As described in Section 4.3.1, the European CEN (Comité Européen de Normalisation) standard operates with an odor concentration unit, $ou_E$ $m^{-3}$, where the odor concentration at the detection threshold is defined as equal to 1 $ou_E$ $m^{-3}$. In countries where the CEN standard is not in use (e.g., the United States), the corresponding dilution-to-threshold value $(D/T)$ without an assigned specific unit is typically used.

In addition to the dynamic olfactometry procedure, it should be mentioned that a sensor array named the "electronic nose" is a rapid and relatively simple technique that can be used for indication of wastewater odors (Stuetz et al. 2000). The electronic nose uses sensors of varying affinities to characterize an odor without reference to its specific chemical composition.

## 10.2   METHODS FOR DETERMINATION OF SUBSTANCES AND PARAMETERS FOR SEWER PROCESS MODELING

The methods dealt with in Section 10.1 are generally intended to provide information on the nature of sewer processes, often in terms of their dependency on external conditions. These methods are, however, also useful in more specific cases, for example, in their ability to determine process-related parameters. To some extent, in contrast to these procedures, the methods dealt with in this section are more specifically designed to determine the actual values for substances and parameters in sewer process models, for example, the WATS model.

Monitoring, sampling, and different types of experiments are generally required for determination of substances and parameters for sewer process modeling (cf. Section 9.4.3). Explicit determination of model compounds and parameters are preferred to indirect and implicit methods. Determination of OUR-versus-time curves is as an important example of an experimental procedure that can be used for determination of parameters for a sewer process model (cf. Section 10.1.3). However, model calibration is to some extent normally needed to establish an acceptable balance between the process details of a model and the possibilities for direct experimental determination of model parameters. The procedures described refer basically to the WATS model as described and formulated in Chapter 9. However, the procedures are generally valid for sewer process models including wastewater compounds characterized by their biodegradability.

Methodologies for determination of substances and parameters for sewer process modeling can be categorized in different ways. The following grouping of methods is an attempt of a systematic application of OUR-based procedures for characterization of organic matter and its transformations under sewer conditions:

(1) Methods for determination of central model parameters (cf. Section 10.2.1). OUR-versus-time measurements of incoming wastewater to a sewer system modified by addition of readily biodegradable substrate (e.g., acetate).
(2) Determination of the biodegradability of wastewater organic matter (cf. Section 10.2.2). OUR-versus-time measurements of incoming wastewater to a sewer system.
(3) Determination of model parameters by iterative simulation of the OUR curve for the incoming wastewater (cf. Section 10.2.3).
(4) Calibration and validation of the sewer process model (cf. Section 10.2.4). OUR measurements of corresponding upstream and downstream wastewater samples followed by a simulation (calibration) with the sewer process model.

These four procedures are outlined in Figure 10.8. In order to achieve optimal parameter estimation, it is recommended that the procedures are carried out in the sequence shown and that a final verification of the sewer process model is done.

When designing a new sewer system, procedure number 4 is, of course, not relevant. Kinetic parameters for the sewer biofilm should therefore be selected and assessed based on information from similar sewer systems.

It is important to understand that wastewater is subject to considerable variability in terms of its substances and processes. Procedures 1 to 4, therefore, correspond to a typical analytical methodology for the determination of the characteristic values of wastewater substances and the stoichiometric and kinetic parameters. Cases where the procedures described in Sections 10.2.1 through 10.2.4 is either difficult or not feasible to follow may exist. A detailed knowledge on wastewater characteristics and experience from laboratory and modeling studies are crucial in such situations for finding alternative variants of the procedures 1 to 4.

The procedures described in Sections 10.2.1 through 10.2.4 refer to the WATS model formulation for aerobic, heterotrophic transformations (cf. Section 9.2.1 and

Procedure
no.

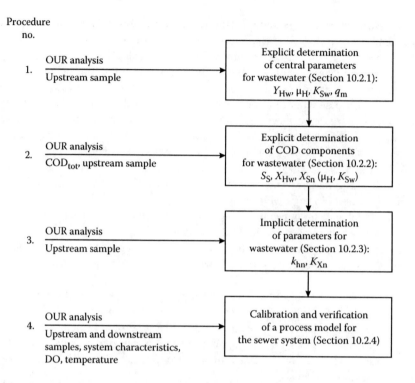

FIGURE 10.8 Overview of OUR-based procedures for determination of wastewater compounds and parameters for sewer process modeling.

Table 9.1). Section 10.2.5 deals with the methods applied for determination of model parameters to include transformations of wastewater compounds under anaerobic conditions (cf. Sections 9.2.3 and 9.2.4).

## 10.2.1 DETERMINATION OF CENTRAL MODEL PARAMETERS

An OUR-versus-time experiment can be carried out on a wastewater sample from the influent to a sewer system. The sample is normally modified by addition of readily biodegradable substrate (e.g., acetate or glucose). The following four process parameters are explicitly determined based on this experimental approach:

1. $\mu_H$, maximum specific growth rate
2. $K_S$, saturation constant for readily biodegradable substrate
3. $Y_{Hw}$, yield constant
4. $q_m$, maintenance energy requirement rate constant

The experiment, or preferably a number of parallel experiments, is carried out until the originally available readily biodegradable substrate and the fast hydrolyzable substrate are depleted. For typical domestic wastewater, this usually takes less than 1 day. At this point in the experiment, it is assumed that the lack of readily biodegradable substrate suppresses the growth process. Readily biodegradable substrate (typically acetate or glucose) is then added to the wastewater, and the experiment continues until this substrate is also depleted, and the OUR value slightly declines. The impact of the substrate added to the microbial system can thereby be determined.

The idea behind the experiment is to let the biomass growth rate change from zero to its maximum value. By adding a known amount of substrate under controlled conditions, the interpretation of the OUR response can be described by the WATS model concept (cf. Table 10.1). The four central process parameters can thereby be determined. The following two conditions are important for a successful outcome of the experiment:

1. Before the readily biodegradable substrate is added, the maintenance energy requirement of the active biomass should ideally correspond to the amount of readily biodegradable substrate produced by hydrolysis of the slowly biodegradable COD fraction. Equilibrium corresponding to an almost constant OUR value should therefore be seen.
2. The active biomass that is available must react directly with the added amount of readily biodegradable substrate, resulting in an immediate exponential growth rate response.

These two conditions are crucial for a correct outcome of the experiment. If the experiment is not successful while using acetate or glucose, a more broad-spectral, but well-defined type of substrate can be selected (e.g., yeast extract). If the experiment is still not successful, a calibration procedure, i.e., an extended procedure number 4, remains as an option.

Figure 10.9 shows an example with six parallel OUR-versus-time experiments. For illustration, the acetate is added in varying amounts when the readily biodegradable and the fast hydrolyzable substrate have been depleted, and a rather constant, albeit slowly decreasing, OUR value is reached. This stage of the system corresponds to a situation where the biomass maintenance energy requirement is supported by the hydrolysis rate of the slowly hydrolyzable substrate. After the acetate is added, an immediate increase in the OUR value is seen, corresponding to the start of non-limited growth of the active biomass. Further exponentially increased growth of the biomass takes place until the added amount of acetate is almost depleted, and the growth rate drops to a level determined by the available slowly hydrolyzable substrate.

It is essential that the determination of the four parameters from the experiment be based on expressions that can be derived from the process concept (cf. Table 10.1). Furthermore, it should be noted that a successful agreement between a measured and a simulated OUR curve is seen. If this is not the case, the validity of the formulated concept for microbial wastewater transformations including the effect of the added substrate is not sufficiently attained.

Experiments have shown that this agreement is normally observed when applying the methodology described. This fact is a strong indication of the validity of the concept for heterotrophic transformations of wastewater as it occurs in sewer networks. Furthermore, it is also crucial for the understanding of the active biomass

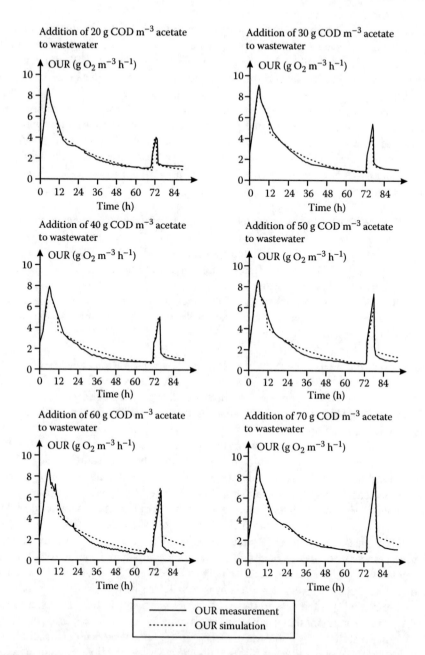

**FIGURE 10.9** OUR-versus-time curves of six parallel experiments with varying amounts of acetate added. Simulation is done with a WATS model version corresponding to formulation in Table 10.1.

as the central compound for the concept. This central position of the biomass, $X_{Hw}$, is reflected by the fact that it is proportional to the OUR value (cf. Equation 10.5 in Section 10.2.2). The success of the concept and the corresponding procedures for the determination of the process parameters are central for the use of the concept in terms of a sewer process model that can be used for design and management purposes.

Details concerning the derivation of expressions for the four process parameters mentioned, based on the model formulation shown in Table 10.1, are found in the work of Vollertsen and Hvitved-Jacobsen (1999). The final expressions for determination of the parameters are as follows:

$$Y_{Hw} = \frac{\Delta S_{S,add} - \Delta S_{O,growth}}{\Delta S_{S,add}} \tag{10.1}$$

$$\mu_H = \frac{\ln\left(\dfrac{OUR(t)}{OUR(t_0)}\right)}{t - t_0} \tag{10.2}$$

$K_{Sw}$ is determined from the slope of the decline for the OUR curve when the added $S_S$ is being depleted.

$$q_m = \frac{\mu_H \dfrac{1 - Y_{Hw}}{Y_{Hw}} \Delta S_{O,maint}}{\Delta S_{O,growth}} \tag{10.3}$$

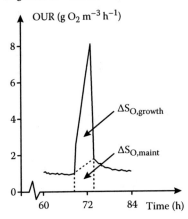

70 g COD m$^{-3}$ of acetate to wastewater

**FIGURE 10.10** Interpretation of the oxygen uptake, $\Delta S_O$, by addition of readily biodegradable substrate corresponding to exponential growth of the biomass. Oxygen uptake is divided into two parts: an uptake for growth and a corresponding uptake for maintenance of the biomass.

Except for $S_{S,add}$, $S_{O,growth}$, and $S_{O,maint}$, which are interpreted in the following discussion, the symbols used in Equations 10.1 through 10.3 are explained in Appendix A. The interpretation of the oxygen uptake, $\Delta S_O$, from an experiment with addition of acetate in a given amount, $\Delta S_{S,add}$, is illustrated in Example 10.1. Figure 10.10 in this example originates from one of the experiments shown in Figure 10.9. As seen in Figure 10.10, $\Delta S_O$ is divided into two parts, a growth-related oxygen uptake part, $\Delta S_{O,growth}$, and an oxygen uptake part, $\Delta S_{O,maint}$, corresponding to the maintenance energy requirement of the biomass.

### Example 10.1: Derivation of the Expression for the Maximum Specific Growth Rate, $\mu_H$, for the heterotrophic biomass, Equation 10.2

The derivation of $\mu_H$ (Equation 10.2) is selected as an example of how Equations 10.1 through 10.3 can be derived based on the conceptual model shown in Table 10.1. The experimental basis for determination of $\mu_H$ is shown in Figure 10.10 based on one of the experiments shown in Figure 10.9.

Referring to Table 10.1, the OUR value at time $t$ is as follows:

$$OUR(t) = \frac{\partial S_{O,growth}}{\partial t} + \frac{\partial S_{O,maint}}{\partial t} = \left( \frac{1-Y_{Hw}}{Y_{Hw}} \mu_H \frac{S_S}{K_{Sw} + S_S} + q_m \right) X_{Hw}$$

The growth process takes place under substrate nonlimiting conditions. The following is therefore approximately the case:

$$\frac{S_S}{K_{Sw} + S_S} = 1$$

Therefore,

$$OUR(t) = \left( \frac{1-Y_{Hw}}{Y_{Hw}} \mu_H + q_m \right) X_{Hw}(t)$$

$$OUR(t_0) = \left( \frac{1-Y_{Hw}}{Y_{Hw}} \mu_H + q_m \right) X_{Hw}(t_0)$$

From Table 10.1, it also follows that $dX_{Hw}/dt = \mu_H X_{Hw}$. In an integrated form, this equation is equivalent to the following:

$$X_{Hw}(t) = X_{Hw}(t_0) e^{\mu_H(t-t_0)}$$

$$\mu_H(t-t_0) = \ln \frac{X_{Hw}(t)}{X_{Hw}(t_0)}$$

By substituting the expressions derived for OUR into this equation, it follows that:

$$\mu_H(t - t_0) = \ln \frac{OUR(t)}{OUR(t_0)}$$

This equation equals Equation 10.2.

The process parameters that are calculated based on Equations 10.1 through 10.3 and the procedure for determination of $K_{Sw}$ are shown in Table 10.2 for each of the six experiments depicted in Figure 10.9.

Apart from some inconsistencies for the determination of $q_m$ when adding a small amount of readily biodegradable substrate, Table 10.2 shows a reasonable agreement between the parameters calculated. It is an important finding that the concept, Table 10.1 (used for the derivation of the expressions for determination of the parameters), corresponds with the experimental evidence. It is also in general observed from the simulation results depicted in Figure 10.9. For the verification of the concept, it is crucial that the biomass as a central component is correctly interpreted. This fact is demonstrated in Figure 10.9 by a rather good agreement between the measured and the simulated OUR value immediately before and after the addition of acetate. It relies on the fact that the OUR under substrate nonlimiting growth conditions is proportional with the biomass (cf. Equation 10.5). Figure 10.9 also shows a weakness of the concept by a less correct simulation of the biomass activity when the added substrate has been depleted.

## 10.2.2 DETERMINATION OF THE BIODEGRADABILITY OF WASTEWATER ORGANIC MATTER

The OUR experimental procedure is in the following used for determination of COD fractions for wastewater in terms of biomass and different fractions of substrate.

As shown in Figure 10.4, an OUR experiment of a wastewater sample reflects— depending on the availability of the substrate—the different phases of activity that the heterotrophic biomass is exposed to. If OUR experiments are carried out under both substrate nonlimited and substrate limited conditions, the different COD components

---

**TABLE 10.2**
**Process Parameters for Six Parallel Experiments Shown in Figure 10.9**

| | | | Experiment No. | | | |
|---|---|---|---|---|---|---|
| Parameter (Unit) | 1 | 2 | 3 | 4 | 5 | 6 |
| $Y_{Hw}$ (−) | 0.65 | 0.66 | 0.71 | 0.67 | 0.67 | 0.65 |
| $\mu_H$ (day$^{-1}$) | 5.4 | 5.6 | 4.7 | 5.7 | 5.3 | 5.3 |
| $K_{Sw}$ (g O$_2$ m$^{-3}$) | 0.7 | 0.9 | 0.9 | 1.0 | 1.1 | 0.8 |
| $q_m$ (day$^{-1}$) | 2.13 | 1.02 | 1.02 | 1.15 | 0.87 | 1.02 |

Parameters are calculated using Expressions 10.1 through 10.3 and the procedure outlined for $K_{Sw}$.

can be determined. Such conditions exist during an OUR experiment where a readily biodegradable substrate is available at the beginning of an experiment.

The following COD fractions can be determined:

(1) OUR experiments under substrate nonlimited conditions:
  - $S_S$, readily biodegradable substrate
  - $X_{Hw}$, the heterotrophic biomass
(2) OUR experiment under substrate limited conditions:
  - $X_{Sn}$, hydrolyzable substrates

Furthermore, the maximum specific growth rate, $\mu_H$, can be determined where substrate nonlimited conditions exist, and the saturation constant, $K_{Sw}$, can be found where such conditions are just being terminated. Determination of $\mu_H$ and $K_{Sw}$ follows the principles described in Section 10.2.1.

The determination of the COD compounds depends on the fact that the substrate uptake can be experimentally related to the OUR curve. The heterotrophic yield constant, $Y_{Hw}$, that is experimentally determined from procedure number 1 (Section 10.2.1) relates the oxygen uptake to the consumption of readily biodegradable substrate regardless of its origin, being either directly available or continuously produced from hydrolyzable COD fractions.

Details concerning the derivation of the expressions for the COD fractions are found in the work Vollertsen and Hvitved-Jacobsen (2002). The following expressions, Equations 10.4 through 10.6, for the determination of the compounds in wastewater at $t_0 = 0$ are derived with two fractions of hydrolyzable organic matter. The interpretation and determination of the two fractions of oxygen uptake, $\Delta S_{O1}$ and $\Delta S_{O2}$, are shown in Figure 10.11:

$$S_S = \frac{\Delta S_{O1}}{1 - Y_{Hw}} \tag{10.4}$$

$$X_{Hw} = \frac{OUR}{\dfrac{1 - Y_{Hw}}{Y_{Hw}} \mu_H + q_m} \tag{10.5}$$

$$X_{S,fast} = \frac{\Delta S_{O2}}{1 - Y_{Hw}} \tag{10.6}$$

The COD fractions can be determined by an OUR experiment typically carried out during 0.5–2 days. The slowly hydrolyzable fraction of COD, $X_{S,slow}$, cannot be determined from the oxygen uptake because the degradation of this fraction takes considerable time and also interferes with the degradation of already produced biomass. A determination based on a COD mass balance is, however, possible:

$$X_{S,slow} = COD_{tot} - (X_{Hw} + S_S + X_{S,fast}) \tag{10.7}$$

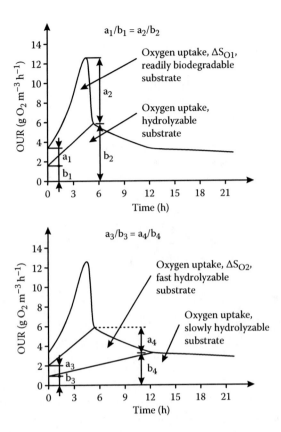

**FIGURE 10.11** Interpretation of oxygen uptake related to determination of $S_S$ and $X_{S,fast}$. The lines dividing different areas under OUR curve are estimated to be straight lines, although theoretically, they are exponential. At relatively high concentrations of $S_S$, it is of minor importance.

The procedure for determination of $X_{Hw}(t_0)$ requires the availability of readily biodegradable substrate at $t = 0$, ideally not less than 10–15 g COD m$^{-3}$, to ensure nonlimited growth conditions. If this is not the case, $X_{Hw}(t_0)$ can approximately be estimated using Equation 10.5, assuming that $\mu_H = 0$. Alternatively, $X_{Hw}$ can also be determined by adding readily biodegradable substrate, for example, acetate, to the wastewater.

The COD fractions as indicated by Equations 10.4 through 10.6 can also be determined by iterative simulation methodologies based on a WATS model corresponding to the matrix formulation in Table 10.1. The kinetic and stoichiometric parameters for the WATS modeling is determined from procedure 1 (cf. Section 10.2.1). However, a successful use of this methodology requires not only theoretical insight into sewer processes but also experience in calibration techniques.

## 10.2.3   DETERMINATION OF MODEL PARAMETERS BY ITERATIVE SIMULATION

Procedure numbers 1 and 2, described in Sections 10.2.1 and 10.2.2, respectively, show that the COD compounds and central kinetic and stoichiometric parameters for the WATS sewer process model can be explicitly determined by means of rather simple OUR experiments. However, some of the process parameters for hydrolysis still need to be determined. These are as follows:

- $k_{hn}$, hydrolysis rate constants
- $K_{Xn}$, saturation constant for hydrolysis

Calibration procedures, i.e., iterative simulation, with the WATS model of the OUR curve can be used to determine these two constants when results from procedures 1 and 2 are available. The WATS model used for the determination of these parameters is based on the matrix shown in Table 10.1. The values shown for $k_{hn}$ and $K_{Xn}$ in Table 9.6 can be used as a starting point for this iteration procedure.

## 10.2.4   CALIBRATION AND VERIFICATION OF THE WATS SEWER PROCESS MODEL

The general characteristics for calibration and verification of sewer process models are described in Section 8.1.1. This section focuses on an exemplification of procedures relevant for calibration of sewer process models for simulation of aerobic organic matter transformations.

Procedure numbers 1 to 3 described in Sections 10.2.1 through 10.2.3 have typically been implemented on wastewater samples at an upstream point of a sewer. The objective of these procedures is therefore typically to characterize the incoming wastewater to the sewer system in terms of COD fractions and process parameters.

In contrast, the objective of procedure 4 is to determine in-sewer characteristics including biofilm processes and reaeration. The characteristics of the water phase considered in procedures 1 to 3 are hereby extended to include all major processes relevant for the aerobic microbial transformations in gravity sewers. Further detailed characterization that is needed when including the anaerobic transformations will be dealt with in Section 10.2.5.

The COD compounds of a downstream wastewater sample from a sewer can be compared with the corresponding values of an upstream sample taking into account the transport time between the two locations. Considering approximate plug flow in the sewer, the difference between corresponding values of the COD fractions of these two wastewater samples reflects the result of the microbial processes during transport. These processes proceed in the wastewater phase as well as in the biofilm—and possibly also in the sewer deposits—all influenced by the reaeration.

The OUR-versus-time curves as described in this chapter constitute for the two wastewater samples, the analytical basis for determination of the compounds, and the relevant process parameters of the wastewater (cf. procedure numbers 1 to 3). The mathematical description of the processes in the sewer is shown in Table 9.1. The description of these processes requires that actual sewer and flow characteristics are available (cf. Section 9.4.1).

Simulation procedure 4 constitutes basically a calibration of the WATS sewer process model for aerobic microbial transformations as described by the matrix formulation in Table 9.1. The biofilm processes and the reaeration are thereby also included. Initial values for the compounds and process parameters for this simulation originate from the sample taken at the upstream position in the sewer. When simulated values of the downstream COD compounds are acceptable—i.e., approaching the corresponding measured values—the calibration procedure is successfully completed. The major model biofilm-related parameters to be included in the calibration process are $k_{1/2}$ and $K_{Sf}$. After having completed the calibration stage, the model is ready for a successive verification process and use in practice.

As previously mentioned, a rather simple 1/2-order flux model for the biofilm processes was selected, although more detailed formulated models are known, for example, the model described by Gujer and Wanner (1990). The reason is that a simple and sound parameter estimation and calibration procedure that can be easily carried out is emphasized.

For aerobic gravity sewers, procedure 4 is the ultimate calibration of the sewer process model. This calibration relies on procedures 1 to 4 using information from upstream and downstream wastewater samples and by including local sewer systems and flow characteristics, temperature, and DO concentrations of the wastewater in the sewer. Example 10.2 outlines the results of calibration and verification of a 5-km intercepting sewer line.

## Example 10.2: Calibration and Verification
## of the WATS Sewer Process Model

Equation 10.2 for determination of $\mu_H$ was in Example 10.1 selected to describe how the different expressions for COD compounds and sewer process parameters can be derived based on OUR-versus-time curves. In this example, the focus is on calibration and verification of the WATS sewer process model, where these compounds and parameters are central. The example concerns the microbial processes under aerobic and dry weather conditions. In order to demonstrate the validity of the sewer process model as shown in a matrix formulation in Table 9.1, the experimental basis is in this case rather extensive.

The field site used for calibration and verification of the WATS sewer process model was a 5.2-km intercepting gravity sewer located between the city of Dronninglund and the wastewater treatment plant in Asaa in the northern part of Jutland, Denmark (cf. Figure 10.12). A series of 29 samples at station 1 and a corresponding number of samples at station 4 were taken according to procedure 4 and exposed to OUR measurements and analysis procedures 1 to 3. Six of the samples (three originating from a summer period and three from a winter period) were used for determination of parameters that were considered universal for all 29 samples. The rest of the samples (11 from a summer period and 12 from a winter period) were used for verification of the WATS model.

The total load of wastewater to the sewer system in terms of person equivalents (PE) is 4350 PE with 3525 PE originating from the city of Dronninglund. Almost no discharges from industries took place. A sanitary sewer network that serves the city and the sewers in the three small villages are predominantly combined. The

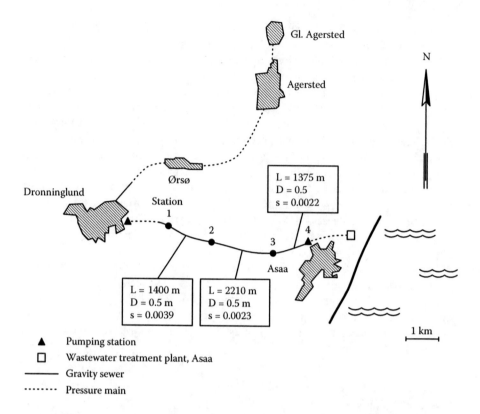

**FIGURE 10.12** Outline of investigated intercepting gravity sewer. Samples for OUR measurements were taken from manholes in stations 1 and 4. $L$, length (m); $D$, diameter (m); $s$, slope (–).

wastewater from Dronninglund is transported 1.2 km in a pressure main followed by a 5.2-km concrete intercepting gravity sewer to the treatment plant in Asaa. Only a negligible amount of wastewater is discharged into the sewer at this stretch.

The average wastewater flow during the 29 events varied between 0.013 and 0.016 $m^3\ s^{-1}$ with a residence time between the two stations (stations 1 and 4) of 2.96–3.11 h. The average summer and winter temperature of the wastewater was 15.2°C and 8.2°C in station 1 and 12.7°C and 7.5°C in station 4, respectively (all are average values with a standard deviation between 0.5 and 1.0). The DO concentration varied typically between 1.0 and 3.0 g $O_2\ m^{-3}$ in all stations. The average value and standard deviation for $COD_{tot}$ were 670 and 145 g COD $m^{-3}$ during both summer and winter, respectively.

Parameter values that were considered universal for all 29 events were determined for the six events that were selected for calibration. These universal parameter values are shown in Table 10.3. The remaining parameters—dependent on the actual event—were determined separately.

The result of the model simulation is validated by the ability of the model to predict quality changes in wastewater organic matter fractions during transport in the sewer from stations 1 to 4. These quality changes are defined in terms of the COD fractions $X_{Hw}$, $S_S$, and $X_{S,fast}$. Figure 10.13 shows measured and simulated

**TABLE 10.3**
**WATS Model Parameter Values Considered Constant for All 29 Events**

| Parameter | Unit | Value |
|---|---|---|
| $K_O$ | g $O_2$ $m^{-3}$ | 0.5 |
| $K_{Sw}$ | g COD $m^{-3}$ | 1.0 |
| $K_{Sf}$ | g COD $m^{-3}$ | 5 |
| $K_{1/2}$ | g $O_2^{0.5}$ $m^{-0.5}$ $day^{-1}$ | 6 |
| $q_m$ | $day^{-1}$ | 1.0 |
| $X_{Hf}$ | g COD $m^{-2}$ | 10.0 |
| $Y_{Hf}$ | g COD biomass (g COD substrate)$^{-1}$ | 0.55 |
| $Y_{Hw}$ | g COD biomass (g COD substrate)$^{-1}$ | 0.55 |
| $\alpha_r$ | – | 1.024 |
| $\alpha_f$ | – | 1.05 |
| $\alpha_w$ | – | 1.07 |
| $\varepsilon$ | – | 0.15 |

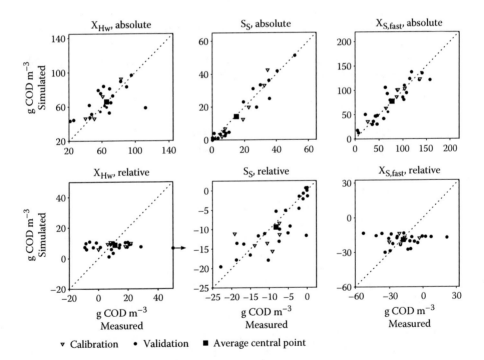

▽ Calibration   • Validation   ■ Average central point

**FIGURE 10.13** Verification of WATS sewer process model for prediction of wastewater quality changes. Measured and simulated absolute values and changes in COD fractions for 29 dry weather events are compared for wastewater during transport in a 5.2-km gravity sewer line.

values for these three main COD fractions as absolute values at station 4 and as changes during transport. The agreement between the absolute values is not considered important for the verification but is shown for general information. However, it is—also for practical use—crucial that the model is able to predict a reliable average value of the changes in COD fractions during dry weather. It is in this respect important to remember that the short-term variability of wastewater compounds in sewers can never be simulated because of nonpredictable impacts.

The results from this study confirm what has been dealt with throughout the book and summarized in Chapter 9: the concept for sewer processes—expressed in terms of the WATS model—is efficiently able to predict wastewater quality changes in a sewer. The example shows that the WATS sewer process model therefore can be used for simulation of the average dry weather process-related performance of the sewer and thereby serves as a tool for process design and management.

It should be noted that a sewer, including its connection with the subsequent treatment plant, in terms of its chemical and biological processes is normally considered during dry weather conditions. Both sanitary sewers and combined sewer networks—taking into account a wet weather functioning of the combined system—are as far as wastewater quality changes are concerned designed to operate under "average" dry weather conditions without being subject to a short-term control. Because such is the case, the WATS sewer process model is a tool for prediction of the transformations under "average" conditions for wastewater quality changes in the sewer. This information is important to remember and also relevant because wet weather is, in terms of the chemical and biological wastewater changes and impacts, just a kind of a short "break." When dealing with the hydraulic impacts, the situation is quite different and basically the opposite.

### 10.2.5 ESTIMATION OF MODEL PARAMETERS FOR ANAEROBIC TRANSFORMATIONS IN SEWERS

Anaerobic processes in wastewater of sewer systems in terms of both the organic matter transformations and the sulfur cycle have been dealt with in Chapter 6. The WATS model including the aerobic–anaerobic sewer processes and the corresponding descriptions of compounds and model parameters is focused on in Chapter 8.

As far as organic matter transformations are concerned, the process rates are significantly slower compared with aerobic transformations. This fact significantly affects the character as well as the magnitude of the process parameters. As an example, readily biodegradable organic matter is preserved and even, to some extent, produced under anaerobic conditions in contrast to the situation when aerobic conditions occur.

The sulfur cycle as depicted in Figure 4.7 shows all major relations and processes relevant for sewer networks. The details in this respect are discussed in Chapters 6 and 8, and the corresponding sulfur compounds and process parameters are thereby defined and discussed. In particular, the outline of the anaerobic, heterotrophic processes and the processes related to the sulfur cycle as applied in the WATS model is shown in Tables 9.3 and 9.4.

The methods for determination of the parameters concerning the anaerobic processes are in general less structured compared with the procedures that have been described in Sections 10.2.1 through 10.2.4. The following points give an overview of which approaches are typically applied for site-specific estimation of compounds and parameters for calibration and verification of the WATS sewer process model:

- Determination of the volatile fatty acids (VFAs) to complete the description of the fermentation process in terms of fermentable, readily biodegradable substrate, $S_F$, and fermentation products, $S_A$.
- Determination of sulfide and a sulfide formation rate. Sampling for analysis of sulfide in the water phase and monitoring of $H_2S$ with sensors in the sewer atmosphere are relevant.
- Determination of the formation rate of readily biodegradable substrate (anaerobic hydrolysis).
- In addition to these three methods, estimation of parameters for calibration of the WATS model may also be based on experience from similar sewer systems.

The first three methods can be considered more or less explicit methods for the determination of compounds and parameters. The four groups of methods for assessment of anaerobic transformations will be outlined in the following discussion.

### 10.2.5.1 Volatile Fatty Acids

The VFAs, primarily formate, acetate, propionate, *n*-butyrate, and isobutyrate, can be determined analytically on an ion chromatograph with a conductivity detector, for example, according to APHA–AWWA–WEF (2005). Determination of fermentable, readily biodegradable substrate, $S_F$, and fermentation products, $S_A$, in units of COD requires that the VFA components be converted to this unit. The following example using formate demonstrates this:

$$HCOO^- + \frac{1}{2}O_2 \rightarrow CO_2 + OH^- \tag{10.8}$$

The stoichiometry of Equation 10.8 shows that the COD/formate ratio is as follows:

$$\frac{16}{45} = 0.36 \text{ g COD (g formate)}^{-1}$$

Table 10.4 outlines the corresponding stoichiometric relations for other VFA components.

The relationship between readily biodegradable substrate ($S_S$), fermentable, readily biodegradable substrate ($S_F$), and fermentation products ($S_A$) is defined as:

$$S_S = S_F + S_A \tag{10.9}$$

**TABLE 10.4**

**COD/Mass Ratios for Selected VFA Compounds**

| VFA Component | COD/Mass Ratio (g COD g$^{-1}$) |
|---|---|
| Formate | 0.36 |
| Acetate | 1.08 |
| Propionate | 1.53 |
| Lactate | 1.08 |
| Butyrate | 1.84 |

### 10.2.5.2 Sulfide and Sulfide Formation Rate

Potential sulfide problems are in practice particularly relevant to assess and therefore often a major subject of WATS model simulations. The sulfur compounds and related process parameters are therefore important. Local information on temperature, pH, and concentration levels of sulfur compounds are therefore both important and also relatively easily monitored. The sewer gas concentration of $H_2S$ can be monitored with sensor instruments, whereas a reliable concentration of sulfide in the water phase requires sampling and analysis. Total sulfide in wastewater can be measured photometrically with the methylene blue method (Cline 1969; APHA–AWWA–WEF 2005).

Because sulfide is a very reactive compound that is easily oxidized or reacts with heavy metals to produce precipitates, and because hydrogen sulfide is also easily emitted from the water phase, precautions must be taken under sampling, and the interpretation of analytical results must be carefully made.

Laboratory experiments for the determination of sulfide production rates from anaerobic biofilms grown on wastewater can be carried out in both simple and more complex operating biofilm reactors with or without a rotating drum (Nielsen 1987; Norsker et al. 1995). Mixed laboratory and field experiments to determine sulfide production rates with biofilm growth under field conditions followed by reactor experiments in the laboratory are described by Bjerre et al. (1998). Sulfide production rates can also be determined in real systems, taking corresponding upstream and downstream samples for analysis and knowing flow and system characteristics (Nielsen et al. 1998). Sulfide production rates have been measured by circulating wastewater in a pilot plant, full-flowing sewer system (Tanaka and Hvitved-Jacobsen 2000).

Field and pilot-plant studies for the determination of sulfide production rates may be feasible in pressure mains; however, under gravity sewer conditions, such methods are less useful because of partial emission and oxidation of sulfide.

### 10.2.5.3 Determination of the Formation Rate for Readily Biodegradable Substrate in Wastewater under Anaerobic Conditions

Preservation and formation of readily biodegradable organic substrate, $S_S$, is characteristic for wastewater under anaerobic conditions (cf. Section 6.7.1). Different processes contribute to the overall result typically seen as a net production of $S_S$. Anaerobic hydrolysis produces $S_S$, but to a minor extent, the sulfate-reducing biomass and the fermenting biomass may also add to a net buildup of specific low molecular organics (cf. Sections 3.2.2 and 3.2.3). However, as also seen in Figure 6.10, it is the anaerobic hydrolysis that governs the production of $S_S$. Furthermore, under anaerobic conditions, the utilization of $S_S$ is limited because of the relatively low growth rate of the heterotrophic biomass.

The formation of readily biodegradable substrate in wastewater is, under anaerobic conditions, investigated in an experimental setup described in Example 10.3.

## Example 10.3: Formation of Readily Biodegradable Substrate by Anaerobic Hydrolysis

This example describes an experimental procedure that can be applied for determination of readily biodegradable substrate production by anaerobic hydrolysis. The procedure is crucial for estimation of an anaerobic hydrolysis rate (cf. Figure 6.10).

The experimental procedure relies on a comparison between an OUR curve of a wastewater sample and an OUR curve carried out on a corresponding wastewater sample, although under changing aerobic and anaerobic conditions (cf. Section 10.1.3). The experimental setup is basically identical with a regular equipment used for an OUR experiment (Tanaka and Hvitved-Jacobsen 1998; Vollertsen and Hvitved-Jacobsen 1999). The first experiment (Experiment 1) is therefore a normal OUR experiment. The other experiment (Experiment 2) is carried out with one or two anaerobic periods, each lasting a few hours. An example of such experiments carried out in parallel is shown in Figure 10.14.

The net production of readily biodegradable substrate, $S_S$, in each of the two experiments is determined from the oxygen uptake (OU), in units of g COD m$^{-3}$.

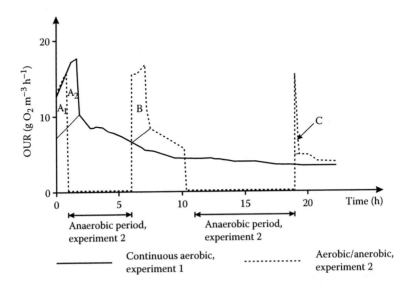

**FIGURE 10.14**  Results from two OUR experiments performed in parallel on two samples of the same wastewater. Experiment 1 is carried out as a normal OUR experiment; Experiment 2 is completed under changing aerobic and anaerobic conditions.

The oxygen uptake, $\Delta S_O$, related to the available amount of readily biodegradable substrate is equal to an area under an OUR curve as depicted in Figure 10.14 (cf. Section 10.1.3 and Figure 10.10). The corresponding $S_S$ is determined taking into account the yield constant, $Y_{Hw}$, for the heterotrophic biomass growth (cf. Equation 10.4). Based on other experiments, a value of $Y_{Hw} = 0.55$ g COD g COD$^{-1}$ is used in the determination of $S_S$.

$$S_S = \frac{\Delta S_O}{1 - Y_{Hw}}$$

If a net production of readily biodegradable substrate occurs during anaerobic periods, it must be seen as a higher total OU ($\Delta S_O$) in the aerobic/anaerobic experiment (Experiment 2) compared with the traditionally performed OUR experiment (Experiment 1).

The initially available readily biodegradable substrate, $S_{S,i}$, is determined based on the results from a traditionally performed OUR experiment. Referring to Figure 10.13, it is calculated by measuring the sum of the two areas, $A_1$ and $A_2$ (cf. Figure 10.10):

$$S_{S,i} = \frac{A_1 + A_2}{1 - Y_{Hw}} = 28.9 \text{ g COD m}^{-3}$$

The initially available readily biodegradable substrate and the potentially generated substrate, $S_{S,i+g}$, produced by anaerobic hydrolysis during the first anaerobic period shown in Figure 10.14 is:

$$S_{S,i+g} = \frac{A_1 + B}{1 - Y_{Hw}} = 35.3 \text{ g COD m}^{-3}$$

It should be noted that part of the initially available substrate, $A_2$, is considered preserved under anaerobic conditions and degraded under aerobic conditions, and is therefore included in the OU of area B.

The $S_S$ determined with this methodology is basically a net production. It is expected that the consumption of substrate for the production of biomass is relatively low under anaerobic conditions, and the net production is therefore approximately equal to the total amount produced. It has also been confirmed that a biofilm has not been generated and no sulfide production takes place (cf. Section 6.7.1). The difference, $\Delta S_{S1}$, between the OU values determined for the two experiments is therefore the estimated amount of readily biodegradable substrate produced by anaerobic hydrolysis:

$$\Delta S_{S1} = 35.3 - 28.9 = 6.4 \text{ g COD m}^{-3}$$

The production of $S_S$ depends on the duration of the anaerobic period—in this experiment, 5 h. An average production rate can therefore be calculated:

$$dS_S/dt = 6.4/5 = 1.3 \text{ g COD m}^{-3} \text{ h}^{-1}$$

In Figure 10.14, it is also seen that the production of readily biodegradable organic matter also occurs during the second anaerobic period from time 10.2 h to time 18.8 h. The produced amount of $S_S$ is:

$$\Delta S_{S2} = \frac{C}{1 - Y_{Hw}} = 4.3 \text{ g COD m}^{-3}$$

The production rate, 0.5 g COD m$^{-3}$ h$^{-1}$, is—as expected—reduced compared with the corresponding result from the first anaerobic period because of a relatively reduced amount of fast hydrolyzable substrate, $X_{S1}$ (cf. Figure 6.9).

The activity of the aerobic heterotrophic biomass, $X_{Hw}$, is maintained during anaerobic conditions. It is seen in Figure 10.14 by comparing the level of the two OUR values (about 16 g O$_2$ m$^{-3}$ h$^{-1}$) immediately before and immediately after the first anaerobic period. This observation has been confirmed for at least up to about 24 h of anaerobic conditions (Tanaka and Hvitved-Jacobsen 1999).

Example 10.3 shows that the (aerobic) biodegradability of the organic matter in wastewater reflects the formation rate of $S_S$ under the preceding anaerobic period. It is furthermore seen that the biodegradability of the hydrolyzable substrate is higher in the first anaerobic period compared with the second anaerobic period. Further investigations shown in Figure 10.15 confirm this finding (Tanaka and Hvitved-Jacobsen 1998). These results also correspond to the observation that a soluble COD value (COD$_S$) below 50 g COD m$^{-3}$ has no real potential for sulfide formation (cf. Equation 4 in Table 6.6). It is of course a rather pragmatic observation as it may reflect the fact that a low COD value often corresponds to a wastewater quality with a relatively high amount of nonbiodegradable and slowly biodegradable, soluble organic matter.

When consumption of $S_S$ under anaerobic conditions by the sulfate-reducing biomass and the fermenting biomass is important, Equation 10.10 expresses the total anaerobic hydrolysis rate. This equation is based on the assumption that methane

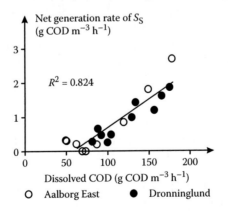

**FIGURE 10.15** Net generation rates of readily biodegradable substrate, $S_S$, under anaerobic conditions in wastewater. Results originate from 19 experiments carried out on wastewater at the inflow to two wastewater treatment plants.

formation in sewers without permanent deposits normally can be neglected (cf. Section 3.2.2).

$$r_{S,tot} = r_{S,net} + 2_{rS,s} + r_{S,ferm} \qquad (10.10)$$

where:

$r_{S,tot}$ = total anaerobic hydrolysis rate (g COD m$^{-3}$ h$^{-1}$)
$r_{S,net}$ = net measured anaerobic production rate of $S_S$ (g COD m$^{-3}$ h$^{-1}$)
$r_{S,s}$ = production rate of sulfide (g S m$^{-3}$ h$^{-1}$)
$r_{S,ferm}$ = substrate-consuming rate by the fermenting biomass (g COD m$^{-3}$ h$^{-1}$)

Equation 10.10 is based on the assumption that Equation 6.4 is an acceptable description of the relation between organic matter consumption and sulfide production, i.e., that 2 mol of $CH_2O$ with a COD value of 32 g COD mol$^{-1}$ is consumed by the production of 1 mol of $H_2S$-S with a molar weight of 32 g S mol$^{-1}$.

Tanaka and Hvitved-Jacobsen (1999) have experimentally confirmed that a relatively high linear correlation coefficient between the production of $S_S$ and $CO_2$ exists under anaerobic conditions. They also found that 50% of $CO_2$ was typically produced by the sulfate-reducing bacteria, the other half by the fermenting biomass. However, the net production rate of $S_S$ was typically about 70% of the total produced $S_S$ by anaerobic hydrolysis (cf. Equation 10.10). Hence, this equation may, even in a reduced form, be valuable for the estimation of the production of readily biodegradable substrate under anaerobic conditions.

## 10.3 FINAL REMARKS

This chapter exemplifies how field, pilot-plant, and bench-scale methodologies can be used to quantify central compounds and parameters for the WATS sewer process model. In addition to these methodologies, it is furthermore important that several other experimental procedures presented throughout this book have contributed to a quantitative understanding and parameter estimation of the governing sewer processes. Without this information, the use of in-sewer process models such as WATS would be irrelevant. These methods for sewer process studies are, in addition to the kinetic and stoichiometric formulation in mathematical terms of the sewer processes, the backbone for the simulation models and the corresponding process design procedure for management of sewers. On top of this general information included in WATS, it is important that local measurements (e.g., structural characteristics, water flow, sulfide concentrations, pH, and temperature) can provide a sound basis for model calibration and verification under specific constraints. The calibration and verification procedure is invaluable as a means for adjustment of the general observed values of model parameters in a specific case.

Sewer process modeling is a rather new discipline, and limited information is available for the practitioner in terms of experience. The methodologies and investigations referred to in this text have formed an important basis for not only the sewer process concept but also for the determination of the corresponding model parameters. However, the variability of wastewater characteristics is considerable, and the types of systems in which the microbial processes proceed are legion. Care must therefore be taken when applying available information on the sewer processes. In this respect, the methodologies for sewer process studies become crucial in acquiring new and site-specific knowledge and information.

Investigations in sewers take place in a nonattractive environment that is often difficult to get into. This fact may have had substantial influence on the rather limited knowledge on the sewer processes. From a microbiological and chemical process point of view, however, the sewer is a highly interesting environment with high variability and diversity. Hopefully, this fact, highly supported by the evidence gained from practical applications, will lead to improved methods for sewer process studies in the future.

## REFERENCES

APHA–AWWA–WEF (2005), *Standard Methods for the Examination of Water and Wastewater*, 21st edition, APHA (American Public Health Association), AWWA (American Water Works Association), WEF (Water Environment Federation), Washington, DC, USA.

ASCE (1983), *Existing Sewer Evaluation and Rehabilitation*, ASCE (American Society of Civil Engineers) *Manual and Report on Engineering Practice 62; WPCF* (Water Pollution Control Federation) *Manual of Practice* FD-6, p. 106.

Ashley, R.M. and M.A. Verbanck (1998), Physical processes in sewers, Congress on Water Management in Conurbations, Bottrop, Germany, June 19–20, 1997. In Emschergenossenschaft: Materialien zum Umbau des Emscher-Systems, *Heft, 9*, 26–47.

Ashley, R.M., J.-L. Bertrand-Krajewski, T. Hvitved-Jacobsen, and M. Verbanck (eds.) (2004), *Solids in Sewers, Scientific and Technical Report No. 14*, IWA (International Water Association) Publishing, London, p. 340.

ASTM (1991), E679-91, *Standard Practice for Determination of Odor and Taste Threshold by a Forced-Choice Ascending Concentration Series Method of Limits*, American Society for Testing and Materials, Philadelphia, PA, USA.

Bertrand-Krajewski, J.-L., D. Laplace, C. Joannis, and G. Chebbo (2000), *Mesures en hydrologie urbaine et réseau d'assainissement*, Tec et Doc, Paris, p. 808.

Bjerre, H.L., T. Hvitved-Jacobsen, B. Teichgräber, and D. te Heesen (1995), Experimental procedures characterizing transformations of wastewater organic matter in the Emscher river, Germany, *Water Sci. Technol., 31*(7), 201–212.

Bjerre, H.L., T. Hvitved-Jacobsen, S. Schlegel, and B. Teichgräber (1998), Biological activity of biofilm and sediment in the Emscher river, Germany, *Water Sci. Technol., 37*(1), 9–16.

CEN (2003), EN 13725, Air quality—determination of odor concentration by dynamic olfactometry, Comité Européen de Normalisation (European Committee for Standardization), London, UK.

Cline, D.J. (1969), Spectrophotometric determinations of hydrogen sulfide in natural waters, *Limnol. Oceanogr., 14*, 454–458.

Dold, P.L., G.A. Ekama, and G. v. R. Marais (1980), A general model for the activated sludge process, *Prog. Water Technol., 12*, 47–77.

Ekama, G.A. and G. v. R. Marais (1978), The dynamic behaviour of the activated sludge process, Research report No. W 27, Department of Civil Engineering, University of Cape Town.

Fenner, R.A. and R.M. Stuetz (2000), Methods of odour measurement and assessment, in: P.N.L. Lens and L.H. Pol (eds.), *Environmental Technologies to Treat Sulfur Pollution*, IWA Publishing, London, pp. 305–328.

Genereux, D.P. and H.F. Hemond (1992), Determination of gas exchange rate constants for a small stream on Walker Branch Watershed, Tennessee, *Water Resour. Res.*, 28(9), 2365–2374.

Gudjonsson, G., J. Vollertsen, and T. Hvitved-Jacobsen (2002), Dissolved oxygen in gravity sewers—measurement and simulation, *Water Sci. Technol.*, 45(3), 35–44.

Gujer, W. and O. Wanner (1990), Modeling mixed population biofilms, in: W.G. Characklis and K.C. Marshall (eds.), *Biofilms*, John Wiley & Sons, New York, pp. 397–443.

Huisman, J.L., C. Gienal, M. Kühni, P. Krebs, and W. Gujer (1999), Oxygen mass transfer and biofilm respiration rate measurement in a long sewer, evaluated with a redundant oxygen balance, in I.B. Joliffe and J.E. Ball (eds.), *Proceedings from the 8th International Urban Storm Drainage Conference,* Sydney, Australia, August 30–September 3, 1999, vol. 1, pp. 306–314.

Jensen, N.Aa. and T. Hvitved-Jacobsen (1991), Method for measurement of reaeration in gravity sewers using radiotracers, *Res. J. WPCF,* 63(5), 758–767.

Madsen, H.I., T. Hvitved-Jacobsen, and J. Vollertsen (2006), Gas phase transport in gravity sewers—a methodology for determination of horizontal gas transport and ventilation, *Water Environ. Res.*, 78(11), 2203–2209.

Nielsen, P.H. (1987), Biofilm dynamics and kinetics during high-rate sulfate reduction under anaerobic conditions, *Appl. Environ. Microbiol.*, 53(1), 27–32.

Nielsen, P.H., K. Raunkjaer, and T. Hvitved-Jacobsen (1998), Sulfide production and wastewater quality in pressure mains, *Water Sci. Technol.*, 37(1), 97–104.

Nielsen, A.H., J. Vollertsen, H.S. Jensen, T. Wium-Andersen, and T. Hvitved-Jacobsen (2008), Influence of pipe material and surfaces on sulfide related odor and corrosion in sewers, *Water Res.*, 42, 4206–4214.

Norsker, N.H., P.H. Nielsen, and T. Hvitved-Jacobsen (1995), Influence of oxygen on biofilm growth and potential sulfate reduction in gravity sewer biofilm, *Water Sci. Technol.,* 31(7), 159–167.

Parkhurst, J.D. and R.D. Pomeroy (1972), Oxygen absorption in streams, *ASCE, J. Sanit. Eng. Div.,* 98(SAI), 101.

Raunkjaer, K., P.H. Nielsen, and T. Hvitved-Jacobsen (1997), Acetate removal in sewer biofilms under aerobic conditions, *Water Res.,* 31(11), 2727–2736.

Sneath, R.W. and C. Clarkson (2000), Odour measurement: a code of practice, *Water Sci. Tech.,* 41(6), 25–31.

Spanjers, H., P.A. Vanrolleghem, G. Olsson, and P.L. Dold (1998), *Respirometry in Control of the Activated Sludge Process: Principles, IAWQ Scientific and Technical Report No. 7,* IAWQ, London, p. 48.

Stuetz, R.M., R.A. Fenner, S. J. Hall, I. Stratful, and D. Loke (2000), Monitoring of wastewater odours using an electronic nose, *Water Sci. Technol.,* 14(6), 41–47.

Stuetz, R. and F.-B. Frechen (eds.) (2001), *Odours in Wastewater Treatment—Measurement, Modelling and Control,* IWA Publishing, London, p. 437.

Tanaka, N. and T. Hvitved-Jacobsen (1998), Transformations of wastewater organic matter in sewers under changing aerobic/anaerobic conditions, *Water Sci. Technol.,* 37(1), 105–113.

Tanaka, N. and T. Hvitved-Jacobsen (1999), Anaerobic transformations of wastewater organic matter under sewer conditions, in: I.B. Joliffe and J.E. Ball (eds), *Proceedings of the 8th International Conference on Urban Storm Drainage,* Sydney, Australia, August 30–September 3, 1999, pp. 288–296.

Tanaka, N. and T. Hvitved-Jacobsen (2000), Sulfide production and wastewater quality—investigations in a pilot plant pressure sewer, *Proceedings of the 1st World Water Congress of the International Water Association* (IWA), vol. 5, pp. 192–199.

Tsivoglou, E.C. and L.A. Neal (1976), Tracer measurement of reaeration: III. Predicting the reaeration capacity of inland streams, *J. Water Pollut. Control Fed.,* 48(12), 2669–2689.

Tsivoglou, E.C., J.B. Cohen, S.D. Shearer, and P.J. Godsil (1968), Tracer measurements of stream reaeration: II. Field studies, *J. Water Pollut. Control Fed.,* 40(2), 285–305.

Tsivoglou, E.C., R.L. O'Connell, M.C. Walter, P.J. Godsil, and G.S. Logsdon (1965), Tracer measurements of atmospheric reaeration: I. Laboratory studies, *J. Water Pollut. Control Fed.,* 37(10), 1343–1363.

Vollertsen, J., A.H. Nielsen, H.S. Jensen, T. Wium-Andersen, and T. Hvitved-Jacobsen (2008), Corrosion of concrete sewers—the kinetics of hydrogen sulfide oxidation, *Sci. Total Environ.,* 394, 162–170.

Vollertsen, J. and T. Hvitved-Jacobsen (1999), Stoichiometric and kinetic model parameters for microbial transformations of suspended solids in combined sewer systems, *Water Res.,* 33(14), 3127–3141.

Vollertsen, J. and T. Hvitved-Jacobsen (2002), Biodegradability of wastewater—a method for COD-fractionation, *Water Sci. Technol.,* 45(3), 25–34.

WEF (2004), *Control of Odors and Emissions from Wastewater Treatment Plants, Water Environment Federation (WEF), WEF Manual of Practice No. 25,* WEF, Alexandria, VA, p. 537.

# 11 Applications
## *Sewer Process Design and Perspectives*

This book is basically devoted to give an overall understanding of in-sewer chemical and microbial processes including the conditions required for these processes to proceed. However, as a natural continuation of previous discussions and on top of this understanding, it is crucial to give both general principles and examples on how this fundamental knowledge can be applied to improve the urban wastewater collection system. It is crucial to remember that process engineering of the sewer network is the ultimate goal of this book and also in this respect parallel to how we consider and manage the wastewater treatment plants.

This book is rich in examples on how knowledge-based process engineering of the wastewater collection system can enhance its performance. It should be widely recognized that the sewer is an infrastructural installation that effectively prevents wastewater from becoming a vehicle for the dissemination of infectious diseases. This functioning of the collection system must be continuously maintained. The importance of an environmentally safe and economically efficient transport of sewage is undeniable. A sound knowledge on the sewer processes is, in this respect, fundamental and central.

Knowledge on sewer processes and tools for prediction and control of the in-sewer processes can today be applied with the aim of assessing process-based impacts in the management of sewer networks. The contents of Chapters 7 through 10, which focus on how to predict the critical impacts of sewer systems and how to control these adverse effects based on sound sewer process models, are topics that can be defined as "applications." The prediction and assessment of hydrogen sulfide and odor-related problems and their control are important examples from these Chapters 7 through 10.

In addition to these very important problems, there are other equally compelling reasons for the management of sewers. These aspects will briefly be dealt with in this chapter, and corresponding examples are also given. The examples are to some extent selected to show the potential of different types of tools that are all based on a sound understanding of sewer processes.

The WATS (Wastewater Aerobic/anaerobic Transformations in Sewers) model, which was presented in Chapter 9, is the ultimate tool when assessing impacts and control of sewer processes. However, it must be kept in mind that empirical equations are often very useful, for example, for screening purposes. Such rather simple tools have also been discussed in the preceding chapters of this book.

## 11.1   WASTEWATER DESIGN: AN INTEGRATED APPROACH FOR WASTEWATER TREATMENT

In-sewer processes may result in adverse effects on the sewer itself and the surrounding environment. Well-known examples are health, corrosion, and odor problems, which are related to the formation of sulfide and volatile organic compounds (VOCs). Other examples are related to the interactions between the sewer and the treatment plant in terms of the quality changes of the wastewater that occur in the sewer network.

The traditional approach for wastewater treatment plant design is to relate the treatment requirements to the flow and quality of the wastewater that exists at the inlet to the plant. This approach basically ignores the fact that "untreated" wastewater as it appears at the inlet to a treatment plant can be managed during transport in a sewer network, to some extent, to comply with the treatment processes.

Contrary to the traditional approach, an integrated process design concept considers wastewater as being subject to quality transformations during transport in the collection system. The fundamental difference between this concept and the traditional approach is that, under the integrated process, these transformations can be managed by the designer and the operator. For the treatment plant design and operation, this means that the "untreated" wastewater as it occurs at the inlet to the plant can be "designed" to comply with the treatment processes. Figure 11.1 gives an overview of the two different approaches for wastewater management and thereby outlines the concept of "wastewater design."

The preceding chapters have described several methods for the management of wastewater in sewers, here expressed as "design." In Section 11.2, different types of structural and operational ways to generate these quality changes of wastewater in sewers will be outlined.

Traditional approach:

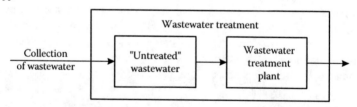

Integrated process approach:

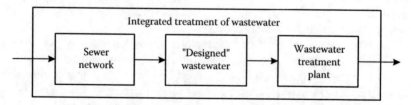

**FIGURE 11.1**   Comparison between the traditional approach for wastewater treatment plant design and operation and the corresponding integrated process concept.

## 11.2 SEWER STRUCTURAL AND OPERATIONAL IMPACTS ON WASTEWATER QUALITY

The process conditions in terms of availability of the electron acceptor are crucial for the type of process—aerobic, anoxic, or anaerobic—that will occur, cf. for example, Table 11.1. This table also shows that the type of sewer has a considerable influence on the type of process that is favored. This is interesting because it can be used actively and in the "opposite direction" by selecting the sewer structure based on which type of wastewater quality is preferred. The sewer network designer and operator thereby have a guideline for which structural and operational conditions should exist to enhance the microbial and chemical processes that are preferred.

Table 11.1 gives an overview of the sewer structural and operational measures that affect the type and course of the in-sewer dry weather processes. The table exemplifies how process design and operation of sewer networks can be implemented. In other words, the table gives an overview of the methods for "wastewater design."

The impact of the different structural or operational measures on the sewer processes and wastewater quality characteristics can be assessed by model simulation, for example, by applying the WATS sewer process model. The following two examples, Examples 11.1 and 11.2, illustrate how structural measures, primarily related to A4 in Table 11.1, affect the in-sewer processes.

---

**TABLE 11.1**

**Examples of Structural (A) and Operational (B) Measures that Affect In-Sewer Processes**

| Structural or Operational Measures Applied in Sewer Process Management | Process or Phenomenon Affected |
|---|---|
| A1: Change of flow regime and degree of turbulence, e.g., slope of sewer line, pipe diameter, and degree of pipe uniformity | Exchange of volatile compounds across the air–water interface, e.g., oxygen (reaeration that affects aerobic or anaerobic conditions) and release of odorous substances |
| A2: Ventilation, e.g., airtight manhole lids or forced ventilation | Release of odorous substances to the urban atmosphere and change of reaeration due to a lower oxygen concentration in the sewer atmosphere |
| A3: Sewer capacity, i.e., residence time of the wastewater | Extent of sewer processes |
| A4: Relative capacity, i.e., water depth/pipe diameter ratio | Reaeration and relative importance of microbial transformations in the water phase and in the biofilm |
| A5: Flow velocity, i.e., shear stress | Development of sewer biofilm and sediments having an effect on the corresponding processes |
| B1: Source control, households (e.g., water consumption, urine separation, and use of kitchen grinders), discharges from industry and reduced infiltration | Sewer upstream quality that affects the in-sewer processes and the downstream quality of the wastewater |
| B2: Injection of oxygen or nitrate | Impact on process conditions (aerobic, anoxic, or anaerobic) |

## Example 11.1: Relative Importance of the Biofilm to Bulk Water Ratio on the DO Consumption in a Gravity Sewer

The total in-sewer respiration process of the wastewater determined as a DO (dissolved oxygen) removal rate originates from the biofilm and the bulk water microbial processes when the contribution from deposits can be neglected. The total respiration rate is, therefore, expressed as follows:

$$r_{tot} = \frac{A}{V} r_f + r_w$$

where:

$r_{tot}$ = total rate of DO consuming processes in a sewer (g $O_2$ m$^{-3}$ h$^{-1}$)
$A$ = area of the biofilm (m²)
$V$ = volume of bulk water (m³)
$r_f$ = rate of DO consuming processes in the biofilm (g $O_2$ m$^{-2}$ h$^{-1}$)
$r_w$ = rate of DO consuming processes in the bulk water phase (g $O_2$ m$^{-3}$ h$^{-1}$)

Figure 11.2 illustrates the DO consumption rate of the biofilm, $r_f$, in percentage of the total DO consumption rate, $r_{tot}$, versus the $A/V$ ratio. The calculations are based on DO and substrate nonlimiting conditions. The value of $r_f$ is considered constant and equal to 1.0 g $O_2$ m$^{-2}$ h$^{-1}$, whereas $r_w$ varies from 2 to 20 g $O_2$ m$^{-3}$ h$^{-1}$.

In a pipe with a diameter of 500 mm, a daily variability of the water depth between 100 and 250 mm results in $A/V$ ratios of 17 and 8 m$^{-1}$, respectively. Figure 11.2 shows that under the conditions given, the corresponding DO consumption rate of the biofilm in percentage of the total rate is about 60% and 40%, respectively, if the bulk water respiration rate is 10 g $O_2$ m$^{-3}$ h$^{-1}$.

The relative importance of the biofilm compared with the bulk water transformations of the wastewater is not just a question of flow conditions in terms of

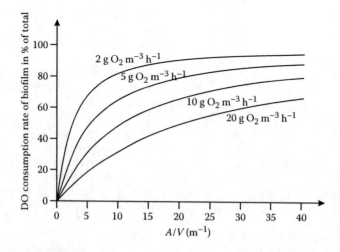

**FIGURE 11.2** Biofilm DO consumption rate in % of total consumption rate versus $A/V$ (area of biofilm/bulk water volume) ratio.

a daily variation. It is, as shown in this example, also a question of which pipe dimension is selected.

## Example 11.2: Reaeration and DO Variability of a Gravity Sewer Pipe

This example illustrates how the flow conditions of a sewer pipe affect the reaeration and the resulting DO concentration in the wastewater. A gravity sewer with a pipe diameter of 500 mm and a slope $s = 0.003$ m m$^{-1}$ is selected as an example. The sewer is without deposits and with a biofilm on the wetted perimeter. The DO consumption rate of the bulk water phase, $r_w$, is at 10°C assumed to have a maximum value equal to 5 g O$_2$ m$^{-3}$ h$^{-1}$, but is however, limited by the reaeration. The DO consumption rate of the biofilm, $r_f$, is—according to Parkhurst and Pomeroy (1972) and for simplicity reasons—described as a first-order process in the DO concentration (cf. Equation 5.7).

Referring to Figure 11.3, the oxygen transfer coefficient ($K_L a$), the flow velocity ($u$), the bulk water DO concentration, and the DO consumption rate of the biofilm ($r_f$) are under steady-state conditions in a gravity sewer pipe plotted versus the flow, Q.

It is shown in Figure 11.3 that the sewer is full flowing at a flow rate of 775 m$^{-3}$ h$^{-1}$ (215 L s$^{-1}$). The figure also shows that the reaeration and the DO concentration

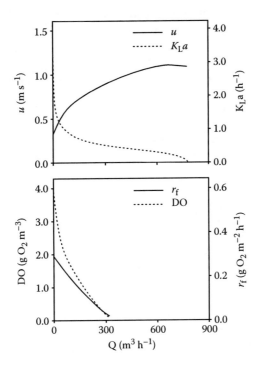

**FIGURE 11.3**  WATS model simulated values of oxygen transfer coefficient ($K_L a$), flow velocity ($u$), bulk water DO concentration, and DO consumption rate for biofilm ($r_f$), versus flow ($Q$) for a gravity sewer pipe. Flow conditions are steady state, and the temperature is 10°C.

vary considerably with the flow conditions. At relatively low flow rates, the DO concentration is about 2–4 g $O_2$ $m^{-3}$, a level that is significantly reduced even at flow rates that are below those corresponding to a half-full flowing pipe.

In this example, it is shown that it is possible to control the magnitude of the aerobic transformation of the wastewater by the selected level of the flow compared with the capacity of the sewer.

Referring to Table 11.1, it is interesting to note that a specific structural change of a sewer network may have an impact on more than just one process, and that it can also exert a total resulting impact of a change in different directions. As an example, turbulence in the water phase may increase the reaeration rate of the water phase. However, by potentially increasing the emission rate of, for example, hydrogen sulfide and odorous substances, the turbulence may intensify the negative effects of anaerobic conditions. This situation may, for example, occur in a gravity sewer pipe located downstream a pressure main with sulfide production.

Other aspects concern the complex temperature impacts of the in-sewer processes. Sewer processes, microbial as well as physicochemical processes such as reaeration, are temperature dependent. The different processes interact, and the overall temperature dependency of a specific phenomenon may therefore not necessarily exert the preferred effects. Changes in the pH value may also result in a corresponding overall effect that is not directly and easily predictable. The use of a sewer process model like the WATS model is therefore basically the only way for assessing such complex impacts of parameter changes.

## 11.3 SEWER PROCESSES: FINAL COMMENTS AND PERSPECTIVES

It is indisputable that the major importance of understanding the sewer microbial and chemical processes lies in the potential for prediction and assessment of impacts and control measures for sulfide and VOCs, i.e., problems related to anaerobic conditions in sewers. It is in this respect essential that this understanding be transferred to the WATS sewer process model and is thereby available for the sewer network managers. Throughout this book, several types of simulation results of the WATS model have been presented.

In addition to these actual aspects, it is important to take a look into further applications of the knowledge gained on sewer processes. Although such uses are today not always clearly formulated, they are at any rate relevant for discussion and consideration. The following subsections will deal with these perspectives.

### 11.3.1 WASTEWATER PROCESSES IN GENERAL

Although this book concerns in-sewer processes applied on sewer networks, other perspectives are also considered, because wastewater occurs in other systems, too. This book, therefore, might not have been titled "sewer processes," but might been considered as a text on "wastewater processes." The authors could have done so at the general level but could not have included a corresponding specific use in all cases. At the end of the book, it is, however, important to mention that the fundamental knowledge dealt with, particularly in Chapters 1 through 6, is not limited to sewer networks alone. Wastewater occurs not just in pipes and open channels but may also occur in different types of wastewater treatment ponds. Taking wastewater processes into account, there is a perspective for further improvement of such relative simple treatment systems.

In several cases, it has been stressed that wastewater and activated sludge, as far as microbial processes are concerned, perform differently. When dealing with microbial and physicochemical processes, it is always important to remember this fact, i.e., to consider wastewater as wastewater and activated sludge as activated sludge. There are, of course, cases where a wastewater type of system occurs between these two "extremes." However, activated sludge is basically a kind of concentrated mix of bacteria and other microorganisms and particulates. This mix of microorganisms is a result of a "processing" whereby it has obtained specific characteristics useful for a treatment. This type of system is fundamentally different from "wastewater."

Although wastewater occurs as different types, the contents of human excreta are generally dominating and—what is also important—it is in its contents varying in a way that is not a result of deliberate planning. Regardless of how, where, and when this situation occurs, the processes discussed in this text are considered a relevant basis for the description of the transformations that occur and the impacts associated with these changes. The sewer is an important reactor, but it is not the only one.

## 11.3.2 IN-SEWER PROCESSES AND WET WEATHER DISCHARGES OF WASTEWATER

Wet weather processes related to wastewater in sewers have, in general, been excluded in this book, because they are based on a different process concept and result in quite different impacts. Microbial and physicochemical processes are dominating in sewers during the dry weather transport of wastewater. In contrast, the physical processes, i.e., the hydraulics, have a dominant role in sewers during wet weather. When dealing with combined sewer networks in terms of pollutant loads during overflow events, dry weather deposits of solids and correspondingly produced biofilms are exposed to erosion and resuspension. Solids transport of inflowing substances during high-flow events is, in addition to the rainfall/runoff hydraulics, central occurring physical in-sewer processes. Widely different process approaches are, therefore, required to describe the dry weather and the wet weather performance of the sewer network.

The most important quality aspect of wet weather sewer performance is related to the discharges of wastewater, i.e., the combined sewer overflows (CSOs), and their impacts on adjacent receiving waters. In addition to the stormwater runoff that transports pollutants from rainwater and urban surfaces during rainfall events, the CSOs furthermore include pollutants that originate from the diluted wastewater and the eroded solids in the combined sewer. When dealing with in-sewer microbial and chemical processes the stormwater sewers are therefore basically not relevant. In contrast, the combined sewers are, of course, during low rainfall events—to some extent—subject to "sewer processes," as it is understood in this book.

The dominance of the wet weather physical processes related to CSO discharges does not mean that microbial and physicochemical processes in sewers play no role at all. However, the nature of these processes is different compared with what is observed during dry weather. The transformation of the organic substances during transport in the sewer is typically of less importance. It is important, however, that the sewer biofilm and deposits can be eroded and resuspended during high-flow events. The following microbial processes are therefore of interest:

- The biofilm growth during the antecedent dry weather period whereby the potential for detachment of microorganisms during wet weather conditions is possible
- The biodegradation characteristics of particularly sewer solids (biofilm and deposits) that are eroded, resuspended, and potentially discharged into adjacent receiving waters during CSO events

Relatively little experimentally based knowledge is available on these types of phenomena. It is therefore interesting that Sakrabani et al. (2009) carried out OUR (oxygen uptake rate)-versus-time experiments on samples collected from a combined sewer during rain events. These measurements revealed that the highest biodegradability of the water occurred during the initial part of the rain event, indicating that the contents of detached biofilm and eroded deposits played a central role. This information can be used to better manage CSO discharges that otherwise could result in the depletion of DO in receiving waters.

The in-sewer heterotrophic microbial process description included in the WATS model can be extended with the two types of wet weather phenomena shown above. Although the microbial processes play a minor in-sewer role, they are important in terms of the receiving water impacts from CSO discharges. This aspect is relevant for those impacts of the CSO discharges that are related to biodegradation of the organic matter, for example, oxygen depletion and growth of heterotrophic biofilms in the receiving waters. The traditional description of receiving water impacts from CSO discharges related to soluble COD and particulate fractions of COD are in such situations an approach with limitations. The soluble fraction includes substances with different biodegradation characteristics, i.e., inert, readily biodegradable, and fast hydrolyzable COD, and the particulate fraction consists of different types of solids with varying adsorption characteristics, settling rates, and biodegradability (Vollertsen et al. 1999). Basic microbial characteristics, for example, biomass growth and corresponding substrate utilization, are not taken into account in the traditional wet weather approach for impacts of CSO discharges. Furthermore, the description is at present based on loads at the sewer–receiving water interface and has normally no process links to solids erosion and washout phenomena in the sewer. A detailed description of this traditional way of dealing with CSO discharges and corresponding receiving water impacts are given by Hvitved-Jacobsen et al. (2010).

An improved concept for CSO impact assessment must include both physical and microbial characteristics and processes. As far as the microbial heterotrophic transformations are concerned, intensive investigations have shown that suspended particles originating from sewer sediments follow the concept for wastewater as depicted in Figure 5.5 (Vollertsen and Hvitved-Jacobsen 1998, 1999; Vollertsen et al. 1999). This finding is important, because it shows that the concept and corresponding model developed for transformations of wastewater in sewers have potential for being expanded to include transformations of the CSO discharges into receiving waters. The concept seems to be valid across the interface between the sewer and the receiving water system. Figure 11.4 defines the corresponding integrated framework for transport and transformation.

A successful implementation of the concept depicted in Figure 11.4 depends on a description of the pathways for sewer solids. A number of steps should be known in this respect, for example, erosion of the sewer sediment and biofilm, solids transport through the sewer and the CSO structure, final degradation in the receiving water, and physicochemical adsorption or sedimentation of specific solids fractions in the receiving waters. Only comprehensive and well-designed field investigations carried out in sewers and receiving waters will provide the experimental evidence for the complex processes that the sewer solids may undergo.

Wet weather processes are subject to high variability. A simple deterministic model result in terms of the impacts on the water quality is not relevant. From a modeling point of view, a stochastic description is, however, a realistic solution for producing relevant results. Furthermore, an approach based on a historical rainfall series as model input is needed for the description of extreme event statistics for a critical CSO impact. A description of the impact in this way is considered central for assessment according to a water quality criterion. This aspect is a key point to include when designing structures for CSO discharges, taking into account the water quality impacts onto the receiving water.

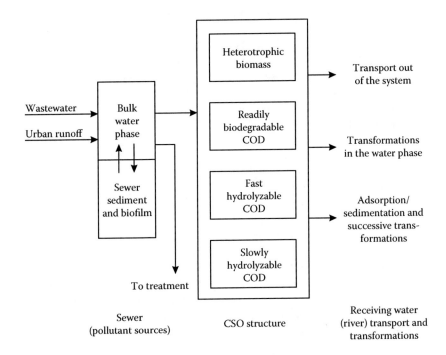

**FIGURE 11.4** A framework including pollutant sources, pathways, and transformations in surface waters receiving CSOs.

## 11.3.3    IN-SEWER PROCESSES AND SUSTAINABLE URBAN
##            WASTEWATER MANAGEMENT

The basic structure of the urban wastewater system as we know it today dates back to about 1850 in most parts of the developed world (cf. Section 1.2.4). The growth of cities during the industrial development of Europe, North America, and Australia required a technical solution for urban water management that could cope with the high quantities of water needed and the associated amount of wastewater produced. A large-scale system for supply of water and collection of wastewater was technologically developed as a pipe system with flow of potable water into and outflow of sewage from the city. With a later addition of end-of-pipe solutions for treatment, this system has efficiently reduced previous serious epidemic diseases and detrimental environmental effects (cf. Section 1.2.5).

Among scientists and practitioners, the question has been raised if today's urban water system is sustainable. It is of course not a simple question to discuss, and the answer also depends on which alternatives are realistic. It is, however, a fact that sanitation got the top vote as a medical breakthrough since 1840, ahead of such important developments as antibiotics, anesthesia, vaccines, and discovery of the DNA structure (cf. Section 1.2.5). Without the present centralized system that separates clean water from the sewage, millions of people worldwide would have died.

The concept of sustainability is, however, a new challenge for urban water managers that must be seriously considered. During the past two decades, the large-scale pipe system out of the city, i.e., the sewer network, and the associated treatment system for wastewater has been proclaimed a nonsustainable solution for management of the wastewater. The existing centralized method of urban wastewater management is difficult to defend if sustainability is considered just met by small-scale and local solutions. Such solutions are, however, only more sustainable compared with a centralized solution, i.e., if they are, in fact, winning. The authors are not sure that this comparison is appropriate if local solutions are proclaimed sustainable just because they are small-scale systems.

It must be realized that urban wastewater management is difficult to cope with in a sustainable manner, because the entire idea is to use resources and pollute them! In spite of this or maybe because of this, the development of the urban infrastructure in a sustainable direction must be seriously considered, particularly because the centralized idea behind the urban water cycle, without doubt, is expected to exist for an indeterminate length of time.

Except for continued improvement and upgrading of the urban wastewater system that is considered a general trend of a technological development, only a few technical solutions have contributed to an improved sustainability. The attempt to reduce water consumption and make the use of water more efficient in households and industry is correct in terms of sustainability as long as a high-level health standard of the population is maintained. In addition to a changing attitude among inhabitants, a number of water-conserving technologies have also contributed to this development. However, reduced water consumption and renovation of the sewer network to reduce the infiltration may also create adverse effects for the existing sewer network. Such problems will, as an example, occur caused by an increased amount of solids deposits in gravity sewers and a prolonged anaerobic residence time in pressure mains. Planners, designers, and operators are facing a tremendous task of finding both nonstructural and technically sustainable solutions to sustain the future development of sewer networks.

It is the authors' opinion that an extended integrated performance of the urban wastewater infrastructure—internally as well as with the surroundings—is fundamental for improved sustainability. The concept of the sewer as a bioreactor has an overall objective to deal with integrated microbial and chemical interactions across the sewer–treatment plant interface, and it seeks to adopt in-sewer process interactions with the surroundings. In this respect, the sewer process concept is claimed as a sustainable approach. Contrary to the "hard" engineering solutions typically applied when implementing solutions for the collection of wastewater, the process concept tends to give the sewer a "value" as a process reactor integrated with its surroundings. This is a new approach for those who manage sewer networks today.

Briefly expressed, the sustainable approach of the sewer process concept can be interpreted by changing wastewater management from an "end-of-pipe treatment" to a "pipe and plant treatment." Why should it not be possible to integrate chemical and microbial processes in wastewater, regardless of which side of the sewer–treatment plant interface it occurs? It is not the only way of approaching a sustainable solution for an urban wastewater system but it is an approach. Finally, it is certainly true that the sewer process concept tends to put much more focus on the dry weather performance of the sewer than is normally done.

## REFERENCES

Hvitved-Jacobsen, T., J. Vollertsen, and A.H. Nielsen (2010), *Urban and Highway Stormwater Pollution: Concepts and Engineering*, CRC Press, Boca Raton, FL, p. 347.

Parkhurst, J.D. and R.D. Pomeroy (1972), Oxygen absorption in streams, *J. Sanit. Eng. Div.*, ASCE, 98(SA1), 121–124.

Sakrabani, R., J. Vollertsen, R.M. Ashley, and T. Hvitved-Jacobsen (2009), Biodegradability of organic matter associated with sewer sediments during first flush, *Sci. Total Environ.*, 407(8), 2989–2995.

Vollertsen, J. and T. Hvitved-Jacobsen (1998), Aerobic microbial transformations of resuspended sediments in combined sewers—a conceptual model, *Water Sci. Technol.*, 37(1), 69–76.

Vollertsen, J. and T. Hvitved-Jacobsen (1999), Stoichiometric and kinetic model parameters for microbial transformations of suspended solids in combined sewer systems, *Water Res.*, 33(14), 3127–3141.

Vollertsen, J., T. Hvitved-Jacobsen, I. McGregor, and R. Ashley (1999), Aerobic microbial transformations of pipe and silt trap sediments from combined sewers, *Water Sci. Technol.*, 39(2), 234–241.

# Appendix A
## *Units and Nomenclature*

### A.1 UNITS

The units used in this book follows the International Systems of Units (le Système international d'unités), also abbreviated as SI. The SI includes the following five base units:

| Name | Unit Symbol | Quantity | Symbol |
|---|---|---|---|
| Meter | m | length | l |
| Kilogram | kg | mass | m |
| Second | s | time | t |
| Kelvin | K | thermodynamic temperature | T |
| Mole | mol[a] | amount of substance | n |

[a] The SI unit symbol is mol and not mole, the name of the quantity.

Electric current (ampere) and luminous intensity (candela) are also SI base units, but are not relevant for this book. In addition to these base units, SI includes numerous derived units.

For traditional reasons, the SI base units are, in some cases, substituted with derived units. As an example, the derived unit, degree Celsius (°C), is in practice typically used as a substitute for Kelvin (K). Another important example is to use $g\ m^{-3}$ or $mg\ L^{-1}$ for concentration and not always M ($mol\ L^{-1}$).

The US Customary System is in general omitted in this text. However, there are specific cases exemplified with the so-called Z-formula for assessment of hydrogen sulfide problems in gravity sewers, where such units by tradition are partly in use (cf. Equation 6.7).

It is furthermore important to note that constants applied in empirical equations—because of the more or less accidental use of units for specific parameters in these expressions—can be assigned corresponding irregular units. Although the use of units for specific constants can be discussed, it is always clearly indicated which units for the parameters in the empirical equations should be selected.

### A.2 NOMENCLATURE

#### A.2.1 GENERAL OVERVIEW

In this book, our aim is not to use a symbol—including an associated index—for more than one descriptor and not to apply a use that may cause confusion. This statement is, however, not always possible to follow. The symbols are therefore clearly specified in the text and if used with different meanings, it is in general done where confusion should not be possible. As an example—and according to SI—g is the symbol for gram. However, and following normal use, g is also symbol for gravitational acceleration with the SI derived unit m s$^{-2}$. Another example is the symbol $A$, which is used to denote area and alkalinity.

Numerous substances, process parameters, and system characteristics are needed for description of processes and phenomena. As a consequence, a systematic and hierarchical use of nomenclature is correspondingly convenient to make the descriptions clear and easy to understand. An index $(x)$ is therefore often added to a symbol $(y)$ giving further details and information to the "derived symbol" $(y_x)$. In specific cases, two indices—or sometimes, even more—can be added to a symbol, for example, $y_{x,z}$.

Whenever possible and convenient, the following symbols and indices are generally used in this text:

## General used Symbols

| | |
|---|---|
| $S$ | Dissolved substance[a] |
| $X$ | Particulate substance |
| $K$ | Equilibrium constant; e.g., saturation constant in a Monod expression, partitioning coefficient, or acid/base constant |
| $k$ | Process rate constant or interfacial mass transfer coefficient |

[a] Capital S is also a symbol for elemental sulfur.

## Selected Indices

In general, indices are used to denote specific characteristics associated with a parameter. A chemical element or a compound can be used as index, e.g., $C_{H_2S}$ as the symbol for a hydrogen sulfide concentration in a water phase. Another specific use is, for example, $S^0_{fast}$ and $S^0_{slow}$, i.e., elemental sulfur occurring at a sewer wall in a fast biodegradable and a slowly biodegradable form, respectively.

The following list shows indices that are often used in the text.

| | |
|---|---|
| $f$ | Biofilm |
| $F$ | Fermentation or fermentable |
| $h$ | Hydrolysis or hydrolyzable |
| $H$ | Heterotrophic |
| $M$ | Methane producing |
| $n$ | Fraction number for hydrolyzable substrate |
| | • $n = 1$: fast, $n = 2$: slow |
| | • $n = 1$: fast, $n = 2$: medium, $n = 3$: slow |
| $O_2$ | Aerobic |
| $NO_3$ or an | Anoxic |
| ana | Anaerobic |
| $r$ | Reaeration |
| $s$ | Sulfide production |
| $w$ | Water phase |

## A.2.2  Tables of Symbols

The following tables provide an overview of selected symbols for compounds, parameters, constants, and system characteristics generally used in this text. The tables include the nomenclature for aerobic transformations. This nomenclature is relevant for the WATS sewer process model and is therefore central for Chapters 8 and 9.

In addition to the symbols for parameters shown in the following tables, the WATS (Wastewater Aerobic/anaerobic Transformations in Sewers) sewer process model may include parameters particularly related to nitrogen and sulfur transformations. Corresponding symbols for compounds, model parameters, constants, etc., are only partly included in the following tables. It is found appropriate that the definitions of the symbols for such parameters are closely associated with the process rate expressions where they typically occur. At the end of the following tables, the specific sections of this text where these model parameters appear are referred to.

### Dissolved Compounds

| | |
|---|---|
| $S_A$ | Fermentation products (g COD m$^{-3}$) |
| $S_{ALK}$ | Bicarbonate alkalinity (mol HCO$_3^-$ m$^{-3}$) |
| $S_F$ | Fermentable, readily biodegradable substrate (g COD m$^{-3}$) |
| $S_O$ | Dissolved oxygen (g O$_2$ m$^{-3}$) |
| $S_{OS}$ | Dissolved oxygen saturation concentration (g O$_2$ m$^{-3}$) |
| $S_S$ | Readily biodegradable substrate (g COD m$^{-3}$) |
| $S_{H_2S}$ | Total sulfide (g S m$^{-3}$) |

### Particulate Compounds

| | |
|---|---|
| $X_H$ | Heterotrophic active biomass (g COD m$^{-3}$ or g COD m$^{-2}$) |
| | • $X_{Hf}$ = heterotrophic biomass in the biofilm (g COD m$^{-2}$) |
| | • $X_{Hw}$ = heterotrophic biomass in the water phase (g COD m$^{-3}$) |
| $X_{Sn}$ | Hydrolyzable substrate, fraction $n$ (g COD m$^{-3}$) |
| | • $n = 1$: fast, $n = 2$: slow |
| | • $n = 1$: fast, $n = 2$: medium, $n = 3$: slow |

### Gas Phase Compounds

The symbols (g) and (aq) are generally used to indicate a gas phase compound and an aqueous phase compound, respectively, e.g., H$_2$S (g) as gaseous hydrogen sulfide in contrast to H$_2$S (aq).

### Selected Stoichiometric and Kinetic Constants

| | |
|---|---|
| $k_{hn}$ | Hydrolysis rate constant for fraction $n$ (day$^{-1}$) |
| | • $n = 1$: fast, $n = 2$: slow |
| | • $n = 1$: fast, $n = 2$: medium, $n = 3$: slow |
| $k_{H_2S}$ | Hydrogen sulfide formation rate constant (h$^{-1}$) |
| $k_{1/2}$ | 1/2-order rate constant (g O$_2^{0.5}$ m$^{-0.5}$ day$^{-1}$) |

| | |
|---|---|
| $K_{fe}$ | Saturation constant for fermentation (g COD m$^{-3}$) |
| | • $K_{fef}$ in the biofilm |
| | • $K_{few}$ in the water phase |
| $K_A$ | Saturation constant for $S_A$ (g COD m$^{-3}$) |
| | • $K_{Af}$ in the biofilm |
| | • $K_{Aw}$ in the water phase |
| $K_O$ | Saturation constant for dissolved oxygen (g O$_2$ m$^{-3}$) |
| $K_S$ | Saturation constant for readily biodegradable substrate (g COD m$^{-3}$) |
| | • $K_{Sf}$ in the biofilm |
| | • $K_{Sw}$ in the water phase |
| $K_{Xn}$ | Saturation constant for hydrolysis, fraction $n$ (g COD g COD$^{-1}$) |
| | • $n = 1$: fast, $n = 2$: slow |
| | • $n = 1$: fast, $n = 2$: medium, $n = 3$: slow |
| $q_{fe}$ | Fermentation rate constant (day$^{-1}$) |
| | • $q_{fef}$ in the biofilm |
| | • $q_{few}$ in the water phase |
| $q_m$ | Maintenance energy requirement rate constant (day$^{-1}$) |
| $Y_H$ | Yield constant for heterotrophic biomass [g COD, biomass (g COD, substrate)$^{-1}$] |
| | • $Y_{Hf}$ in the biofilm |
| | • $Y_{Hw}$ in the water phase |
| $\mu_H$ | Maximum specific growth rate for heterotrophic biomass (day$^{-1}$) |
| $\varepsilon$ | Efficiency constant for the biofilm biomass (–) |
| | • $\varepsilon_A$ aerobic conditions |
| | • $\varepsilon_{An}$ anaerobic conditions |
| $\eta_{h,anox}$ | Efficiency constant for anoxic hydrolysis (–) |
| $\eta_{h,anae}$ | Efficiency constant for anaerobic hydrolysis (–) |

## Selected Process Rate Expressions

| | |
|---|---|
| $r_{grw}$ | Growth rate of heterotrophic biomass in suspension (g COD m$^{-3}$ day$^{-1}$) |
| $r_{grf}$ | Growth rate of heterotrophic biomass in a biofilm (g COD m$^{-3}$ day$^{-1}$) |
| $r_{maint}$ | Rate of maintenance energy requirement for biomass in suspension (g COD m$^{-3}$ day$^{-1}$) |
| $r_{hydr}$ | Rate of hydrolysis (g COD m$^{-3}$ day$^{-1}$) |
| $F$ | Rate of oxygen transfer (g O$_2$ m$^{-3}$ s$^{-1}$) |

## Flow Related and Sewer System Parameters

| | |
|---|---|
| $A/V$ | Ratio of biofilm area to bulk water volume, i.e., $R^{-1}$ (m$^{-1}$) |
| $g$ | Gravitational acceleration (m s$^{-2}$) |
| $d_m$ [a] | Hydraulic mean depth (m) |
| $K_L a$ | Air–water oxygen transfer coefficient, reaeration constant (s$^{-1}$, h$^{-1}$, or day$^{-1}$) |
| $R$ [b] | Hydraulic radius (m) |
| $s$ | Slope (m m$^{-1}$) |
| $u$ | Mean flow velocity (m s$^{-1}$) |

[a] The cross-sectional area of the water volume divided by the water surface width.
[b] The cross-sectional area of the water volume divided by the wetted perimeter.

## General Environmental Parameters

| | |
|---|---|
| $T$ | Temperature (°C) |
| $t$ | Time (s, h, or day) |
| $\alpha$ | Temperature coefficient (–) |
| | • $\alpha_f$ biofilm |
| | • $\alpha_r$ reaeration |
| | • $\alpha_s$ sulfide production |
| | • $\alpha_w$ water phase |

## Sections with Nomenclature for Parameters Particularly Related to Nitrogen and Sulfur Transformations

- Anoxic transformations: Section 5.6
- Anaerobic transformations: Section 6.7
- Aerobic sulfide oxidation: Section 5.5
- Concrete corrosion: Section 6.5

# Appendix B
## Definitions and Glossary

This appendix includes selected definitions used in this book. Definitions that mainly refer to a specific chapter where they are further explained and discussed will in general not be included in this list but can be found in the relevant chapters or via the index list. Commonly used terms that are well known are also excluded from this list.

**Collecting sewer:** A sewer that collects wastewater (and runoff water) from different sources (residences, commercial buildings, industry, and institutions) and to which no other sewer tributary exists. A collecting sewer is also named a lateral sewer.

**Combined sewer:** A sewer (in contrast to a separate sewer) designed to receive both wastewater and runoff water from impervious areas such as streets, roads, and roofs.

**Combined sewer overflows (CSO):** The mixed untreated flow of wastewater and runoff water from impervious urban surfaces and roads that are discharged from an overflow structure in a combined sewer network. The discharge takes place when the capacity of the downstream network including treatment plant is exceeded.

**Force main:** A sewer (in contrast to a gravity sewer) that conveys wastewater under pressure to raise it from one level to another. A force main is also named a pressure main or pressure sewer.

**Intercepting sewer:** A large sewer in a separate or combined sewer system that conveys the flow of wastewater to a wastewater treatment plant. An intercepting sewer collects the flow from small sewers and trunk sewers and has typically few connections.

**Lateral sewer:** See Collecting sewer.

**Mitigation:** Mitigation, in this book, is defined as the act of reducing adverse effects caused by chemical and biological processes in sewers. In particular, it refers to the reduction of formation or air phase occurrence of volatile substances such as hydrogen sulfide ($H_2S$) and volatile organic compounds (VOCs).

**Pollutant:** The word pollutant is not well defined and, to some extent, can also give a wrong signal. In spite of this, here it is used identically with the words "compound" and "substance."

**Pressure sewer:** See Force main.

**Separate sewer:** A sewer (in contrast to a combined sewer) receiving sanitary wastewater.

**Trunk sewer:** A sewer that collects wastewater or stormwater from small sewers (collecting sewers and lateral sewers) and typically conveys the flow to an intercepting sewer.

# Appendix C

## *Acronyms*

This appendix features selected acronyms used in the text.

**BOD:** Biochemical oxygen demand
**COD:** Chemical oxygen demand
**CSO:** Combined sewer overflow
**DO:** Dissolved oxygen
**EPS:** Extracellular polymers
**NUR:** Nitrate uptake rate
**OM:** Organic matter
**OUR:** Oxygen uptake rate
**PAH:** Polycyclic aromatic hydrocarbon
**SOB:** Sulfide oxidizing bacteria
**SOD:** Sediment oxygen demand
**SRB:** Sulfate-reducing biomass
**SS:** Suspended Solids
**SWR:** Stormwater runoff
**TKN:** Total Kjeldahl nitrogen
**TN:** Total nitrogen
**TOC:** Total organic carbon
**TS:** Total solids
**TP:** Total phosphorous
**TSS:** Total suspended solids
**VFA:** Volatile fatty acid
**VOC:** Volatile organic compound
**VS:** Volatile solids
**VSC:** Volatile sulfur compounds
**VSS:** Volatile suspended solids
**WWTP:** Wastewater treatment plant

# Index

Page numbers followed by *f* and *t* indicate figures and tables, respectively.